P9-CCG-859

ACIDITY FUNCTIONS

ORGANIC CHEMISTRY

A SERIES OF MONOGRAPHS

Edited by

ALFRED T. BLOMQUIST

Department of Chemistry, Cornell University, Ithaca, New York

1. Wolfgang Kirmse. CARBENE CHEMISTRY, 1964
2. Brandes H. Smith. BRIDGED AROMATIC COMPOUNDS, 1964
3. Michael Hanack. CONFORMATION THEORY, 1965
4. Donald J. Cram. FUNDAMENTALS OF CARBANION CHEMISTRY, 1965
5. Kenneth B. Wiberg (Editor). OXIDATION IN ORGANIC CHEMISTRY, PART A, 1965; PART B, *In preparation*
6. R. F. Hudson. STRUCTURE AND MECHANISM IN ORGANO-PHOSPHORUS CHEMISTRY, 1965
7. A. William Johnson. YLID CHEMISTRY, 1966
8. Jan Hamer (Editor). 1,4-CYCLOADDITION REACTIONS, 1967
9. Henri Ulrich. CYCLOADDITION REACTIONS OF HETEROCUMULENES, 1967
10. M. P. Cava and M. J. Mitchell. CYCLOBUTADIENE AND RELATED COMPOUNDS, 1967
11. Reinhard W. Hoffman. DEHYDROBENZENE AND CYCLOALKYNES, 1967
12. Stanley R. Sandler and Wolf Karo. ORGANIC FUNCTIONAL GROUP PREPARATIONS, 1968
13. Robert J. Cotter and Markus Matzner. RING-FORMING POLYMERIZATIONS, PART A, 1969; PART B, *In preparation*
14. R. H. DeWolfe. CARBOXYLIC ORTHO ACID DERIVATIVES, 1970
15. R. Foster. ORGANIC CHARGE-TRANSFER COMPLEXES, 1969
16. James P. Snyder (Editor). NONBENZENOID AROMATICS, I, 1970
17. C. H. Rochester. ACIDITY FUNCTIONS, 1970

ACIDITY FUNCTIONS

COLIN H. ROCHESTER

Department of Chemistry

University of Nottingham, England

1970

ACADEMIC PRESS

London and New York

ACADEMIC PRESS INC. (LONDON) LTD
Berkeley Square House
Berkeley Square
London, W1X 6BA

U.S. Edition published by
ACADEMIC PRESS INC.
111 Fifth Avenue
New York, New York 10003

Library of Congress Catalog Card Number: 76-117126
SBN: 12-590850-4

PRINTED IN GREAT BRITAIN BY
SPOTTISWOODE, BALLANTYNE & CO. LTD
LONDON AND COLCHESTER

Preface

This book is concerned with the thermodynamic and kinetic behaviour of organic solutes in concentrated solutions of acids and bases. The interpretation of the results of studies in this section of Physical-organic Chemistry has largely relied on the acidity function concept first proposed by L. P. Hammett and A. J. Deyrup in 1932. A landmark in the study of acidity functions was provided by two authoritative reviews published by F. A. Long and M. A. Paul in 1957. These laid the basis for much of the later work on acidity functions. Since 1957 many aspects of the subject have been drastically altered particularly by the observed lack of generality of acidity function behaviour. The present status of the usefulness and limitations of acidity functions is critically examined here.

The relationship between the pH scale for dilute aqueous solutions and other acidity function scales for concentrated solutions is emphasized. Tabulations of acidity function scales are listed and are discussed in detail for aqueous solutions of many strong and weak inorganic and organic acids and bases. The results of kinetic studies of acid and base catalysed reactions in concentrated aqueous solutions of acids and bases are described and have been discussed primarily from the point of view of using acidity function correlations as criteria for the diagnosis of the mechanism of such reactions. The acidity function approach for solutions of acids and bases in mixed aqueous and non-aqueous solutions is critically reviewed. The measurement of the acidity of oxide surfaces by the adsorption of typical acidity function indicators on the surfaces has not been mentioned as it would be out of place in a book otherwise completely devoted to Solution Chemistry.

My greatest debts of gratitude are to my father who first aroused and encouraged my interest in chemistry and to Professor V. Gold, who inspired my interest in acidity functions during the years I spent at King's College, London. I gratefully acknowledge the advice, helpful discussion and encouragement given by many members of the Chemistry Department at the University of Nottingham, and in particular by Professor D. D. Eley, OBE, FRS, and Drs M. F. A. Dove and J. C. Roberts. I also wish to thank Mrs Marjorie Smalley who typed the manuscript, Mr Ian Goddard who drew the figures and Mr John Symonds for help with the indexing. Tables 1.2 and 1.3 from BS1647: 1961 pH scale are reproduced by permission of the

British Standards Institution, 2 Park Street, London W1Y 4AA from whom copies of the complete standard may be obtained. Finally, I am indebted to my wife for help both with the indexing and in many other ways.

COLIN H. ROCHESTER

February, 1970

Contents

CHAPTER 4 The rates of reactions in concentrated aqueous acid solutions. Theoretical approaches

CHAPTER 5 The rates of reactions in concentrated aqueous acid solutions. Experimental results

CHAPTER 1

The Definition and Measurement of pH and the Hammett Acidity Function H_0 for Aqueous Solutions

The definition of acids and bases in terms of proton transfer equilibria was proposed by J. N. Brønsted (1923). An acid is a species which can lose a proton and a base is a species which can accept a proton. It follows that the ionization of a neutral molecule HA as an acid may be written

$$HA \rightleftharpoons A^- + H^+ \tag{1.1}$$

For the forward reaction HA is behaving as an acid and for the backward reaction A^- is behaving as a base on the Brønsted definition. It is a general result that the ionization of an acid always leads to a base which differs from the parent acid by one proton H^+. An acid and base which differ only by one proton are referred to as a conjugate acid–base pair. This concept of acidity is equally applicable to positively charged, negatively charged and neutral acids. Of course an acid in the Brønsted sense must contain at least one hydrogen atom.

1.1. The Acidity of Dilute Aqueous Solutions

An acid in solution only ionizes to an appreciable extent according to equation (1.1) if there is a base present to accept the proton which is being lost. It is fortunate that many common solvents are themselves capable of acting as acids or bases. Thus, for example, water is capable of acting as a base and the ionization of an acid HA in water is more correctly represented (Brønsted 1926) by

$$HA + H_2O \rightleftharpoons H_3O^+ + A^- \tag{1.2}$$

It follows that the ionization of any acid in water leads to an increase in the concentration of H_3O^+ ions in the solution.

The ionization of a base in water is represented by equation (1.3) in which the solvent acts as an acid.

$$B + H_2O \rightleftharpoons BH^+ + OH^- \tag{1.3}$$

Because of the dual acidic and basic (amphiprotic) character of water it undergoes self-ionization according to equation (1.4).

$$2H_2O \rightleftharpoons H_3O^+ + OH^- \qquad (1.4)$$

Consideration of the equilibria (1.3) and (1.4) shows that the ionization of a base in water leads to a decrease in the concentration of H_3O^+ ions in the solution.

The extent of ionization of acids and bases in water is directly dependent, via the equilibrium constants for equations (1.2), (1.3) and (1.4), on the activity of H_3O^+ ions in solution.

The activity of hydrogen ions in a dilute aqueous solution may be identified with the "acidity" of that solution. However it is unsatisfactory to define an acidity scale directly in terms of hydrogen ion activity because individual ion activities cannot be measured experimentally. Sørensen (1909) suggested a logarithmic scale of acidity based on measurement of the e.m.f. of electrochemical cells. The modern definition of the pH scale of acidity for aqueous solutions is based on Sørensen's idea.

1.2. The Definition and Measurement of pH

The experimental determination of the pH of a solution requires measurement of the e.m.f. of a cell of the type

$$\text{Pt, } H_2(g)|\text{acid solution} \vdots \text{KCl(aq)}|\text{AgCl,Ag} \qquad (1.5)$$

A glass or quinhydrone electrode may be used in place of the hydrogen electrode although the latter enables a more rigorous definition of pH. The calomel half cell (Hg_2Cl_2, Hg) is commonly used as an alternative reference electrode. The potassium chloride concentration is usually 3·5 M or saturated (at 25°C, 4·16 M). The e.m.f. of cell (1.5) is given by equation (1.6) in which E^0 is the

$$E = E^0 - \frac{RT}{F}\ln\frac{a_{Cl^-}a_{H^+}}{p^{1/2}} + E_j \qquad (1.6)$$

standard e.m.f. of the cell, a_{Cl^-} is the activity of chloride ions in the potassium chloride solution, a_{H^+} is the activity of hydrogen ions in the acid solution, p is the hydrogen pressure at the hydrogen electrode, and E_j is the potential at the liquid junction between the two solutions. The e.m.f. of the cell is measured for two different acid solutions X and S with the pressure of hydrogen and concentration of potassium chloride unchanged. The difference between the e.m.f. values is given by equation (1.7) providing it is assumed that the difference between the liquid junction potentials E_j(X) and E_j(S), is negligible.

$$E(\text{X}) - E(\text{S}) = -\frac{RT}{F}\ln[a_{H^+}(\text{X})/a_{H^+}(\text{S})] \qquad (1.7)$$

The pH of a solution X of unknown acidity is defined by equation (1.8) in which pH(S) refers to a standard solution S. Comparison of equations (1.7)

$$\mathrm{pH(X)} - \mathrm{pH(S)} = \frac{F[E(\mathrm{X}) - E(\mathrm{S})]}{2{\cdot}303RT} \tag{1.8}$$

and (1.8) shows that effectively pH has been equated with $-\log_{10} a_{\mathrm{H}^+}$. However the hydrogen ion activity cannot be independently measured and therefore the equality $\mathrm{pH} = -\log_{10} a_{\mathrm{H}^+}$ can never be definitely established for any solution. The problem then arises as to how to assign to the standard solution a pH which is as far as can be ascertained equal to minus the logarithm of the hydrogen ion activity.

The American (NBS) scale of pH is based on a series of standard solutions the pH of which is estimated in the following way (Bates, 1962). A suitable buffer solution is chosen and the e.m.f. E of the cell (1.9) (without liquid

$$\mathrm{Pt, H_2(g)\,|\,Buffer\ solution, Cl^-\,|\,AgCl,Ag} \tag{1.9}$$

junction) is measured with a series of concentrations of chloride ion added to the buffer. The function $-\log_{10}(a_{\mathrm{H}^+}\gamma_{\mathrm{Cl}^-})$ is evaluated via equation (1.10) and the values are extrapolated to zero chloride ion concentration to give $-\log_{10}(a_{\mathrm{H}^+}\gamma_{\mathrm{Cl}^-})^0$.

$$-\log_{10}(a_{\mathrm{H}^+}\gamma_{\mathrm{Cl}^-}) = \frac{(E - E^0)F}{2{\cdot}303RT} + \log_{10} m_{\mathrm{HCl}} \tag{1.10}$$

The standard pH which is assigned to the solution is then calculated using equation (1.11) in which the activity coefficient of the chloride ion at infinite

$$\mathrm{pH(S)} = -\log_{10}(a_{\mathrm{H}^+}) = -\log_{10}(a_{\mathrm{H}^+}\gamma_{\mathrm{Cl}^-})^0 - \frac{AI^{1/2}}{1 + 1{\cdot}5I^{1/2}} \tag{1.11}$$

dilution in the buffer solution is predicted by Debye–Hückel theory (Bates and Guggenheim, 1960). A is the Debye–Hückel constant (0·5115 for water at 25°C if I is in molal units) and I is the ionic strength of the solution which equals the sum $\frac{1}{2}\sum m_i z_i^2$ over all species i (charge z_i, molal concentration m_i) in the solution. The pH(S) values deduced in the above way for a series of standards must of course be consistent with equation (1.8) when the e.m.f. of cell (1.5) is compared for any two of the standards.

The accurately determined pH(S) at eighteen temperatures of the five primary standards of the NBS scale are given in Table 1.1.

The British scale of pH is described in the British Standard Institution publication "Specification for pH Scale" (British Standard 1647: 1961). It is based on a single standard which is a 0·05 M solution of potassium hydrogen

TABLE 1.1.

Primary standards of the NBS pH scale at eighteen temperatures (Bates, 1962)

Temperature (°C)	A†	B†	C†	D†	E†
0		4·003	6·984	7·534	9·464
5		3·999	6·951	7·500	9·395
10		3·998	6·923	7·472	9·332
15		3·999	6·900	7·448	9·276
20		4·002	6·881	7·429	9·225
25	3·557	4·008	6·865	7·413	9·180
30	3·552	4·015	6·853	7·400	9·139
35	3·549	4·024	6·844	7·389	9·102
38	3·548	4·030	6·840	7·384	9·081
40	3·547	4·035	6·838	7·380	9·068
45	3·547	4·047	6·834	7·373	9·038
50	3·549	4·060	6·833	7·367	9·011
55	3·554	4·075	6·834		8·985
60	3·560	4·091	6·836		8·962
70	3·580	4·126	6·845		8·921
80	3·609	4·164	6·859		8·885
90	3·650	4·205	6·877		8·850
95	3·674	4·227	6·886		8·833

† Standard solutions are: (A) saturated solution of potassium hydrogen tartrate at 25°; (B) 0·05 *m* potassium hydrogen phthalate; (C) 0·025 *m* potassium dihydrogen phosphate, 0·025 *m* disodium hydrogen phosphate; (D) 0·086 95 *m* potassium dihydrogen phosphate, 0·030 43 *m* disodium hydrogen phosphate; (E) 0·01 *m* borax.

phthalate. The pH of this solution is defined by equation (1.12) for temperatures in the range $0° < t < 55°$ and by equation (1.13) for $55° < t < 95°$.

$$\mathrm{pH} = 4{\cdot}000 + \frac{1}{2}\left(\frac{t-15}{100}\right)^2 \tag{1.12}$$

$$\mathrm{pH} = 4{\cdot}000 + \frac{1}{2}\left(\frac{t-15}{100}\right)^2 - \left(\frac{t-55}{500}\right) \tag{1.13}$$

The pH of the primary standard for $0° < t < 95°$ is given in Table 1.2.

On the British approach the pH of the standard is interpreted as conforming to the experimentally measurable definition of pH (1.14).

$$\mathrm{pH} = -\log(C_{\mathrm{H^+}}\, y_{\pm}) \pm 0{\cdot}02 \tag{1.14}$$

TABLE 1.2

Temperature dependence of defined pH of primary standard (0·05 M potassium hydrogen phthalate) of British pH scale (British Standard 1647: 1961)

Temperature (°C)	pH	Temperature (°C)	pH	Temperature (°C)	pH
0	4·011	35	4·020	70	4·121
5	4·005	40	4·031	75	4·140
10	4·001	45	4·045	80	4·161
15	4·000	50	4·061	85	4·185
20	4·001	55	4·080	90	4·211
25	4·005	60	4·091	95	4·240
30	4·011	65	4·105		

In this equation C_{H^+} is the molar concentration of hydrogen ions and $y_{\pm}$ is the mean molar activity coefficient of any typical 1:1 (uni–univalent) electrolyte in the solution. Except at temperatures $< 5°$ and $> 90°$ the NBS and British scales of pH are consistent with each other to within 0·005 units. Comparison of the pH values for potassium hydrogen phthalate in Tables 1.1 and 1.2 show this.

Because of its many advantages the glass electrode is frequently used in place of the hydrogen electrode for pH measurement. However the glass electrode does not necessarily show the same response to variation in hydrogen ion activity as does the hydrogen electrode particularly over wide ranges of pH. For this reason it is recommended that pH determination with a glass electrode is made with reference to two standards $S_1(pH_1)$ and $S_2(pH_2)$ preferably with pH values on either side, and not too distant, from the unknown. Equation (1.15) is then applicable (Bates, 1962). A set of secondary pH standards have been established for this purpose and are given in Table (1.3.)

$$pH(X) = pH_1 + \frac{[E(X) - E_1](pH_2 - pH_1)}{E_2 - E_1} \tag{1.15}$$

It must be noted that the pH scale will only have any real theoretical significance for solutions of ionic strength less than 0·1 M. At higher ionic strengths the Bates–Guggenheim (1960) activity coefficient expression (equation 1.11) will not be applicable, neither will the assumption made in the British interpretation of pH that the activity coefficients of different uni–univalent electrolytes show similar variations with increasing ionic strength. Thus although e.m.f. measurements can be made and pH deduced via equations (1.8) or (1.14) for any solution, no matter how concentrated, the precise

TABLE 1.3

pH values of secondary standards to be used for calibration of glass electrodes (British Standard 1647: 1961)

Solution	pH 12°C	pH 25°C	pH 38°C	Source of data†
0·1 M $KH_3(C_2O_4)_2.2H_2O$		1·48	1·50	b
0·01 M HCl + 0·09 M KCl		2·07	2·08	b
0·1 M CH_3COOH+ 0·1 M CH_3COONa‡	4·65	4·64	4·65	a, b
0·01 M CH_3COOH + 0·01 M CH_3COONa‡	4·71	4·70	4·72	a
0·025 M KH_2PO_4 + 0·025 M Na_2HPO_4		6·85	6·84	b
0·05 M $Na_2B_4O_7.10H_2O$		9·18	9·07	b
0·025 M $NaHCO_3$ + 0·025 M Na_2CO_3		10·00		c

† Sources of data: (a) MacInnes *et al.* (1938), (b) Hitchcock and Taylor (1938), (c) Bates *et al.* (1950).

‡ It is recommended that these solutions be prepared from acetic acid diluted and half neutralized by sodium hydroxide, and not from sodium acetate.

interpretation of the pH values is lost. The cancellation of liquid junction potentials in deducing equation (1.7) from equation (1.6) also becomes progressively more questionable for comparison of solutions of very high ionic strength with a conventional pH standard with ionic strength in the acceptable range. This limits the usefulness of pH as a measure of acidity in aqueous solution to the approximate range $1 < \text{pH} < 13$.

1.3. The Colorimetric Measurement of pH and Acid Ionization Constants

The colorimetric method is used almost exclusively for the measurement of acidity functions for concentrated solutions. It is therefore relevant to consider this method applied to the ionization of weak acids or bases in solutions for which the pH scale is meaningful.

The ionization of a weak acid indicator HA at low concentration (ca. 10^{-4} M) in a buffer solution of known pH is represented by equation (1.2). Writing the ionization constant K_a for the indicator according to equation (1.16) leads to

$$K_a = (a_{H^+}\, a_{A^-}/a_{HA}) = (a_{H^+}\, m_{A^-}/m_{HA})(\gamma_{A^-}/\gamma_{HA}) \qquad (1.16)$$

$$\text{pH} = \text{p}K_a + \log_{10}(m_{A^-}/m_{HA}) + \log_{10}(\gamma_{A^-}/\gamma_{HA}) \qquad (1.17)$$

equation (1.17) for the relationship between the pH of the buffer solution {corrected for the effect of the small added concentration of HA (Robinson and

Biggs, 1955)}, and $pK_a(=-\log_{10}K_a)$ and the ionization ratio (m_{A^-}/m_{HA}) of the indicator.

In solutions of ionic strength $I < 0{\cdot}1\ m$ the activity coefficient γ_{HA} of the neutral acid may be taken as one. The ionic activity coefficient γ_{A^-} may be calculated via the Guggenheim (1935) extension, equation (1.18), of the Debye–Hückel equation.

$$\log_{10}\gamma_i = -\frac{0{\cdot}5115 z_i^2 I^{1/2}}{1+I^{1/2}} + bI \tag{1.18}$$

For indicators which are fairly large organic molecules a value for b of 0·2 has been found most satisfactory (Robinson and Biggs, 1955; Robinson, 1959). The ionization ratio (m_{A^-}/m_{HA}) may be determined colorimetrically providing A^- and HA have measurably different electronic absorption spectra when present in solution at concentrations of the order of 10^{-4}–10^{-5} M.

Figure 1.1 shows the spectra of 4-*t*-butylphenol at a fixed stoichiometric concentration ($1{\cdot}89 \times 10^{-4}$ mole litre^{-1}) in a series of aqueous solutions of different acidity. Curve (a) is for the phenol in water and is the spectrum of unionized 4-t-butylphenol. Curve (f) for the phenol in 0·06 M sodium hydroxide solution is the spectrum of the 4-t-butylphenoxide anion. Curves (b)–(e) are for the phenol in four $NaHCO_3/NaOH$ buffer solutions (Bates and Bower, 1956) whose pH is such that appreciable concentrations of both the phenol and its conjugate base anion are present in solution. The set of spectra allow calculation of the ionization ratios of the phenol in the four buffer solutions as follows.

The Beer–Lambert law relates the optical density D_i due to a molecular or ionic species at concentration C_i in solution by equation (1.19). The path

$$D_i = \epsilon_i C_i l \tag{1.19}$$

length of the optical cell is l and ϵ_i is the extinction coefficient of the species at the wavelength at which D_i is measured. If for a stoichiometric concentration C of acid the measured optical densities at a fixed wavelength are D_{HA} for the neutral acid, D_{A^-} for its conjugate base, and D for an equilibrium mixture of the two forms then equations (1.20)–(1.23) are applicable. Here ϵ_{HA} and ϵ_{A^-} are the extinction coefficients of the acid and its conjugate base respectively at the wavelength to which the optical densities refer and C_{HA} and C_{A^-} are the concentrations of the two forms in the equilibrium mixture. Combination of

$$D_{HA} = \epsilon_{HA} Cl \tag{1.20}$$

$$D_{A^-} = \epsilon_{A^-} Cl \tag{1.21}$$

$$D = \epsilon_{HA} C_{HA} l + \epsilon_{A^-} C_{A^-} l \tag{1.22}$$

$$C = C_{HA} + C_{A^-} \tag{1.23}$$

these equations leads to equation (1.24) which enables direct calculation of the ionization ratio of the indicator. The result should be independent of the wave-length chosen for study.

$$\left(\frac{m_{A^-}}{m_{HA}}\right) = \left(\frac{C_{A^-}}{C_{HA}}\right) = \left(\frac{D - D_{HA}}{D_{A^-} - D}\right) \tag{1.24}$$

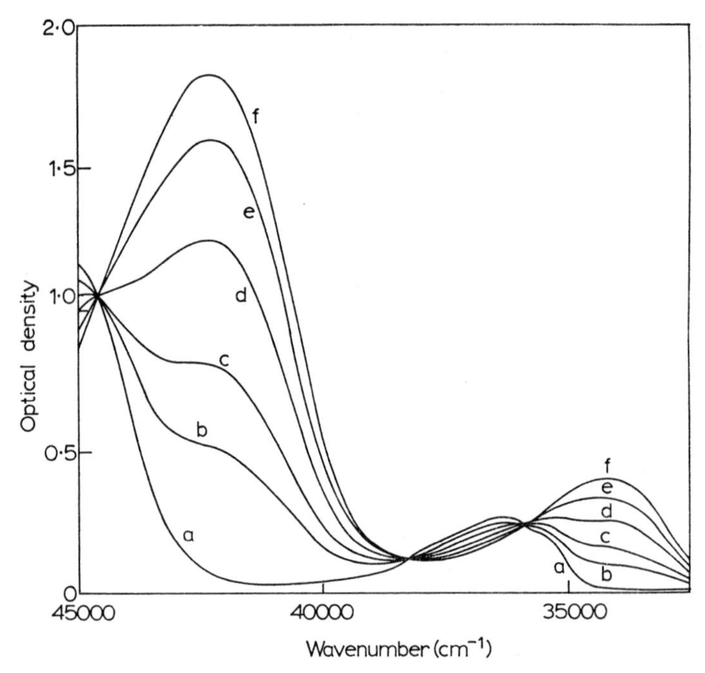

FIG. 1.1. Spectra of 4-t-butylphenol ($1{\cdot}89 \times 10^{-4}$ M; 1 cm cells) in (a) water; (b)–(e) $NaHCO_3$/NaOH buffer solutions of pH: (b) 9·70, (c) 9·97, (d) 10·38, (e) 10·96; (f) 0·06 M NaOH (Rochester, 1965).

The ionization ratio (m_{A^-}/m_{HA}) thus measured is combined in equation (1.17) with γ_{A^-} estimated from equation (1.18) and the pH of the buffer solution (corrected for the effect of the small concentration of indicator and based on e.m.f. data in accord with the definition of pH) to enable calculation of pK_a for the acid HA. Providing the ionic strengths of the buffered systems are in the range for which pH is a meaningful measure of acidity and equation (1.18) is applicable the pK_a deduced should be independent of the composition and ionic strength of the buffer solutions. Robinson and Biggs (1955) have studied the ionization of 4-nitrophenol in four phosphate buffer mixtures with ionic strengths in the range $0{\cdot}0063\ m < I < 0{\cdot}66\ m$. Below $I = 0{\cdot}1\ m$ the pK_a deduced from their spectrophotometric results was a constant within acceptable experimental error. However at greater ionic strengths large deviations occurred and the method is therefore unreliable.

A significant factor which to some extent tests the reliability of the use of equation (1.24) for the deduction of indicator ionization ratios is the presence of isosbestic points in the set of observed spectra. In Fig. 1.1 there are three such points which occur at wavelengths at which ϵ_{A^-} and ϵ_{HA} are identical. It follows from equations (1.22) and (1.23) that the recorded optical density D is independent of the ratio (C_{HA}/C_{A^-}) for such wavelengths. It may be concluded that the extinction coefficients, frequencies and half band widths of the absorption maxima shown in Fig. 1.1 are not subject to change due to solvent or general medium effects within the series of solutions studied. If such effects become important isosbestic points would not be shown and as will be discussed later (Section 1.4) this complicates and makes more unreliable the evaluation of ionization ratios from the spectra.

The calculated pK_a values should be independent of the wavelength chosen for the measurement of optical density. Thus, for example, Robinson and Biggs (1957) determined the pK_a of n-propyl 4-aminobenzoate in water at 25°C by the spectrophotometric method. The observed optical density changes at 286 nm gave $pK_a = 2{\cdot}488$ whereas those at 300 nm gave $pK_a = 2{\cdot}486$. The ionization constants should also be the same as those determined by conductivity, electromeric titration, or other methods (Hammett *et al.* 1934).

The spectrophotometric measurement of pH is analagous to the above method except it is the pH and not pK_a which is the unknown factor in equation (1.17). Thus measurement of the extent of ionization of a weak acid of known pK_a in a solution of unknown pH enables calculation of the pH by an identical procedure to that described above. The same limitations apply. In particular the ionic strength must be less than 0·1 *m* and the spectra must preferably be free from medium effects.

1.4. Concentrated Acid Solutions. The Definition and Measurement of the Hammett Acidity Function

The breakdown of pH as a useful measure of acidity for solutions whose ionic strength is greater than ca. 0·1 *m* leads to the necessity for some other quantitative scale which expresses the acidity of more concentrated solutions. With this in mind Hammett and Deyrup (1932) suggested that a convenient measure of acidity would be provided by spectrophotometric determination of the extent of protonation of weakly basic indicators in acid solution. The

$$B + H^+ \rightleftharpoons BH^+ \qquad (1.25)$$

relevant proton transfer equilibrium may be written as equation (1.25) where B is an electrically neutral weak base. In dilute acid solution the extent of protonation of the base will be related to pH by equation (1.26) in which γ_B may

$$pH = pK_{BH^+} - \log_{10}\left(\frac{m_{BH^+}}{m_B}\right) - \log_{10}\left(\frac{\gamma_{BH^+}}{\gamma_B}\right) \tag{1.26}$$

be taken as one and γ_{BH^+} may be calculated using equation (1.18). pK_{BH^+} is the acid dissociation constant of BH^+ which is the conjugate acid of B. As the concentration of acid is increased to and beyond ionic strengths greater than ca. 0·1 *m* equation (1.26) becomes progressively less reliable. Hammett and Deyrup (1932) therefore defined an acidity function H_0 by equation (1.27) and

$$H_0 = pK_{BH^+} - \log_{10}\left(\frac{C_{BH^+}}{C_B}\right) \tag{1.27}$$

proposed that H_0 could be taken as a quantitative measure of the acidity of a solution. Combination of equations (1.27) and (1.28) and introducing

$$pK_{BH^+} = -\log_{10}\left(\frac{a_{H^+}\, C_B\, y_B}{C_{BH^+}\, y_{BH^+}}\right) \tag{1.28}$$

$h_0 = (a_{H^+} y_B / y_{BH^+})$ leads to equation (1.29). It follows that H_0 will be independent of the particular indicator which is used to measure it providing (y_B/y_{BH^+})

$$H_0 = -\log_{10} h_0 = -\log_{10}\left(\frac{a_{H^+}\, y_B}{y_{BH^+}}\right) \tag{1.29}$$

is identical for all bases in the same acid solution. As will be discussed later (Chapter 3) although this identity holds for series of bases of similar structure there are often spectacular deviations between acidity functions defined by bases whose structures differ. At low ionic strengths $y_B \rightarrow 1$ and $y_{H^+} \rightarrow y_{BH^+}$ and therefore $H_0 \rightarrow -\log_{10} C_{H^+} \equiv pcH$. At infinite dilution $y_B = y_{BH^+} = 1$ and $H_0 = -\log_{10} a_{H^+} = pH$. The Hammett acidity function H_0 is therefore referred to infinite dilution as standard state. For pure water, with no added solute, $H_0 = pH = 7{\cdot}00$ at 25°C and for solutions of increasing acidity H_0 decreases.

The evaluation of the H_0 scale for a particular series of solutions of increasing acidity is made using equation (1.27). The spectrophotometric method provides a simple and usually reliable means of measuring the ionization ratios (C_{BH^+}/C_B) of weak bases and the dissociation constants of their conjugate acids. It follows that for a weak base to be useful for setting up an H_0 scale it must have an electronic absorption spectrum measurably different from that of its conjugate acid. Also the extinction coefficient of either B or BH^+ at the wavelength selected for study must be such that the concentration of indicator necessary for measurable absorption changes has negligible effect on the overall acidity of the medium.

It is important to prove that the indicator colour changes refer to a simple protonation of B to BH^+. Hammett and Deyrup (1933) showed that a good criterion of the mode of ionization of an indicator in acid solution was the van't Hoff factor, *i*, deduced from measurement of freezing point depression for solutions of the indicators in 100% sulphuric acid. They used a series of aniline derivatives as bases and found that $i =$ ca. 2. Thus, for example, $i = 2{\cdot}08$ for a 0·064 *m* solution of 2-nitroaniline in 100% H_2SO_4. This is in accord with the desired mode of ionization, equilibrium (1.30). Certain aromatic azo

$$\text{2-}NO_2C_6H_4NH_2 + H_2SO_4 \rightleftharpoons \text{2-}NO_2C_6H_4NH_3^+ + HSO_4^- \qquad (1.30)$$

compounds, ketones and carboxylic acids also gave van't Hoff factors of 2 (Hammett and Deyrup, 1933; Flexser *et al.* 1935; Treffers and Hammett, 1937) suggesting a simple protonation was occurring. However although a series of compounds of related structure (in particular with respect to the group functioning as the basic site) may give the required *i*-factor it is somewhat dangerous to assume that all compounds of related structure will behave

TABLE 1.4

van't Hoff *i*-factors for neutral solute molecules in 100% H_2SO_4. Values are quoted as the nearest whole number to the average of several determinations with different indicator concentrations

Solute	*i*	Reference†
Benzoic acid	2	a, b
2-Methylbenzoic acid	2	a
2,4,5-Trimethylbenzoic acid	2	a
3,4,5-Trimethoxybenzoic acid	2	a
2,6-Dibromo-3,4,5-trimethoxybenzoic acid	2	a
2-Methyl-6-nitrobenzoic acid	2	a
2,4-Dimethylbenzoic acid	2	a
2-Methoxybenzoic acid	4	c
3-Methoxybenzoic acid	4	c
4-Methoxybenzoic acid	4	c
2-Benzoylbenzoic acid	4	b
2,4,6-Trimethylbenzoic acid	4	a
2,3,5,6-Tetramethylbenzoic acid	4	c
Ethyl benzoate	2	b
Methyl 2,4,6-trimethylbenzoate	5	b
Normal methyl 2-benzoylbenzoate	2	b
Pseudo methyl 2-benzoylbenzoate	5	b

† References (a) Treffers and Hammett (1937); (b) Newman *et al.* (1945); (c) Newman and Deno (1951).

similarly. This is exemplified by the values given in Table 1.4 of i found for a series of carboxylic acids and their esters (Treffers and Hammett, 1937; Newman *et al.* 1945; Newman and Deno, 1951).

For the compounds for which $i = 2$ equation (1.31) is applicable and conforms to the basic requirement for a Hammett indictor (R = alkyl or H). When

$$RCOOR + H_2SO_4 \rightleftharpoons RCO_2HR^+ + HSO_4^- \quad (1.31)$$

$i = 4$ a rather different mode of ionization occurs. This is given in equation (1.32) and involves the formation of a carbonium ion by abstraction of OH^- from the neutral solute,

$$RCOOH + 2H_2SO_4 \rightleftharpoons RCO^+ + H_3O^+ + 2HSO_4^- \quad (1.32)$$

The two compounds for which $i = 5$ ionize according to equilibrium (1.33) which is written for methyl 2,4,6-trimethylbenzoate. Clearly any neutral solute species which ionizes in concentrated acid solution according to equilibria (1.32) or (1.33) is not a Hammett base and may not be used as an indicator for definition of H_0 acidity functions.

$$\text{2,4,6-}Me_3C_6H_2COOMe + 3H_2SO_4 \rightleftharpoons \text{2,4,6-}Me_3C_6H_2\overset{+}{C}O + MeHSO_4 + H_3O^+ + 2HSO_4^- \quad (1.33)$$

Some bases, for example 2,6-dimethylbenzoic acid and 3,5-dibromo-2,4,6-trimethylbenzoic acid give freezing point depressions in accord with i-factors between 2 and 4 and not apparently identical with either. This result implies that these solutes are in part ionizing by (1.32) and again means they are unsuitable for H_0 determination. Of course it does not necessarily follow that a solute will ionize in 100% H_2SO_4 in the same way as it does in say aqueous H_2SO_4 or aqueous $HClO_4$. However, the assumption that the mode of ionization is the same despite medium changes of this kind is usually made and seems to be generally applicable.

The evaluation from optical density values at a single wavelength of (C_{BH^+}/C_B) for the base B in a series of acid solutions is straightforward providing a set of spectra are obtained which do not show medium effects. Equation (1.34) is applicable where for a fixed added concentration of B D_B is the optical density in a dilute acid solution, D_{BH^+} is that in a very concentrated acid

$$(C_{BH^+}/C_B) = (D - D_B)/(D_{BH^+} - D) \quad (1.34)$$

solution, and D refers to some intermediate acidity. The presence of isosbestic points is a good test for the reliability of the spectra for the direct calculation of

(C_{BH^+}/C_B) in this way. Unfortunately for concentrated acid solutions differing widely in acidity there are also significant changes in the solvent environment around any solute species. It follows that the spectra quite frequently show serious wavelength shifts with changing solvent composition. This makes the evaluation of (C_{BH^+}/C_B) more difficult and less reliable. Several methods of approaching this problem have been suggested.

Flexser *et al.* (1935) measured the spectra of acetophenone in water and in a series of aqueous sulphuric acid solutions. For sulphuric acid concentrations in the range $0 < [H_2SO_4] < 55\%$ the absorption due to undissociated acetophenone showed a lateral shift in position but a negligible change in intensity. Similarly for $85{\cdot}96\% < [H_2SO_4] < 95{\cdot}99\%$ the spectrum of the conjugate acid of acetophenone showed lateral shifts but no intensity changes. At intermediate concentrations ($55\% < [H_2SO_4] < 85{\cdot}96\%$) the spectra showed large intensity changes characteristic of the protonation equilibrium. However superimposed on these were the lateral shifts due to medium effects. There were consequently no isosbestic points present. It was suggested that as the medium effects only caused a shift in the position and not in the intensity of the absorption maxima they could be corrected for by shifting the curves so that they did all pass through a common point. This was chosen as the intersection point of the two curves which most nearly corresponded to 50% protonation of acetophenone. This requires that the magnitude of the shift caused by the medium effect is independent of wavelength. This is probably reasonable if the wavelength at which optical densities are taken to calculate (C_{BH^+}/C_B) is not very far removed from that of the isosbestic point. Equation (1.34) is now applicable. This method has been widely used and gives results which compare favourably with other methods (for example see Flexser *et al.* 1935; Noyce and Jorgenson, 1962).

Gold and Hawes (1951) used a means of calculation of ionization ratios which required measurement of optical densities at two wavelengths λ_1 and λ_2. They considered first the situation where only one of the indicator forms (B or BH^+) absorbed at λ_1 and λ_2 and the extinction coefficients of the absorbing form were $\epsilon_1{}^0$ and $\epsilon_2{}^0$. If for an equilibrium mixture the observed extinctions are ϵ_1 and ϵ_2 it follows that in the absence of medium effects the fraction α of the absorbing form of the indicator present in solution will be given by equation (1.35). However if medium effects cause wavelength shifts the magnitude of

$$\alpha = (\epsilon_1/\epsilon_1{}^0) = (\epsilon_2/\epsilon_2{}^0) \tag{1.35}$$

the errors in extinction will depend on the rate of change of ϵ^0 with λ $(\partial\epsilon^0/\partial\lambda)$ at each wavelength. The wavelengths λ_1 and λ_2 are chosen such that equation (1.36) is applicable.

$$\frac{1}{\epsilon_1{}^0}\left(\frac{\partial\epsilon_1{}^0}{\partial\lambda}\right) = -\frac{1}{\epsilon_2{}^0}\left(\frac{\partial\epsilon_2{}^0}{\partial\lambda}\right) \tag{1.36}$$

Combination of the measurements at the two wavelengths allows calculation of α via equation (1.37). This method effectively cancels the opposing errors

$$\alpha = \frac{1}{2}\left(\frac{\epsilon_1}{\epsilon_1{}^0} + \frac{\epsilon_2}{\epsilon_2{}^0}\right) \tag{1.37}$$

due to medium effects at the two wavelengths. However the theory assumes that $(\partial\epsilon/\partial\lambda)$ is constant over a wavelength range comparable with the wavelength shifts caused by medium effects. The method is probably only reliable for small solvent effects. Equation (1.38) is the relevant equation for calculation of ionization ratios when both B and BH^+ absorb appreciably at the two wavelengths.

$$\frac{C_{BH^+}}{C_B} = \frac{1}{2}\left\{\left(\frac{\epsilon - \epsilon_B}{\epsilon_{BH^+} - \epsilon}\right)_{\lambda_1} + \left(\frac{\epsilon - \epsilon_B}{\epsilon_{BH^+} - \epsilon}\right)_{\lambda_2}\right\} \tag{1.38}$$

As before λ_1 and λ_2 are chosen such that the solvent effect causes an increase in apparent extinction coefficient at one wavelength and a decrease at the other. The fundamental idea behind the calculation is to exactly cancel the errors in the apparent ionization ratios which would be given by equation (1.34).

An extension of this approach was used by Alder *et al.* (1966) who calculated (C_{BH^+}/C_B) via equation (1.34) for five or six wavelengths symmetrically placed across a convenient absorption maximum, either due to B or BH^+. The mean ionization ratio is given by equation (1.39) where a series of n wavelengths

$$\frac{C_{BH^+}}{C_B} = \frac{1}{n}\sum_{j=1\to n}\left(\frac{\epsilon - \epsilon_B}{\epsilon_{BH^+} - \epsilon}\right)_{\lambda_j} \tag{1.39}$$

$\lambda_1, \lambda_2, \ldots \lambda_j, \ldots \lambda_n$ are studied. It may be noted that equation (1.38) is the particular form of equation (1.39) when $n = 2$. The variance V_1 (the square of the standard deviation) of (C_{BH^+}/C_B) was estimated together with a second variance V_2 estimated from many repeated determinations of the separate extinction coefficients of the unionized base B and its conjugate acid BH^+. In a least squares plot of the calculated (C_{BH^+}/C_B) values against the logarithm of the strong acid concentration the weight of each point was taken as $1/(V_1 + V_2)$. This method has the considerable advantage that it takes into account the progressively decreasing reliability of the calculated ionization ratios as their values become more removed from one. It is applicable either to well behaved sets of spectra showing isosbestic points or to cases where medium effects are significant.

The evaluation of pK_{BH^+} follows directly from knowledge of (C_{BH^+}/C_B) as a function of acid concentration. From the definition of pK_{BH^+} in equation (1.28) it follows for infinite dilution when all activity coefficients equal one that

$$pK_{BH^+} = \log_{10}\left(\frac{C_{BH^+}}{C_B}\right) - \log_{10} C_{H^+} \tag{1.40}$$

equation (1.40) is valid. Thus extrapolation to infinite dilution of a plot of this function against acid concentration should lead to a value for pK_{BH^+}. Furthermore K_{BH^+} will be the correct thermodynamic dissociation constant referred to pure solvent as standard state. Providing the acid concentration is not too high

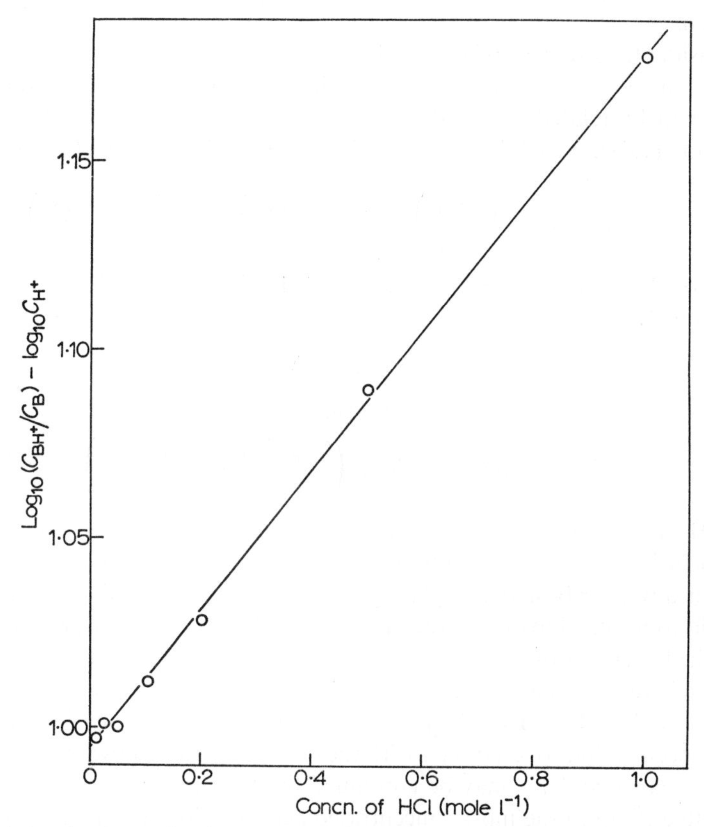

FIG. 1.2. The determination of pK_{BH^+} for 4-nitroaniline in water by extrapolation of equation (1.40) to infinite dilution. Data from Paul (1954).

(< ca. 2 M) the plots are generally linear and therefore the extrapolation is reliable. Paul (1954) measured ionization ratios for 4-nitroaniline in a series of aqueous hydrochloric acid solutions and these data are plotted as an example in Fig. 1.2. The accuracy of the results and the linearity of the curve enabled $pK_{BH^+}(25°C) = 0{\cdot}99$ to be deduced with confidence. This clearly demonstrates the reliability of the extrapolation procedure at low acid concentrations.

For weaker bases than 4-nitroaniline the protonation equilibrium must be studied at higher acid concentrations. As the acid concentration is increased above ca. 2 mole litre^{-1} it is usually found that the function $\log_{10}(C_{BH^+}/C_B) - \log_{10} C_{H^+}$ ceases to vary linearly with acid concentration. The curvature of the plots coupled with the increased distance over which extrapolation must be made makes the above method progressively less satisfactory for a series of bases of decreasing strength. A stepwise comparison procedure is therefore adopted.

Consider the situation where two bases B_1 and B_2 are protonated to measurable extents in the same concentrated acid solution. The observed ionization ratios will be related to the dissociation constants of B_1H^+ and B_2H^+ by equation (1.41). Providing the two bases have similar structure it is not

$$pK_{B_1H^+} - pK_{B_2H^+} = \log_{10}\left(\frac{C_{B_1H^+}}{C_{B_1}}\right) - \log_{10}\left(\frac{C_{B_2H^+}}{C_{B_2}}\right) + \log_{10}\left(\frac{y_{B_1H^+}\, y_{B_2}}{y_{B_1}\, y_{B_2H^+}}\right) \quad (1.41)$$

unreasonable to assume the approximation (1.42) and therefore the validity of equation (1.43).

$$\log_{10}\left(\frac{y_{B_1H^+}\, y_{B_2}}{y_{B_1}\, y_{B_2H^+}}\right) \approx 0 \quad (1.42)$$

$$pK_{B_1H^+} - pK_{B_2H^+} = \log_{10}\left(\frac{C_{B_1H^+}}{C_{B_1}}\right) - \log_{10}\left(\frac{C_{B_2H^+}}{C_{B_2}}\right) \quad (1.43)$$

If B_1H^+ is a comparatively strong base then $pK_{B_1H^+}$ can be measured by extrapolation of equation (1.40) to zero acid concentration. Suppose now that B_2H^+ is a weaker base than B_1H^+ ($pK_{B_2H^+} pK_{B_1H^+}$) and that the higher end of the acid concentration range over which ($C_{B_1H^+}/C_{B_1}$) can be measured coincides with the lower end of the range over which ($C_{B_2H^+}/C_{B_2}$) is measurable. In the region of overlap equation (1.43) should be applicable and allows evaluation of $pK_{B_2H^+}$. A third base B_3 ($pK_{B_3H^+} < pK_{B_2H^+}$) for which protonation occurs at acid concentrations overlapping with those for B_2H^+ may now be studied and $pK_{B_3H^+}$ evaluated. This may be continued for a series of bases of decreasing strength and up to the most concentrated acid solutions available. Providing the fundamental assumption (1.42) is valid thermodynamic dissociation constants referred to pure solvent standard state are obtained.

Hammett and Deyrup (1932) in their original acidity function study measured the ionization of five primary amine bases in aqueous perchloric acid. The results are plotted in Fig. 1.3. Where two curves for successive indicators overlap they are parallel. This confirms the applicability of equations (1.42) and (1.43) for the particular set of aniline bases used. With pK_{BH^+}(4-nitroaniline) = 0·99 (Fig. 1.2) the stepwise deduction of $pK_{BH^+}(25°)$ for the other bases gave –0·31 for 2-nitroaniline, –1·03 for 4-chloro-2-nitroaniline, –3·30 for 2,4-dichloro-6-nitroaniline and –4·55 for 2,4-dinitroaniline. More recent measurements by Yates and Wai (1964) using eleven primary amine indicators give an excellent illustration of parallel variation of $\log_{10}(C_{BH^+}/C_B)$ for structurally similar bases in overlapping ranges of $HClO_4$ concentration.

When (C_{BH^+}/C_B) have been measured for a set of bases in a series of acid solutions and pK_{BH^+} has been deduced for each indicator the defined Hammett acidity function is evaluated by substitution in equation (1.27). The H_0 values deduced from Hammett and Deyrup's (1932) results for aqueous perchloric

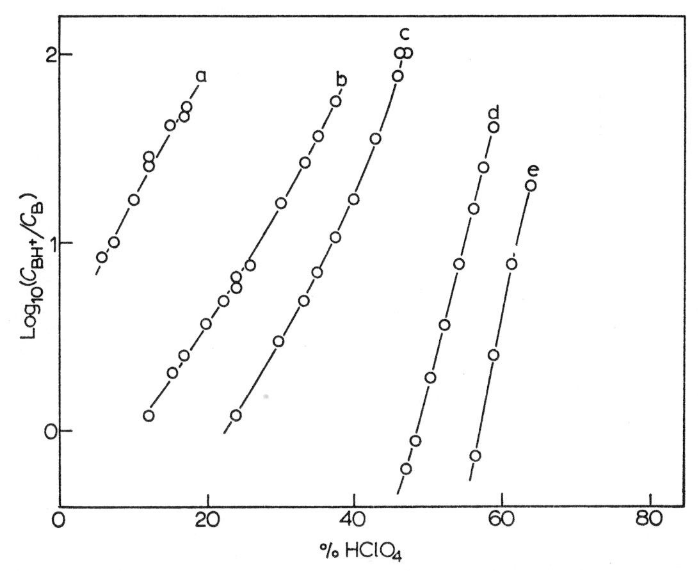

FIG. 1.3. Ionization ratios for (a) 4-nitroaniline, (b) 2-nitroaniline, (c) 4-chloro-2-nitroaniline, (d) 2,4-dichloro-6-nitroaniline, (e) 2,4-dinitroaniline in aqueous perchloric acid. Data from Hammett and Deyrup (1932).

acid at 25° are shown in Fig. 1.4. The smooth curve drawn through the five sets of points defines the H_0 acidity function. It must be noted that the scale differs by −0·41 units from that of Hammett and Deyrup because they took pK_{BH^+} (4-nitroaniline) = 1·40, a value which was later proved wrong.

The deduction of H_0 as described above depends largely upon the necessity that successive indicator curves should be parallel in the region where they overlap. Bonner and Phillips (1966) have studied the protonation of benzophenone and twelve of its derivatives in aqueous sulphuric acid from 40% to 90% H_2SO_4. Plots of $\log_{10}(C_{BH^+}/C_B)$ against % H_2SO_4 were linear within experimental error. The slopes $\{d\log_{10}(C_{BH^+}/C_B)/d(\%H_2SO_4)\}$ are given in Table 1.5 together with range of sulphuric acid concentrations for which each indicator was studied. The bases are listed in the order of decreasing strength. These results show that quite large deviations in slope can occur even for bases of related structure which are protonated in solutions of similar acidity. In particular 3-chloro-4-methoxybenzophenone ($pK_{BH^+} = -5{\cdot}63$) and benzophenone ($pK_{BH^+} = -5{\cdot}70$) which were both protonated in 65–80% H_2SO_4 gave slopes differing by ca. 23%. This suggests that in this case equation (1.42) is becoming rather more approximate than is satisfactory for definition of an H_0 scale. When bases of dissimilar structure are compared large differences in

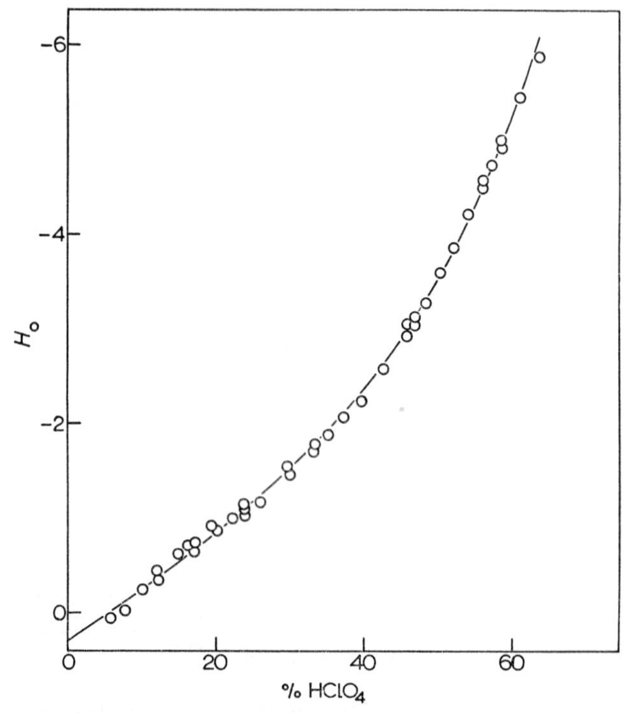

FIG. 1.4. The Hammett acidity function H_0 for aqueous perchloric acid at 25°C deduced from measurements for five primary amine indicators (Hammett and Deyrup, 1932).

slope sometimes occur. Neither equation (1.42) nor equation (1.43) are then applicable. This will be discussed in detail later (Chapter 3).

The stepwise deduction of pK_{BH^+} and hence H_0 requires comparison of $\log_{10}(C_{BH^+}/C_B)$ at low values for one base with $\log_{10}(C_{BH^+}/C_B)$ at high values for another base. This makes the method dependent on the least reliable (C_{BH^+}/C_B) data. Alder *et al.* (1966) have therefore suggested an improved way of interpreting the results which they applied to measurements of the protonation of nine bases by seven different acids in sulpholane solvent. Linear least-square plots of $\log_{10}(C_{BH^+}/C_B)$ against $\log_{10} C_A$ (where C_A = acid concentration) were calculated such that each point was given a weight appropriate to

TABLE 1.5

Rate of change of $\log_{10}(C_{BH^+}/C_B)$ with sulphuric acid concentration for a series of substituted benzophenones (Bonner and Phillips, 1966)

Substituent	% H_2SO_4	$\dfrac{d\log_{10}(C_{BH^+}/C_B)}{d(\%\,H_2SO_4)}$	$-pK_{BH^+}$
2,2′,4,4′-Tetramethoxy	40–55	0·111	3·34
2,4,4′-Trimethoxy	45–60	0·116	3·60
4,4′-Dimethoxy	50–65	0·088	4·41
4-Hydroxy-3,3′,4′-trimethoxy	50–65	0·089	4·50
2,4-Dihydroxy	55–70	0·082	4·82
4-Methoxy	60–75	0·104	4·93
4-Hydroxy	60–75	0·094	5·03
4-Chloro-4′-methoxy	60–75	0·092	5·25
3-Chloro-4-methoxy	65–80	0·109	5·63
Unsubstituted	65–80	0·086	5·70
4-Chloro	70–85	0·105	6·16
3-Chloro	70–85	0·114	6·27
4,4′-Dichloro	75–90	0·115	6·46

its accuracy. Providing the regression lines for different bases had similar slopes the weighted points were then fitted to a series of parallel lines. From knowledge of pK_{BH^+} for one indicator the other dissociation constants were therefore calculable. This method is only applicable when $\log_{10}(C_{BH^+}/C_B)$ varies linearly with some function of the acid concentration. For each system studied by Alder *et al.* the plots of $\log_{10}(C_{BH^+}/C_B)$ against $\log_{10} C_A$ were linear and parallel for all the bases. It follows that the H_0 scale deduced for each acid was also linear in $\log_{10} C_A$.

A detailed description of the calculation of H_0 by a multiple regression technique has been given by Mörikofer *et al.* (1959). Their method does not require $\log_{10}(C_{BH^+}/C_B)$ and $\log_{10} C_A$ to be linearly related and therefore should be more generally applicable.

REFERENCES

Alder, R. W., Chalkley, G. R., and Whiting, M. C. (1966). *Chem. Comm.* 405.
Bates, R. G. (1962). *J. Res. Natl. Bur. Std.* **66A**, 179.
Bates, R. G., and Bower, V. E. (1956). *Anal. Chem.* **28**, 1322.
Bates, R. G., and Guggenheim, E. A. (1960). *Pure Appl. Chem.* **1**, 163.
Bates, R. G., Pinching, G. D., and Smith, E. R. (1950). *J. Res. Natl. Bur. Std.* **45**, 418.
Bonner, T. G., and Phillips, J. (1966). *J. Chem. Soc. B.* 650.
Brønsted, J. N. (1923). *Rec. Trav. Chim.* **42**, 719.
Brønsted, J. N. (1926). *J. Phys. Chem.* **30**, 777.
Flexser, L. A., Hammett, L. P., and Dingwall, A. (1935). *J. Amer. Chem. Soc.* **57**, 2103.

Gold, V., and Hawes, B. W. V. (1951). *J. Chem. Soc.* 2102.
Guggenheim, E. A. (1935). *Phil. Mag.* **19**, 588.
Hammett, L. P., and Deyrup, A. J. (1932). *J. Amer. Chem. Soc.* **54**, 2721.
Hammett, L. P., and Deyrup, A. J. (1933). *J. Amer. Chem. Soc.* **55**, 1901.
Hammett, L. P., Dingwall, A., and Flexser, L. (1934). *J. Amer. Chem. Soc.* **56**, 2010.
Hitchcock, D. I., and Taylor, A. C. (1938). *J. Amer. Chem. Soc.* **60**, 2710.
MacInnes, D. A., Belcher, D., and Shedlovsky, T. (1938). *J. Amer. Chem. Soc.* **60**, 1094.
Mörikofer, A., Simon, W., and Heilbronner, E. (1959). *Helv. Chim. Acta*, **42**, 1737.
Newman, M. S., and Deno, N. C. (1951). *J. Amer. Chem. Soc.* **73**, 3651.
Newman, M. S., Kuivila, H. G., and Garrett, A. B. (1945). *J. Amer. Chem. Soc.* **67**, 704.
Noyce, D. S., and Jorgenson, M. J. (1962). *J. Amer. Chem. Soc.* **84**, 4312.
Paul, M. A. (1954). *J. Amer. Chem. Soc.* **76**, 3236.
Robinson, R. A. (1959). "The Structure of Electrolytic Solutions" (W. J. Hamer, ed.), p. 253. Wiley, New York.
Robinson, R. A., and Biggs, A. I. (1955). *Trans. Faraday Soc.* **51**, 901.
Robinson, R. A., and Biggs, A. I. (1957). *Aust. J. Chem.* **10**, 128.
Rochester, C. H. (1965). *J. Chem. Soc.* 4603. (Spectra previously unpublished.)
Sørensen, S. P. L. (1909). *Biochem. Z.* **21**, 131, 201.
Treffers, H. P., and Hammett, L. P. (1937). *J. Amer. Chem. Soc.* **59**, 1708.
Yates, K., and Wai, H. (1964). *J. Amer. Chem. Soc.* **76**, 5408.

CHAPTER 2

The Hammett Acidity Function for Aqueous Solutions of Strong and Weak Acids

2.1. The Hammett Acidity Function

The original hope of the acidity function concept was that there might be a unique H_0 acidity function for a particular series of solutions of changing acidity. For example, for aqueous perchloric acid it was thought that a single H_0 scale might exist which could be expressed as a function of $HClO_4$ concentration. The assumption inherent in this idea is that having measured H_0 with a set of weak bases of different strengths then the extent of ionization of any other weak base in the acid solutions should be in accord with equation (1.27). A plot of $\log(C_{BH^+}/C_B)$ against H_0 should be linear with slope $-1{\cdot}00$ for all neutral bases. If this were so the H_0 scale would be only dependent on the acid solution, and would be completely independent of the structure of the indicators used to measure it. Equation (1.29) shows that for this to occur the value for (y_B/y_{BH^+}) must be the same for all neutral bases in a particular acid solution. Also $d(y_B/y_{BH^+})/d$(acid concentration) must be independent of the base.

That these assumptions are not generally valid is now well known. Groups of indicators with different structures, and especially different basic sites, often show significant deviations in acidity function behaviour. Thus, for example, for the protonation of 11 amides in aqueous sulphuric acid $\log_{10}(C_B/C_{BH^+})$ plotted against an H_0 scale defined using amine bases as indicators gave linear plots with slopes in the range 0·39–0·77 for 10 of the amides (Katritzky *et al.* 1963). The slope of 1·00 required for equation (1.27) to be valid was only given by one of the amide bases. Similarly, and perhaps at first sight a little more surprising, the ionization of primary and tertiary amines in aqueous sulphuric acid conforms to different acidity function scales (Arnett and Mach, 1964). These observations mean that a numerical H_0 scale will only be applicable to the ionization of bases for which $\log_{10}(C_{BH^+}/C_B)$ accurately parallels that particular scale. Furthermore comparisons of the acidity of solutions of different strong acids can only be unambiguously made by comparing the H_0

acidity functions which have been measured for each acid using the same set of base indicators.

In their original measurements of H_0 functions for aqueous perchloric acid and aqueous sulphuric acid solutions Hammett and Deyrup (1932) used fifteen indicators of which seven were primary amines. It has become standard nomenclature to give the name "Hammett acidity function" to any H_0 scale defined from measurements of the ionization of primary amines. Weak bases for which $\log(C_{BH^+}/C_B)$ plotted against the Hammett acidity function gives a linear plot with unit slope are then called Hammett bases. Hammett bases may have any structure and need not necessarily be amines. The criterion is that all Hammett bases conform to the same acidity function behaviour for which primary amines provide the standard. The Hammett acidity function retains the symbol H_0. Acidity functions defined by non-Hammett bases are usually given a different symbol so that there is a clear distinction between them and H_0. In this chapter the Hammett acidity function for aqueous solutions of several strong and weak acids will be presented and discussed. Consideration of the deviations between acidity function behaviour for Hammett and non-Hammett bases will be deferred until later.

2.2. Experimental H_0 Scales for Aqueous Acids

2.2.1. *Sulphuric Acid* (*and Oleum*)

The results of four separate sets of measurements of the H_0 scale for aqueous sulphuric acid solutions are listed in Tables 2.1–2.4. The variations of H_0 with sulphuric acid concentration are compared in Fig. 2.1. The acidity function (Table 2.1) quoted by Paul and Long (1957) is based on the pioneering measurements made by Hammett and Deyrup (1932). The original scale given by Hammett and Deyrup differs from the present one by a constant amount (0·41 units) in accord with the difference between $pK_{BH^+} = 1{\cdot}40$ for 4-nitroaniline used by them and the more correct $pK_{BH^+} = 0{\cdot}99$ which has been measured by the extrapolation to infinite dilution method. Figure 2.1 shows that Hammett and Deyrup's H_0 scale was remarkably accurate up to 60% by weight sulphuric acid despite the fact that that had to use a colorimetry technique. The highly accurate spectrophotometers of modern times were not then available. However above 60% H_2SO_4 their scale shows large differences (up to 1·1 units) from the more recent results of Jorgenson and Hartter (1963) and Ryabova *et al.* (1966). Apart from the two most weakly basic indicators used by Ryabova *et al.* the acidity functions given in Tables 2.3 and 2.4 were measured using primary amines as indicators. Hammett and Deyrup (1932) used as indicators several other classes of base which have since proved unsuitable for definition of H_0. Thus the ketones benzalacetophenone, β-benzoylnaphthalene, 4-benzoyl biphenyl, and anthraquinone all probably conform to a different acidity function dependence than do primary amines (Bonner and Phillips, 1966). The tertiary amine *N,N*-dimethyl-2,4,6-trinitroaniline is also not a Hammett base (Arnett and Mach, 1964). Bascombe and Bell (1959) found that the spectra of 4-nitrodiphenylamine and its conjugate

TABLE 2.1

The Hammett acidity function H_0 for aqueous sulphuric acid at 25°C (Paul and Long, 1957)

wt % H_2SO_4	$-H_0$	wt % H_2SO_4	$-H_0$	wt % H_2SO_4	$-H_0$
10	0·31	55	3·91	96	8·98
15	0·66	60	4·46	97	9·14
20	1·01	65	5·04	98	9·36
25	1·37	70	5·65	99	9·74
30	1·72	75	6·30	99·3	9·89
35	2·06	80	6·97	99·5	10·03
40	2·41	85	7·66	99·7	10·22
45	2·85	90	8·27	99·8	10·36
50	3·38	95	8·86	99·9	10·59
				100·0	11·10

This scale is based on the measurements made by Hammett and Deyrup (1932) and is unreliable for > 60% H_2SO_4 (see text). Indicators used were (pK_{BH^+} in brackets): 4-nitroaniline (0·99); 2-nitroaniline (–0·29); 4-chloro-2-nitroaniline (–1·03); 4-nitrodiphenylamine (–2·48); 2,4-dichloro-6-nitroaniline (–3·32); 4-nitroazobenzene (–3·47); 2,6-dinitro-4-methylaniline (–4·44); 2,4-dinitroaniline (–4·53); *N,N*-dimethyl-2,4,6-trinitroaniline (–4·81); benzalacetophenone (–5·73); *β*-benzoylnaphthalene (–6·04); 4-benzoylbiphenyl (–6·31); 6-bromo-2,4-dinitroaniline (–6·71); anthraquinone (–8·27); 2,4,6-trinitroaniline (–9·41). The pK_{BH^+} data are the "best" values quoted by Paul and Long.

acid in aqueous sulphuric acid were subject to medium effects which made 4-nitrodiphenylamine unreliable as an indicator for H_0 determination.

The H_0 acidity function given in Table 2.4 covers a range of sulphuric acid concentrations from 1–100% and appears to be a reliable scale. The H_0 data were in part based on earlier work by Vinnik and Ryabova (1964). The values given in 1964 had to be adjusted by a constant factor of 0·49 units because an inaccurate H_0 scale had been used to deduce pK_{BH^+} for 2,4,6-trinitroaniline. For 10–40% sulphuric acid Bascombe and Bell's (1959) results agree very closely with those of Ryabova *et al.* Below 10% sulphuric acid there is a small discrepancy (up to 0·15 units) between the two scales. At low sulphuric acid concentrations Bascombe and Bell's H_0 function is based on measurements with three indicators for which pK_{BH^+} were deduced independently by extrapolation to infinite dilution. The data in Table 2.2 are the most detailed and probably the most reliable for low percentages of sulphuric acid.

TABLE 2.2

The Hammett acidity function H_0 for aqueous sulphuric acid at 25°C (Bascombe and Bell, 1959)

wt % H_2SO_4	H_0	wt % H_2SO_4	H_0	wt % H_2SO_4	H_0
0·0049	3·03	0·612	1·06	17·8	−0·90
0·0057	2·94	2·26	0·48	19·1	−1·08
0·0058	2·93	2·67	0·41	22·2	−1·21
0·0066	2·82	2·69	0·40	22·6	−1·25
0·0096	2·77	3·39	0·28	24·6	−1·38
0·0116	2·67	5·10	0·09	25·1	−1·41
0·0131	2·51	8·3	−0·23	28·7	−1·69
0·0234	2·35	9·1	−0·27	33·6	−2·03
0·0248	2·33	10·75	−0·42	33·8	−2·16
0·0401	2·14	10·9	−0·44	35·2	−2·15
0·0425	2·12	11·1	−0·45	35·6	−2·22
0·0470	2·08	13·4	−0·62	37·1	−2·37
0·0590	1·99	13·8	−0·64	38·6	−2·45
0·0602	1·98	14·1	−0·68		
0·0734	1·90	14·2	−0·66		
0·0890	1·83	14·75	−0·71		
0·136	1·64	16·3	−0·83		
0·248	1·43	16·4	−0·83		
0·325	1·32	16·8	−0·88		
0·400	1·22	17·1	−0·86		

Indicators used to determine this scale were (pK_{BH^+} in brackets): 4-amino-azobenzene (2·82); 3-nitroaniline (2·50); 4-nitroaniline (1·02); 2-nitroaniline (−0·29); *N*,*N*-dimethyl-2,4-dinitroaniline (−1·00); 4-chloro-2-nitroaniline (−1·02). For the first three pK_{BH^+} was measured directly by extrapolation to infinite dilution, and for the remainder the stepwise comparison method was used.

Above 60% sulphuric acid the Paul and Long (1957) H_0 scale may be disregarded. The acidity functions measured by Jorgenson and Hartter (1963) and by Ryabova *et al.* (1966) are however in fairly good agreement at least up to 98% H_2SO_4 where they deviate by only 0·14 units. Above 98% H_2SO_4 the deviation becomes greater and is 0·32 units for 99·44% sulphuric acid. The reason for this is not obvious from a comparion of the measured ionization constants of the weak bases used in the setting up of the two scales. Thus in general the pK_{BH^+} values agree to within 0·12 units, a magnitude of error which is acceptable in view of the possible cumulative addition of errors which can occur during the stepwise deduction of dissociation constants. The only notable exception is 6-bromo-2,4-dinitroaniline for which a discrepancy $pK_{BH^+} = 0{\cdot}41$ exists. Ryabova *et al.* (1966) found that the ionization ratio C_{BH^+}/C_B for this indicator was a function of the wavelength chosen for study. They evaluated C_{BH^+}/C_B from areas under the spectral curves and concluded

TABLE 2.3

The Hammett acidity function H_0 for aqueous sulphuric acid (Jorgenson and Hartter, 1963). This scale is for an unspecified ambient laboratory temperature.

wt % H_2SO_4	$-H_0$	wt % H_2SO_4	$-H_0$	wt % H_2SO_4	$-H_0$
60	4·46	78	7·03	94	9·68
62	4·70	80	7·34	95	9·85
64	4·95	82	7·66	96	10·03
66	5·20	84	7·97	97	10·21
68	5·50	85	8·14	98	10·41
70	5·80	86	8·29	98·82	10·62
72	6·10	88	8·61	99·27	10·92
74	6·41	90	8·92	99·44	11·12
76	6·71	92	9·29		

Indicators used were (pK_{BH^+} in brackets): 2,4-dinitroaniline (–4·53); 2,6-dinitroaniline (–5·54); 4-chloro-2,6-dinitroaniline (–6·14); 2-bromo-4,6-dinitroaniline (–6·68); 3-methyl-2,4,6-trinitroaniline (–8·22); 3-bromo-2,4,6-trinitroaniline (–9·46); 3-chloro-2,4,6-trinitroaniline (–9·71); 2,4,6-trinitroaniline (–10·10). Reference to water standard state was made via measurements with: 2-nitroaniline (–0·29); 4-chloro-2-nitroaniline (–1·03); 2,5-dichloro-4-nitroaniline (–1·78); 2-chloro-6-nitroaniline (–2·43); 2,6-dichloro-4-nitroaniline (–3·27).

that the results were therefore probably more reliable than those of Jorgenson and Hartter (1963). However the figure of –6·68 for pK_{BH^+} of 6-bromo-2,4-dinitroaniline (Jorgenson and Hartter, 1963) is in much closer agreement with $pK_{BH^+}(17°C) = -6{\cdot}59$ deduced by Milyaeva (1958) and by Gel'bshtein *et al.* (1956a, 1956b). A medium effect was also reported by Ryabova *et al.* for 2,4-dichloro-6-nitroaniline and for this base (C_{BH^+}/C_B) was deduced from measured areas under absorption curves.

Ryabova *et al.* (1966) found that $d(C_B/C_{BH^+})/d(H_2SO_4)$ for 3-bromo-2,4,6-trinitroaniline and 3-chloro-2,4,6-trinitroaniline was smaller (more negative) than for 3-methyl-2,4,6-trinitroaniline and 2,4,6-trinitroaniline. The former two bases were not therefore used to establish the acidity function given in Table 2.4. Jorgenson and Hartter (1963) used all four amines to define their scale (Table 2.3) and no discrepancy between the acidity function behaviour of the two pairs of bases was apparent from their results.

TABLE 2.4

The Hammett acidity function H_0 for aqueous sulphuric acid at 25°C (Ryabova *et al.* 1966)

wt % H_2SO_4	$-H_0$	wt % H_2SO_4	$-H_0$	wt % H_2SO_4	$-H_0$
1	−0·84	42	2·69	87	8·60
2	−0·31	45	2·95	90	9·03
5	0·02	47	3·13	92	9·33
8	0·28	50	3·41	94	9·59
10	0·43	52	3·60	96	9·88
12	0·58	55	3·91	98	10·27
14	0·73	57	4·15	99	10·57
16	0·85	60	4·51	99·1	10·62
18	0·97	62	4·83	99·2	10·66
20	1·10	65	5·18	99·3	10·71
22	1·25	67	5·48	99·4	10·77
25	1·47	70	5·92	99·5	10·83
27	1·61	72	6·23	99·6	10·92
30	1·82	75	6·71	99·7	11·02
32	1·96	77	7·05	99·8	11·18
35	2·19	80	7·52	99·85	11·28
37	2·34	82	7·84	99·90	11·42
40	2·54	85	8·29	99·95	11·64
				100·00	11·94

Indicators used (pK_{BH^+} in brackets): 4-nitroaniline (1·0); 2-nitroaniline (−0·33); 5-chloro-2-nitroaniline (−1·55); 2,5-dichloro-4-nitroaniline (−1·90); 2-chloro-6-nitroaniline (−2·54); 2,4-dichloro-6-nitroaniline (−3·28); 2,4-dinitroaniline (−4·45); 2,6-dinitroaniline (−5·64); 4-chloro-2,6-dinitroaniline (−6·25); 6-bromo-2,4-dinitroaniline (−7·09); 3-methyl-2,4,6-trinitroaniline (−8·34); 2,4,6-trinitroaniline (−9·98); 4-nitrotoluene (−11·27); 4-chloronitrobenzene (−12·63). pK_{BH^+} for 4-nitroaniline was evaluated by extrapolation to low H_2SO_4 concentrations, the remainder were deduced by stepwise comparison.

The variation of H_0 with temperature has been investigated by Gel'bshtein *et al.* (1956a, 1956b) using 4-nitroaniline, 2-nitroaniline, 4-chloro-2-nitroaniline, 2,4-dichloro-6-nitroaniline, 2,4-dinitroaniline, 6-bromo-2,4-dinitroaniline and 2,4,6-trinitroaniline as indicators. However, their scale for 25°C agrees fairly closely with that of Paul and Long (1957) which has been superceded at high H_2SO_4 concentrations by the measurements of Jorgenson and Hartter (1963) and Ryabova *et al.* (1966). Tickle *et al.* (1970) have also determined the variation of H_0 with temperature and their scale for 25°C agrees much more closely with the acidity functions given in Tables 2.2–2.4.

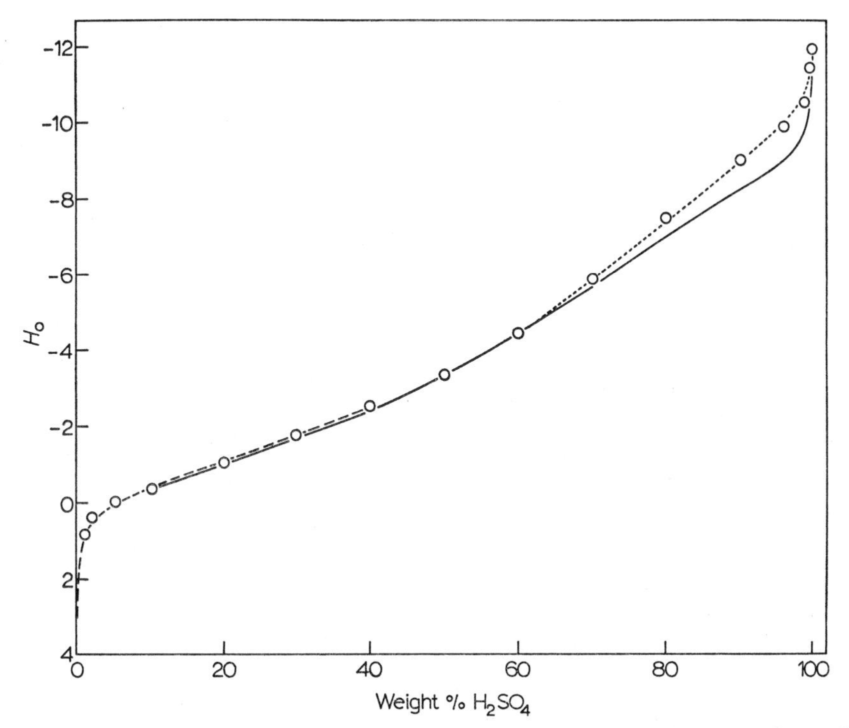

FIG. 2.1. Hammett acidity function for aqueous sulphuric acid. Comparison of measurements by different workers. Full line: Long and Paul (1957), Hammett and Deyrup (1932); Dashed line: Bascombe and Bell (1959); Dotted line: Jorgenson and Hartter (1963); Circles: Ryabova *et al.* (1966).

Table 2.5 contains the H_0 scales for aqueous H_2SO_4 at 15, 25, 35, 45 and 55°C as quoted by Tickle *et al.* (1970). Ten primary amines (aniline derivatives) were used as indicators. The pK_a values for 4-nitroaniline as a function of temperature were in good agreement with the results obtained by Boyd and Wang (1965) for this amine. The variation dH_0/dT of H_0 with temperature for a particular H_2SO_4 concentration was independent of temperature for $15° \leqslant T \leqslant 55°$. This observation is probably applicable for temperatures up to at least 80°C (Gel'bshtein *et al.* 1956a, 1956b). The variation $(dH_0/dC_{H_2SO_4})$ of H_0 with H_2SO_4 concentration becomes less as the temperature is raised.

Thus for $C_{H_2SO_4} > 50\%$ a plot of H_0 (55°) against H_0 (15°) is linear with 0·92 slope. As will be seen later this has importance when we are considering the interpretation of correlations which exist in the literature between equilibrium or rate data at high temperatures and H_0 values for 25°C.

TABLE 2.5

The effect of temperature on the Hammett acidity function H_0 for aqueous sulphuric acid solutions (Tickle *et al.*, 1970).

	$-H_0$				
wt % H_2SO_4	15°	25°	35°	45°	55°
4	−0·22	−0·22	−0·23	−0·23	−0·23
8	0·18	0·18	0·19	0·19	0·19
12	0·45	0·46	0·46	0·46	0·46
16	0·72	0·73	0·73	0·73	0·73
20	0·98	0·99	0·99	0·99	0·99
24	1·24	1·24	1·24	1·24	1·24
28	1·50	1·49	1·49	1·49	1·48
32	1·81	1·79	1·78	1·77	1·75
36	2·11	2·09	2·07	2·05	2·03
40	2·41	2·39	2·37	2·35	2·33
44	2·71	2·69	2·67	2·65	2·63
48	3·02	3·00	2·98	2·96	2·94
52	3·48	3·44	3·40	3·36	3·32
56	3·92	3·87	3·82	3·78	3·72
60	4·40	4·35	4·31	4·26	4·21
64	4·90	4·85	4·79	4·74	4·69
68	5·46	5·39	5·33	5·27	5·21
72	6·05	5·96	5·88	5·80	5·72
76	6·65	6·54	6·43	6·32	6·21
80	7·28	7·17	7·06	6·95	6·84
84	7·91	7·79	7·68	7·57	7·46
88	8·57	8·44	8·31	8·18	8·05
92	9·25	9·11	8·96	8·81	8·66
96	9·99	9·83	9·67	9·51	9·35
98	10·38	10·22	10·06	9·90	9·74

It is relevant at this point to consider whether there is any apparent correlation between the Hammett acidity function and the composition of sulphuric acid/water mixtures. Brand (1950) considered the ionization of a weak base B in concentrated sulphuric acid as

$$B + H_2SO_4 \overset{K}{\rightleftharpoons} BH^+ + HSO_4^- \tag{2.1}$$

and deduced equation (2.2) (in which X represents a mole fraction). Brand

$$H_0 = \mathrm{p}K_a + \mathrm{p}K + \log_{10}\left(\frac{X_{HSO_4^-}}{X_{H_2SO_4}}\right) + \log_{10}\left(\frac{y_{BH^+}y_{HSO_4^-}}{y_B\,y_{H_2SO_4}}\right) \tag{2.2}$$

assumed that in aqueous sulphuric acid for which $C_{H_2SO_4} \gg C_{H_2O}$ the equilibrium (2.3) lies entirely on the HSO_4^-/H_3O^+ side. From the known stoichio-

$$H_2SO_4 + H_2O \rightleftharpoons HSO_4^- + H_3O^+ \tag{2.3}$$

metric concentrations of H_2SO_4 and H_2O he therefore calculated $X_{HSO_4^-}/X_{H_2SO_4}$ and showed that the Hammett and Deyrup (1932) H_0 scale conformed to the empirical equation (2.4) for sulphuric acid concentrations in the range

$$H_0 = \text{const.} + \log_{10}\frac{X_{HSO_4^-}}{X_{H_2SO_4}} \tag{2.4}$$

90–99·8 wt %. Deno and Taft (1954) confirmed and extended this result. They showed that if an equilibrium constant of 50 was assumed for equation (2.3) then equation (2.4) became applicable for sulphuric acid concentrations down to 83%. The constant in equation (2.4) was found to be –8·48 for the revised Hammett and Deyrup H_0 scale given by Paul and Long (1957). It also followed from equilibrium (2.3) and the identity (2.4) that the mole fractions of water and H_3O^+ ions in concentrated sulphuric acid were related to H_0 via equation (2.5). The constants in equation (2.4) and (2.5) differ by $\log_{10} 50$ in accord with the assumed equilibrium constant of 50 for the ionization (2.3). Gel'bshtein

$$H_0 = -6{\cdot}78 + \log_{10}\frac{X_{H_2O}}{X_{H_3O^+}} \tag{2.5}$$

et al. (1956a, 1956b) tested the applicability of equation (2.4) to their H_0 acidity function for aqueous sulphuric acid. They calculated $X_{HSO_4^-}$ and $X_{H_2SO_4}$ according to the method of Brand (1950) and found for 90–98 wt % H_2SO_4 at 20° that equation (2.4) was valid with the constant equal to –8·56. This confirms the results of Brand (1950) and Deno and Taft (1954) and would imply that the term log $(y_{BH^+}y_{HSO_4^-}/y_B y_{H_2SO_4})$ in equation (2.3) is a constant in the appropriate range of water–sulphuric acid concentrations. This is a surprising conclusion and it led Brand (1950) to suggest that the apparent fitting of equation (2.4) might well be fortuitous. The equation is not applicable for the H_0 acidity function at 80° (Gel'bshtein *et al.* 1956b).

The above correlations have been considered for the earlier H_0 scales which more recent measurements have suggested are probably in error, particularly at the high sulphuric acid concentrations relevant to this discussion. Attempts to correlate the H_0 acidity function determined by Jorgenson and Hartter (1963) with $\log_{10}(X_{HSO_4^-}/X_{H_2SO_4})$ were only successful for sulphuric acid concentrations greater than 95%. The calculated constant in equation (2.4) was –9·47, a somewhat different value from that given by the earlier H_0 scales (Paul and Long, 1957; Gel'bshtein *et al.* 1956a, 1956b). Jorgenson and Hartter (1963) found that at sulphuric acid concentrations less than 95%, the acidity function deviated from $\log_{10}(X_{HSO_4^-}/X_{H_2SO_4})$ in a way which could not be accounted for in terms of the Deno and Taft (1954) approach in which a particular value was selected for the equilibrium constant for equation (2.3). Jorgenson and Hartter conclude that the success of the Deno and Taft approach was fortuitous.

Wyatt (1957, 1960) has pointed out that the value of 50 estimated by Deno and Taft for the equilibrium constant for equation (2.3) is not consistent with the observed activities and partial molar heat contents of water in concentrated sulphuric acid solutions (Kunzler and Giaque, 1952; Giaque *et al.* 1960). Wyatt has adopted an approach which considers the existence of the equilibria (2.6) and (2.7) in concentrated (> 80% H_2SO_4) sulphuric acid solutions. The

$$2H_3O^+ + HSO_4^- \rightleftharpoons H_3O^+(H_2O) + H_2SO_4 \tag{2.6}$$

$$H_3O^+ + H_2SO_4 \rightleftharpoons H_3O^+(H_2SO_4) \tag{2.7}$$

formulation $H_3O^+(H_2O)$ represents the hydronium ion solvated by a water molecule and $H_3O^+(H_2SO_4)$ the hydronium ion solvated by a sulphuric acid molecule. The appearance of Raman lines due to molecular sulphuric acid (Young *et al.* 1959; Young and Walrafen, 1961) for aqueous sulphuric acid containing less than 50 mole % is compatible with equation (2.6).

Young and Walrafen (1961) have produced evidence from Raman spectra for the existence of the species $H_5SO_5^+$ (which may be formally written H_3O^+, H_2SO_4) in aqueous sulphuric acid of H_2SO_4 concentration greater than 50 mole %. Wyatt has deduced the following equilibrium constants (25°C) for equations (2.3), (2.6) and (2.7). Calculation has shown that these values of the

$$K(2.3) = \frac{a_{HSO_4^-}\, a_{H_3O^+}}{a_{H_2SO_4}\, a_{H_2O}} = 2550 \tag{2.8}$$

$$K(2.6) = \frac{a_{H_3O^+(H_2O)}\, a_{H_2SO_4}}{a^2_{H_3O^+}\, a_{HSO_4^-}} = 0{\cdot}05 \tag{2.9}$$

$$K(2.7) = \frac{a_{H_3O^+(H_2SO_4)}}{a_{H_3O^+}\, a_{H_2SO_4}} = 1{\cdot}0 \tag{2.10}$$

three constants give a satisfactory account of the observed densities of aqueous sulphuric acid and the activities and partial molar heat contents of water in water–sulphuric acid mixtures (Wyatt, 1960). Also the calculated concentrations (Kaandorp *et al.* 1962) of the various species present in concentrated aqueous sulphuric acid are generally in good agreement with Raman (Young *et al.* 1959; Young and Walrafen, 1961) and p.m.r. (Hood and Reilly, 1957;

Gillespie and White, 1960) data for these solutions. The only significant discrepancy occurs in the region of 50 mole % (84·5 wt %) sulphuric acid. When the stoichiometric molarity ratio of water to sulphuric acid is one, Raman spectra (Young *et al.* 1959) give $[HSO_4^-]/[H_2SO_4]$ is 4·3 whereas the value is 6·8 on Wyatt's treatment. Wyatt (1960) has suggested that the discrepancy might be removed if the species $H_2SO_4.H_2O$ is formed in the solutions. Zarakhani and Vinnik (1963) have found evidence for this species, which may also be formally written $HSO_4^-.H_3O^+$ (Kachurin, 1967), from Raman spectra and freezing point data for aqueous sulphuric acid.

The Hammett acidity function calculated using the equilibrium constants (2.9) and (2.10) and assuming that equilibrium (2.3) lies effectively completely on the right hand side were in agreement with the Paul and Long (1957) H_0 scale for sulphuric acid concentrations in the range 79·7–98·2 wt %. Thus Wyatt's treatment is consistent with the Raman spectra, p.m.r. spectra, thermodynamic properties and H_0 acidity function for mixtures of water and sulphuric acid which contain predominantly (> 50 mole %) sulphuric acid. It would be interesting to calculate what changes in the equilibrium constants (2.9) and (2.10) would be needed to give a satisfactory account of the more recently determined H_0 scales (Jorgenson and Hartter, 1963; Ryabova *et al.* 1966). If Wyatt's ideas are still to be applicable then the new equilibrium constant values must be such that the satisfactory correlations for the other properties of the solutions are maintained.

Equation (2.11) results from combination of the definition of H_0, equation (1.27), and the expression for the equilibrium constant K for equilibrium (2.3).

$$H_0 + \log_{10}\frac{a_{H_2SO_4}}{C_{HSO_4^-}} = \log_{10}\left(\frac{y_{BH^+}\, y_{HSO_4^-}}{y_B}\cdot\frac{1}{K}\right) \tag{2.11}$$

It follows that if H_0 is a linear function of $\log(C_{H_2SO_4}/C_{HSO_4^-})$, as found by Brand (1950), then $(y_{BH^+}y_{HSO_4^-}/y_B y_{H_2SO_4})$ must be a constant. Vinnik and Ryabova (1964) calculated the quantity on the left hand side of equation (2.11) from the Paul and Long (1957) H_0 scale and known sulphuric acid activities $a_{H_2SO_4}$ (Giaque *et al.* 1960). They then suggested that for sulphuric acid solutions of H_2SO_4 content greater than 99 % the total concentration of ion will not be very high and therefore ionic activity coefficients in equation (2.11) might be expected to be consistent with the Debye–Hückel Limiting Law. A graph of $H_0 + \log_{10}(a_{H_2SO_4}/C_{HSO_4^-})$ against $\sqrt{C}$ was linear for solutions containing 99·4–99·8 % sulphuric acid. The concentration C was taken as half of the total concentration of ions in the solutions. The straight line plot fitted equation (2.12).

$$\log\left(\frac{y_{BH^+}\, y_{HSO_4^-}}{y_B}\cdot\frac{1}{K}\right) = -0{\cdot}844 - 0{\cdot}264\sqrt{C} \tag{2.12}$$

The intercept (–0·844) given in this equation is based on the revised H_0 scale given by Ryabova *et al.* (1966). However the slope of the line (–0·264) is not in accord with that (ca. 0·7) excepted on the Debye–Hückel theory for a low concentration of ions in a solvent with dielectric constant of ca. 100 (Brand *et al.* 1952b, 1953; Gillespie and Cole, 1956; Gillespie and White, 1958). This

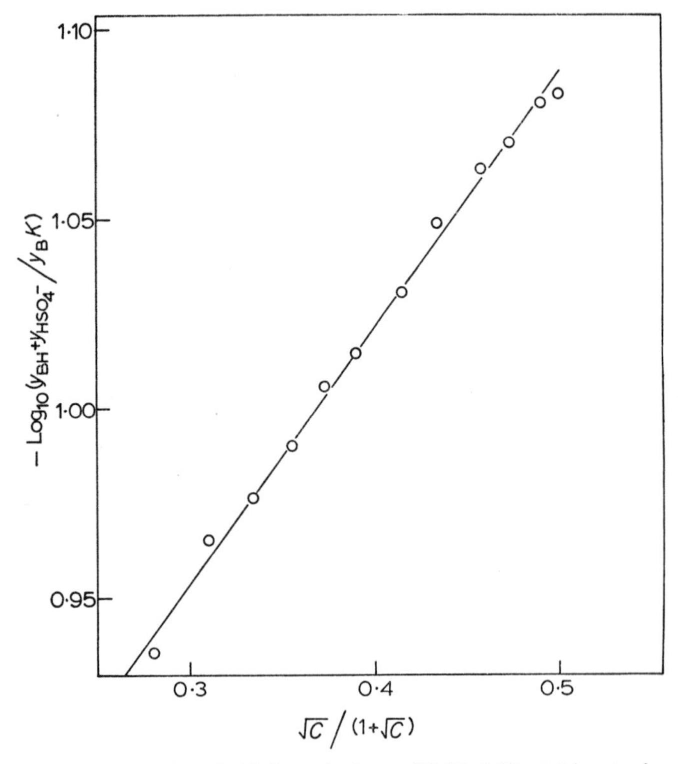

FIG. 2.2. Test of equation (2.13) for solutions of 0·15–1·00 wt % water in sulphuric acid. Data of Vinnik and Ryabova (1964).

probably arises because the ionic strengths of solutions were well in excess of the normally accepted limits for the validity of the Debye–Hückel limiting law. Thus, for example, for solutions in anhydrous sulphuric acid of metal hydrogen sulphates at ionic strengths comparable to those considered here the osmotic coefficients calculated from freezing point data (Bass and Gillespie, 1960) deviated widely from the values predicted by Debye–Hückel Theory (Gillespie 1966).

It is interesting to note that if the $\sqrt{C}$ values of Vinnik and Ryabova are divided by $(1+\sqrt{C})$ then a graph of $H_0+\log_{10}(a_{H_2SO_4}/C_{HSO_4^-})$ against

$\sqrt{C}/(1+\sqrt{C})$ is linear with a slope of $-0{\cdot}68$ (Fig. 2.2). The H_0 values used for this plot were taken from the revised H_0 scale given by Ryabova *et al.* (1966). Equation (2.13) is applicable for sulphuric acid concentrations in the range

$$\log\left(\frac{y_{BH^+}\,y_{HSO_4^-}}{y_B}\cdot\frac{1}{K}\right) = -0{\cdot}749 - \left(\frac{0{\cdot}68\sqrt{C}}{1+\sqrt{C}}\right) \qquad (2.13)$$

99–99·85%. This equation is in much better agreement with Debye–Hückel theory since the slope $-0{\cdot}68$ is very close to the predicted slope of twice the Debye–Hückel A factor for sulphuric acid at 25°. It must be concluded that at the ionic concentrations considered here the Debye–Hückel theory is only applicable in its extended form which takes into account the finite size of the ions (Robinson and Stokes, 1959). The assumption inherent in equation (2.13) of an ion size parameter of ca. 3·5 ångström is probably reasonable. The linearity and slope of the line in Fig. 2.2 suggests that the Hammett acidity function for dilute solutions of water in sulphuric acid is dependent on the variation of the activity coefficients y_{BH^+} and $y_{HSO_4^-}$ with changing ionic strength. Up to quite high ionic strengths ($I \sim 1{\cdot}02$ M in 99% H_2SO_4) these activity coefficients are apparently consistent with equation (2.14) where A is the correct Debye–Hückel factor for sulphuric acid solvent. Since all the ions being considered are univalent this equation is in accord with Debye–Hückel theory.

$$-\log_{10} y_{\pm} = \frac{A\sqrt{C}}{1+\sqrt{C}} \qquad (2.14)$$

The Hammett acidity function has been measured for solutions of sulphur trioxide in sulphuric acid (Brand, 1950; Brand *et al.* 1952a; Vinnik and Ryabova, 1964). Brand *et al.* used a series of mono and dinitro substituted benzene derivatives as indicators. The results of their measurements have been summarized by Paul and Long (1957), and are compared in Table 2.6 with the H_0 scale determined by Vinnik and Ryabova. The latter is based on measurements using 4-nitrotoluene ($pK_{BH^+} = -11{\cdot}27$) and 4-chloronitrobenzene ($pK_{BH^+} = -12{\cdot}63$) as indicators. These pK_{BH^+} values were quoted by Ryabova *et al.* (1966) and necessitate subtraction of a constant factor of 0·49 units from the original H_0 scale for oleum given by Vinnik and Ryabova.

The numerical values of the two acidity functions in Table 2.6 are such that they form a natural extension into SO_3/H_2SO_4 mixtures of the two corresponding H_0 scales (Table 2.1 and Table 2.4) for aqueous sulphuric acid. Comparison shows that the rate of change of H_0 with sulphur trioxide concentration is similar for the two sets of data. Thus in the range 0·2 to 3·1 wt % sulphur trioxide the deviation ΔH_0 (Table 2.6) between the two scales is a constant. Above 3·1% sulphur trioxide the agreement is not quite so good but

TABLE 2.6

The Hammett acidity function for solutions of sulphur trioxide in sulphuric acid (oleum)

wt % H_2SO_4	wt % SO_3	$-H_0$†	$-H_0$‡	ΔH_0
100·045	0·2	12·14	11·19	0·95
100·067	0·3	12·20	11·24	0·96
100·090	0·4	12·26	11·27	0·99
100·112	0·5	12·30	11·31	0·99
100·135	0·6	12·33	11·34	0·99
100·180	0·8	12·38	11·40	0·98
100·225	1·0	12·43	11·44	0·99
100·337	1·5	12·54	11·54	1·00
100·450	2·0	12·62	11·63	0·99
100·562	2·5	12·68	11·69	0·99
100·7	3·11	12·75	11·75	1·00
101·0	4·44	12·86	11·82	1·04
101·5	6·66	13·02	11·95	1·07
102·0	8·9	13·14	12·06	1·08
102·5	11·1	13·27	12·18	1·09
103	13·3		12·28	
104	17·8		12·47	
105	22·2		12·62	
106	26·6		12·74	
107	31·1		12·87	

† Vinnik and Ryabova (1964); Ryabova *et al.* (1966).
‡ Brand (1950); Brand *et al.* (1952a); Paul and Long (1957).

is still probably within experimental error. Despite the good agreement between the two sets of measurements of the variation of H_0 with sulphur trioxide concentration the problem arises as to which scale is most likely to represent the best absolute H_0 values for these solutions.

The acidity functions for oleum are referred to standard state pure water via a stepwise comparison of indicators of decreasing basicity ionized in aqueous sulphuric acid and oleum. The Paul and Long scale for oleum is referred to water standard state via the measurements of Hammett and Deyrup (1932) for aqueous sulphuric acid. The absolute values of H_0 are therefore in error in accord with the fact that Hammett and Deyrup used several bases which do not show the accepted acidity function behaviour of Hammett bases as characterized by the ionization of primary amines. The acidity function for oleum quoted by Vinnik and Ryabova (1964) is however based on measurements for aqueous sulphuric acid using only primary amines as indicator. It must be concluded that the latter scale is probably a better estimate of the correct absolute H_0 acidity function for oleum. However this is by no means certain. Thus large deviations occur between the reported H_0 values for aqueous sulphuric acid in the region of 100% sulphuric acid even where the stepwise reference to water is accomplished using only primary amines as indicators (Gel'bshtein *et al.* 1956a, 1956b; Jorgenson and Hartter, 1963; Ryabova *et al.* 1966).

The acidity function for oleum solutions measured by Lewis and Bigeleisen (1943) and discussed by Coryell and Fix (1955) is not relevant to the present discussion because it is based on use of a positively charged ionic indicator derived from dinitrodibromofluorescein. A true Hammett H_0 acidity function must be based on the ionization behaviour of neutral bases in acid solution. In fact the subscript zero in the symbol H_0 is chosen to represent the charge on the base form of the indicators used to define the acidity function.

Lewis and Bigeleisen (1943) considered that the acidity function for oleum follows $\log_{10} a_{SO_3}$ where a_{SO_3} is the activity of sulphur trioxide in the solutions. However Paul and Long (1957) compared Brand's H_0 scale (Brand, 1950; Brand *et al.* 1952a) with the partial pressures of sulphur trioxide (Brand and Rutherford, 1952) over oleum and found that H_0 did not correlate satisfactorily with $\log_{10} P_{SO_3}$.

The self-ionization equilibria which exist in 100% sulphuric acid are given in equations (2.15)–(2.19) (Bass *et al.* 1960; Wyatt, 1969).

$$2H_2SO_4 \rightleftharpoons H_2O + H_2S_2O_7 \quad (2.15)$$

$$H_2O + H_2SO_4 \rightleftharpoons H_3O^+ + HSO_4^- \quad (2.16)$$

$$H_2S_2O_7 + H_2SO_4 \rightleftharpoons H_3SO_4^+ + HS_2O_7^- \quad (2.17)$$

$$2H_2SO_4 \rightleftharpoons H_3SO_4^+ + HSO_4^- \quad (2.18)$$

$$2H_2SO_4 \rightleftharpoons H_3O^+ + HS_2O_7^- \quad (2.19)$$

The acidity of 100% sulphuric acid and oleum may be attributed to the existence of the protonated sulphuric acid molecule $H_3SO_4^+$ in the solutions. This of course compares with the acidity of dilute aqueous solutions being associated with the H_3O^+ ion. As the sulphur trioxide concentration is increased more $H_2S_2O_7$ is formed and this ionizes according to equation (2.17) to give a higher concentration of $H_3SO_4^+$ ions. Gillespie (1967) has found that equation

$$H_0 = \text{const.} - \log_{10}(H_3SO_4^+) \quad (2.20)$$

(2.20) gives a reasonable fit with the experimental H_0 scale for oleums. The situation is probably complicated by the formation of the polysulphuric acids $H_2S_3O_{10}$ and $H_2S_4O_{13}$ in these solutions (Gillespie and Robinson, 1962; Gillespie and Malhotra, 1967). Addition of $KHSO_4$ to oleum decreases the acidity of the solutions (Brand, 1950). The added $KHSO_4$ is completely ionized

(Gillespie and Oubridge, 1956) and will repress the formation of $H_3SO_4^+$ ions in solution via equation (2.18). The interpretation of acidity in terms of the concentration of $H_3SO_4^+$ ions is therefore consistent with the observed effect of $KHSO_4$ on H_0.

2.2.2. *Deuterosulphuric Acid in Deuterium Oxide*

Högfeldt and Bigeleisen (1960) have determined the Hammett acidity function D_0 for solutions of D_2SO_4 in heavy water D_2O in the concentration range 10^{-4} M < $C_{D_2SO_4}$ < 12 M. They also measured H_0 for aqueous sulphuric acid using the same set of indicators. The H_0 scale agreed very closely with that quoted by Paul and Long (1957). The $H_0(H_2SO_4/H_2O)$ and $D_0(D_2SO_4/\text{-}D_2O)$ acidity functions for acid concentrations < 1 M are compared in Table 2.7. The measured dissociation constants of the conjugate acids of the indicators in H_2O and D_2O are given in Table 2.8.

TABLE 2.7

Comparison of the Hammett acidity functions H_0 for aqueous H_2SO_4 and D_0 for D_2O/D_2SO_4 at 22 ± 2°C (Högfeldt and Bigeleisen, 1960)

$-\log_{10} C_{L_2SO_4}$†	H_0 (H_2SO_4)	D_0 (D_2SO_4)	$-\log_{10} C_{L_2SO_4}$†	H_0 (H_2SO_4)	D_0 (D_2SO_4)
3·80	3·50	3·50	1·80	1·54	1·71
3·60	3·28	3·28	1·60	1·38	1·51
3·40	3·08	3·08	1·40	1·22	1·33
3·20	2·84	2·90	1·20	1·03	1·13
3·00	2·64	2·73	1·00	0·83	0·95
2·80	2·45	3·54	0·80	0·65	0·72
2·60	2·26	2·38	0·60	0·45	0·51
2·40	2·07	2·20	0·40	0·23	0·26
2·20	1·89	2·01	0·20	−0·02	−0·01
2·00	1·72	1·85	0·00	−0·30	−0·30

† $C_{L_2SO_4}$ = concentration of acid (H_2SO_4 or D_2SO_4) in mole litre.

Below $C_{L_2SO_4}$ = ca. 4×10^{-4} mole litre^{-1} (where L = H or D) the H_0 and D_0 scales are identical. However in the approximate concentration range 4×10^{-4} mole litre^{-1} < $C_{L_2SO_4}$ < 0·6 mole litre^{-1} the two functions deviate from each other in the sense $D_0 > H_0$. This difference is in accord with the observation that DSO_4^- is a weaker acid than HSO_4^- (Drucker, 1937). In the further range 0·6 mole litre^{-1} < $C_{L_2SO_4}$ < 12 mole litre^{-1} the H_0 and D_0 acidity functions are again identical. In this range sulphuric acid in water or deuterosulphuric

TABLE 2.8

Comparison of pK_{BH^+} and pK_{BD^+} for the indicators used by Högfeldt and Bigeleisen (1960) to establish the Hammett acidity scales H_0 for aqueous H_2SO_4 and D_0 for D_2O/D_2SO_4

Indicator	pK_{BH^+}	pK_{BD^+}	ΔpK
4-Aminoazobenzene	2·79	3·32	0·53
3-Nitroaniline	2·50†	3·00	0·50
2-Nitroaniline	−0·31	0·30	0·61
4-Chloro-2-nitroaniline	−1·02	−0·46	0·56
4-Nitrodiphenylamine	−2·50	−2·23‡	0·27‡
2,4-Dichloro-6-nitroaniline	−3·16	−2·73	0·43
2,6-Dinitro-4-methylaniline	−4·45	−4·04	0·41
2,4-Dinitroaniline	−4·42	−4·03	0·39
N,N-Dimethyl-2,4,6-Trinitroaniline	−4·98	−4·67	0·31
Benzalacetophenone	−5·68	−5·53	0·15

† Determined by Bascombe and Bell (1959).
‡ Values possibly in error due to medium effects on indicator spectra (Bell and Bascombe, 1959).

acid in heavy water may be assumed to be approximately completely dissociated according to equations (2.21) and (2.22) respectively (de Fabrizio, 1966). Thus the concentration of solvated hydrogen ions (L_3O^+) is effectively

$$H_2SO_4 + H_2O \rightleftharpoons H_3O^+ + HSO_4^- \tag{2.21}$$

$$D_2SO_4 + D_2O \rightleftharpoons D_3O^+ + DSO_4^- \tag{2.22}$$

equal to the stoichiometric acid concentration. From the definition of the acidity function, equation (2.23) (L = D or H), it follows that the identity of the

$$L_0 = pK_{BL^+} - \log_{10}\frac{C_{BL^+}}{C_B} = -\log_{10} C_{L^+} + \log_{10}\frac{y_{BL^+}}{y_B\, y_{L^+}} \tag{2.23}$$

H_0 and D_0 scales suggests the applicability of equation (2.24) for solutions of H_2SO_4 and D_2SO_4 of equal stoichiometric acid concentration. This is in part

$$(y_{BH^+}/y_B\, y_{H^+}) = (y_{BD^+}/y_B\, y_{D^+}) \tag{2.24}$$

due to the very similar dielectric constants of H_2O and D_2O (Malmberg, 1958). Thus effects of ion association on acidity and electrostatic contributions to the activity coefficients will be similar for the two solvents.

Högfeldt and Bigeleisen (1960) used some indicators (Table 2.8) which are not strictly Hammett bases to establish their D_0 scale. However their H_0 scale, measured using the same indicators, agreed very closely with that of

Paul and Long (1957). Furthermore in the range of acid concentrations (<70 wt % H_2SO_4) studied there is negligible deviation between the Paul and Long acidity function and the more recent scales of Jorgenson and Hartter (1963) and Ryabova *et al.* (1966) established using only primary amine (Hammett) bases. The D_0 acidity function may therefore be taken as a reliable measure of the Hammett acidity function for D_2O/D_2SO_4 solutions.

Schubert and Burkett (1956) have measured the extent of ionization of mesitaldehyde in 58·8%, 64%, 65% and 71% sulphuric acid in water and deuterosulphuric acid in D_2O. On the assumption that for the conjugate acid of mesitaldehyde $pK_{BH^+}(H_2O) = pK_{BD^+}(D_2O)$ it was found that $(H_0 - D_0) =$ 0·36 units for the four acid concentrations. However, in accord with the findings of Högfeldt and Bigeleisen (1960), $pK_{BH^+}(H_2O) \neq pK_{BD^+}(D_2O)$ and in the range of acid concentrations 59–71% $D_0 = H_0$. The measured differences between $\log_{10}(C_{BH^+}/C_B)$ in H_2O and $\log_{10}(C_{BD^+}/C_B)$ in D_2O are associated not with differences between the acidity of aqueous H_2SO_4 and D_2O/D_2SO_4 but with the differences in acid strength of the conjugate acids (BH^+ or BD^+) of mesitaldehyde in H_2O and D_2O. The different extents of protonation of 2,4,6-triisopropylbenzaldehyde in aqueous H_2SO_4 and D_2O/D_2SO_4 (Schubert and Myhre, 1958) may similarly be associated with the isotopic change in dissociation constant of the conjugate acid of the aldehyde.

2.2.3. *Hydrochloric Acid*

Prior to 1957 several independent determinations of the H_0 acidity function for aqueous hydrochloric acid had been made (Hammett and Paul, 1934; Braude, 1948; Paul, 1954; Bell *et al.* 1955; Vinnik *et al.* 1956). The results of these studies were summarized by Paul and Long (1957) in terms of an H_0 scale (Table 2.9) which was based on the most reliable pK_{BH^+} values then available for the base indicators which had been used to establish H_0. The indicators were primary amines, in accord with the strict definition of a Hammett H_0 acidity function, apart from diphenylamine (Hammett and Paul, 1934) and 4-nitrodiphenylamine (Bell *et al.* 1955). Hammett and Paul (1934) anchored their measurements to water as standard state, not only with 4-nitroaniline, but also with aminoazobenzene and benzeneazodiphenylamine. In quoting the H_0 scale for aqueous hydrochloric acid Paul and Long (1957) have taken the data for high hydrochloric acid concentrations of Vinnik *et al.* (1956) who used only primary aniline bases to establish H_0.

Milyaeva (1958) has measured an H_0 acidity function for aqueous hydrochloric acid (0·11 mole litre^{-1} $< C_{HCl} <$ 8·61 mole litre^{-1}) at 17°C using 2-nitroaniline and 5-chloro-2-nitroaniline as indicators. The Milyaeva (17°) and Paul and Long (25°) H_0 scales were parallel functions of HCl molarity. This confirms the results of a more detailed investigation of the effect of temperature on H_0 (Gel'bshtein *et al.* 1956a, 1956c). Salomaa (1957) has also reported a

TABLE 2.9

The Hammett acidity function for aqueous hydrochloric acid at 25°C (Paul and Long, 1957)

C_{HCl} (mole litre^{-1})	$-H_0$	C_{HCl} (mole litre^{-1})	$-H_0$	C_{HCl} (mole litre^{-1})	$-H_0$
0·1	−0·98	3·0	1·05	7·0	2·50
0·25	−0·55	3·5	1·23	8·0	2·86
0·5	−0·20	4·0	1·40	9·0	3·22
0·75	0·03	4·5	1·58	10·0	3·59
1·0	0·20	5·0	1·76	11·0	3·99
1·5	0·47	5·5	1·93	12·0	4·41
2·0	0·69	6·0	2·12	13·0	4·82
2·5	0·87	6·5	2·34		

This scale is based on measurements with the following bases (pK_{BH^+} in brackets): 4-nitroaniline (0·99); diphenylamine (0·78); 2-nitroaniline (−0·29); 4-chloro-2-nitroaniline (−1·03); 5-chloro-2-nitroaniline (−1·52); (−2·48); 2,4-dichloro-6-nitroaniline (−3·37); 2,4-dinitroaniline (−4·62).

few H_0 values for aqueous HCl. His results, based on measurements with 4-nitroaniline, are in agreement with the Paul and Long scale.

The variation of H_0 with temperature for aqueous hydrochloric acid has been investigated by Gel'bshtein *et al.* (1956a, 1956c). 4-Nitroaniline, 2-nitro-

TABLE 2.10

The variation of H_0 with temperature for aqueous hydrochloric acid (Gel'bshtein *et al.* 1956a, 1956c)

C_{HCl} (mole litre^{-1})	$-H_0$				$10^3.(dH_0/dT)$
	20°	40°	60°	80°	
0·1	−1·09	−1·10	−1·12		+0·8†
0·4	−0·40	−0·35	−0·30		−2·5†
0·8	−0·08	−0·01	0·07	0·16	−4·0‡
1·0	0·06	0·12	0·18	0·23	−2·8‡
2·0	0·53	0·59	0·65	0·71	−3·0‡
3·0	0·86	0·93	1·00	1·06	−3·3‡
4·0	1·18	1·24	1·30	1·36	−3·0‡
5·0	1·47	1·53	1·59	1·65	−3·0‡
6·0	1·76	1·83	1·89	1·96	−3·3‡
6·5	1·91	1·97	2·04	2·10	−3·2‡

† $(dH_0/dT) = [H_0(60°) - H_0(20°)]/40$
‡ $(dH_0/dT) = [H_0(80°) - H_0(20°)]/60$

aniline and 4-chloro-2-nitroaniline were used as indicators. The results are given in Table 2.10. The absolute H_0 values deviate from those in Table 2.9 by up to ca. 0·4 units for $C_{HCl} = 6{\cdot}5$ mole litre^{-1}. However despite this the variation of H_0 with temperature may be taken as reliable. As for sulphuric acid (dH_0/dT) is apparently a constant for a particular HCl concentration and temperatures in the range $20°C < T < 80°C$. The quantity $[H_0(80°) - H_0(20°)]$/-60 (Table 2.10) therefore gives an indication of the rate of change of acidity function with temperature at each acid concentration. (dH_0/dT) is approximately independent of concentration for 1 mole litre$^{-1} < C_{HCl} < 6{\cdot}5$ mole litre^{-1}. Thus for this concentration range the H_0 acidity functions at different temperatures are parallel to each other.

2.2.4. *Deuterium Chloride in Deuterium Oxide*

The D_0 acidity function for DCl/D_2O mixtures (10^{-4} mole litre$^{-1} < C_{DCl} <$ 1 mole litre^{-1}) has been measured by Högfeldt and Bigeleisen (1960). The variation of D_0 with DCl concentration was the same as that of H_0 with HCl concentration for aqueous hydrochloric acid. For acid concentrations less than 1 mole litre^{-1} the two scales were numerically identical. Also at these low concentrations the acidity function plotted against the logarithm of the acid concentration gave a linear graph with a slope of $-1{\cdot}0$. Equation (2.25) is therefore applicable (and similarly for D_0).

$$H_0 \approx -\log_{10} C_{HCl} \quad (C_{HCl} \approx C_{H^+};\, y_{H^+} y_B / y_{BH^+} \approx 1) \tag{2.25}$$

In view of the validity of this equation for dilute (< 1 mole litre^{-1}) acid solutions the identity of D_0 and H_0 is not surprising. Unfortunately measurements for more concentrated acid solutions have not been made.

2.2.5. *Hydrobromic Acid*

Vinnik *et al.* (1956) have measured an H_0 scale for aqueous hydrobromic acid solutions (Table 2.11). Their measurements were based on $pK_{BH^+} = 1{\cdot}11$ for 4-nitroaniline and have therefore been adjusted by $-0{\cdot}12$ units in accord with the Paul and Long (1952) "best value" of $pK_{BH^+} = 0{\cdot}99$ for 4-nitroaniline.

The value of $pK_{BH^+} = -4{\cdot}44$ for 2,4-dinitroaniline evaluated from measurements with hydrobromic acid solutions differs significantly from that of $pK_{BH^+} = 4{\cdot}62$ (Table 2.9) also measured by Vinnik *et al.* (1956) with hydrochloric acid solutions. The "best value" of Paul and Long (1957) was $-4{\cdot}53$ and was taken from the original results of Hammett and Deyrup (1932) for aqueous perchloric and sulphuric acids. More recent measurements for sulphuric acid have given $-4{\cdot}53$ (Jorgenson and Hartter, 1963) and $-4{\cdot}45$ (Ryabova *et al.* 1966). It is difficult to assess which of these figures is likely to be the most reliable but taking an average the Paul and Long value is still

TABLE 2.11

The Hammett H_0 acidity function for aqueous hydrobromic acid at $24 \pm 2°C$ (Vinnik *et al.* 1956)

m_{HBr} (mole kg^{-1})	H_0	m_{HBr} (mole kg^{-1})	H_0	m_{HBr} (mole kg^{-1})	H_0
0·001	2·91	1·0	−0·19	8·95	−2·92
0·005	2·21	2·0	−0·67	9·71	−3·16
0·01	1·92	3·0	−1·04	10·53	−3·42
0·02	1·63	5·81	−1·93	11·41	−3·69
0·05	1·33	6·36	−2·11	12·36	−4·00
0·1	1·02	6·95	−2·25	13·38	−4·34
0·2	0·70	7·57	−2·49	14·51	−4·69
0·5	0·28	8·24	−2·70	15·73	−4·97

Indicators used were (pK_{BH^+} in brackets): 4-nitroaniline (0·99); 2-nitroaniline (−0·29); 5-chloro-2-nitroaniline (−1·52); 2,4-dichloro-6-nitroaniline (−3·27); 2,4-dinitroaniline (−4·44).

acceptable. The measurements with hydrobromic acid solutions gave $pK_{BH^+} = -3{\cdot}27$ for 2,4-dichloro-6-nitroaniline and this is in very good agreement with the other estimates (Paul and Long, 1957; Ryabova *et al.* 1966).

2.2.6. *Hydrofluoric Acid*

The Hammett H_0 acidity function for aqueous hydrogen fluoride given in Table 2.12 is based on the results of Bell *et al.* (1956), and Hyman *et al.* (1957). The scale is referred to water standard state via $pK_{BH^+} = 0{\cdot}99$ for 4-nitroaniline (Paul and Long, 1957). Hyman *et al.* (1957) deduced H_0 for 98–100 wt % HF from measurements using the very weak base 2,4,6-trinitroaniline for which they took $pK_{BH^+} = -9{\cdot}41$ (Paul and Long, 1957). This figure comes from Hammett and Deyrup's (1932) results and is probably in error because its stepwise deduction was made via ionization data for several non-Hammett bases. More recent determinations using only Hammett bases in aqueous sulphuric acid are −10·10 (Jorgenson and Hartter, 1963) and −9·98 (Ryabova *et al.* 1966). The data in Table 2.12 for $C_{HF} > 98$ wt % has been calculated taking the mean (−10·04) of these two values. The H_0 scale for $75\% < C_{HF} < 95\%$ is based on measurements using anthraquinone as indicator (Hyman *et al.* 1957). Because the protonation of ketones does not follow the Hammett acidity function defined using primary amines as bases (Bonner and Phillips, 1966) the H_0 scale for aqueous HF is probably in error for this concentration range. Anthraquinone is not a Hammett base (Yates and Wai, 1965).

Hydrogen fluoride is a weak acid in water with $K_a = 6{\cdot}7 \times 10^{-4}$ at 25°C

(Broene and De Vries, 1947). Thus the acidity of an aqueous solution of HF is much less than that of a solution of the same molar concentration of either of the strong acids HCl or HBr. However the acidity of aqueous HF increases rapidly as the water content of the solutions becomes small. Thus for anhydrous

TABLE 2.12

Hammett H_0 acidity function for aqueous hydrofluoric acid at 25°C (Bell *et al.* 1956; Hyman *et al.* 1957)

wt % HF	$-H_0$	wt % HF	$-H_0$	wt % HF	$-H_0$
2·5	−1·06	50	3·84	98·13	9·49
5	−0·65	55	5·01	98·38	9·74
10	−0·24	60	5·86	98·40	9·72
15	0·15	65	6·42	98·48	9·75
20	0·51	70	6·81	98·83	9·97
25	0·88	75	7·16§	99·07	9·94
30	1·20	80	7·50§	99·29	10·10
35	1·60	85	7·84§	99·78	10·22
40	2·08	90	8·16§	100†	10·59
45	2·79	95	8·49§	100‡	10·8

This scale is based on measurements with the following indicators (pK_{BH+} in brackets): 4-nitroaniline (0·99); 2-nitroaniline (−0·29); 4-chloro-2-nitroaniline (−1·03); 4-nitrodiphenylamine (−2·50); 2,6-dinitro-4-methylaniline (−4·44); 6-bromo-2,4-dinitroaniline (−6·71); anthraquinone (−8·27); 2,4,6-trinitroaniline (−10·04).

† H_2O content < 0·01 mole litre^{-1}

‡ H_2O content < 0·001 mole litre^{-1}

§ These values are probably unreliable as they have been deduced from measurements with anthraquinone as indicator and the ionization behaviour of anthraquinone is somewhat different from that of true Hammett bases (Yates and Wai, 1965).

hydrogen fluoride (< 0·001 mole litre^{-1} water) $H_0 = -10{\cdot}8$. This high acidity is probably associated with the formation of the ions HF_2^-, $H_2F_3^-$, $H_3F_4^-$, ... $H_{n-1}F_n^-$ in the solutions (Bell *et al.* 1956; Hyman *et al.* 1957; Bell, 1959). The equilibrium constant for equilibrium (2.26) is 3·9 at 25°C (Broene and De

$$HF + F^- \rightleftharpoons HF_2^- \tag{2.26}$$

Vries, 1947). This implies that at quite low concentrations of hydrogen fluoride the fluoride anions in solution are associated with HF molecules and therefore exist predominantly as HF_2^-. Infrared spectra of aqueous HF are consistent with this (Jones and Penneman, 1954) and also provide evidence for the higher "polymers" $H_{n-1}F_n^-$ (where $n > 1$). These anions are derived from the acids H_nF_n ($n > 1$) which have not been detected in aqueous solution. It must be concluded that they are completely ionized strong acids in aqueous HF. The high acidity of concentrated solutions of hydrogen fluoride in water supports this conclusion. Bell (1959) has suggested that a concentrated solution of hydrogen fluoride (concentration C_{HF}) in water behaves like a solution of a strong acid of concentration (C_{HF}/n) where n is the average for all the ions $H_{n-1}F_n^-$ in the solution. It is, however, difficult to test this hypothesis quantitatively (Bell *et al.* 1956).

2.2.7. *Perchloric Acid*

Hammett and Deyrup (1932) measured an H_0 scale for aqueous perchloric acid up to 64% $HClO_4$. Their scale (corrected to the standard $pK_{BH^+} = 0{\cdot}99$ for 4-nitroaniline) is plotted in Fig. 1.4. Bonner and Lockhart (1957) extended

TABLE 2.13

H_0 acidity function for aqueous perchloric acid at 25°C (Yates and Wai, 1964)

wt % $HClO_4$	$-H_0$	wt % $HClO_4$	$-H_0$	wt % $HClO_4$	$-H_0$
4·9	−0·02	50·3	3·50	67·1	6·99
10·1	0·36	51·7	3·73	68·7	7·35
15·1	0·67	52·8	3·86	70·0	7·75
20·4	1·01	54·6	4·22	71·1	8·08
25·4	1·30	55·9	4·40	71·9	8·25
30·4	1·62	57·7	4·78	72·9	8·60
33·4	1·83	59·2	5·06	73·7	8·90
36·3	2·05	61·2	5·54	75·0	9·15
39·6	2·35	62·4	5·77	76·1	9·53
42·6	2·62	63·0	5·97	77·0	9·96
45·6	2·96	64·5	6·39	77·7	10·07
48·7	3·34	65·7	6·63	78·6	10·31

Indicators used were (pK_{BH^+} in brackets): 2-nitroaniline (−0·29); 4-chloro-2-nitroaniline (−1·07); 2,5-dichloro-4-nitroaniline (−1·79); 2-chloro-6-nitroaniline (−2·41); 2,6-dichloro-4-nitroaniline (−3·20); 2,4-dinitroaniline (−4·26); 2,6-dinitroaniline (−5·25); 4-chloro-2,6-dinitroaniline (−6·12); 2-bromo-4,6-dinitroaniline (−6·69); 3-methyl-2,4,6-trinitroaniline (−8·56); 3-bromo-2,4,6-trinitroaniline (−9·77).

the data to 71% perchloric acid using 2,4-dinitro-1-naphthylamine as indicator. More recently Yates and Wai (1964) have determined the H_0 acidity function for 4·9–78·6 wt % $HClO_4$ in water. Their scale (Table 2·13) is based on measurements using only primary amines as indicators. The same bases were chosen as were used by Jorgenson and Hartter (1963) for defining H_0 for aqueous sulphuric acid (Table 2.3).The scale was referred to standard state water via the Paul and Long (1957) value of $pK_{BH^+} = -0{\cdot}29$ for 2-nitroaniline. The Hammett and Deyrup (1932), Paul and Long (1957), Bonner and Lockhart (1957) and Yates and Wai (1964) acidity functions were all in reasonable agreement with each other although the data of Yates and Wai are probably the most reliable.

The pK_{BH^+} of the primary anilines used to establish H_0 for aqueous $HClO_4$ were generally in satisfactory agreement with the corresponding values deduced from a study of the ionization of the same bases in aqueous H_2SO_4 (Jorgenson and Hartter, 1963). Notable exceptions are the figures for 2,4-dinitroaniline and 2,6-dinitroaniline. Yates and Wai (1964) argued that their values were more accurate than those from the results for aqueous sulphuric acid because deduction of the latter were dependent on the original, less reliable colorimetric measurements of Hammett and Deyrup (1932). However, the independent study of aqueous sulphuric acid by Ryabova *et al.* (1966) gave $pK_{BH^+} = -4{\cdot}45$ for 2,4-dinitroaniline and $pK_{BH^+} = -5{\cdot}64$ for 2,6-dinitroaniline. These values are in much closer agreement with those deduced by Jorgenson and Hartter (1963) than those given by Yates and Wai (1964). The magnitude of the deviations between the pK_{BH^+} values for a particular indicator determined by independent studies given a indication of the inaccuracies which are inherent in the determination of pK_{BH^+} by the stepwise comparison of progressively weaker bases.

That perchloric acid is a strong acid is well known. Thus Raman spectra (Heinzinger and Weston, 1965; Covington *et al.* 1965) and proton and chlorine nuclear magnetic resonance (Hood *et al.* 1954; Hood and Reilly, 1960; Akitt *et al.* 1965) of aqueous perchloric acid solutions are consistent with the conclusions that $HClO_4$ is completely ionized in water up to a concentration $C_{HClO_4} = $ ca. 6–8 mole litre^{-1} (45–55 wt %). At $C_{HClO_4} = 12$ mole litre^{-1} the acid is still probably at least 91 % ionized (Akitt *et al.* 1965). These observations are in accord with the H_0 data. Thus comparison of H_0 for aqueous solutions of sulphuric acid and perchloric acid of equal molar concentration shows that if $C_{acid} > 6$ mole litre^{-1} then $H_0(HClO_4) > H_0(H_2SO_4)$. In fact at high concentrations perchloric acid gives more acidic solution than does any other of the common strong acids. It has been estimated that H_0 might be of the order of −16 for aqueous perchloric acid of composition approaching 100 % $HClO_4$ (Yates and Wai, 1964). This is only an approximate guess but the trend of the acidity function with increasing perchloric acid concentration certainly suggests that as 100 % $HClO_4$ is approached acidities will be attained which are much greater than those of anhydrous hydrogen fluoride or sulphuric acid. The attempts by Dawber (1966a, 1966b) to correlate H_0 with degree of ionization for aqueous perchloric acid will be described later (Section 2.3.2).

2.2.8. *Periodic Acid and Iodic Acid*

The ionization of 4-nitroaniline and 2-nitroaniline in aqueous periodic acid and aqueous iodic acid has been studied by Dawber (1965). The relevant H_0 scales are given in Table 2.14.

TABLE 2.14

The Hammett acidity function for aqueous iodic acid and aqueous periodic acid (Dawber, 1965)

Molarity acid	H_0 (iodic acid)	H_0 (periodic acid)	Molarity acid	H_0 (iodic acid)	H_0 (periodic acid)
0·025	1·60		1·75	−0·06	0·14
0·05	1·36	2·00	2·0	−0·13	0·01
0·01	1·10	1·58	2·25	−0·19	−0·10
0·20	0·86	1·22	2·5	−0·25	−0·18
0·35	0·63	1·03	3·0	−0·36	
0·50	0·48	0·91	3·5	−0·46	
0·75	0·30	0·74	4·0	−0·54	
1·00	0·18	0·58	4·5	−0·60	
1·25	0·08	0·43	5·0	−0·66	
1·50	0·00	0·28	5·5	−0·72	

Indicators used were (pK_{BH^+} in brackets as quoted by Paul and Long, 1957); 4-nitroaniline (−0·99); 2-nitroaniline (−0·29).

Näsänen (1954) has deduced a value of 0·028 mole litre^{-1} for the acid ionization constant (25°C) of periodic acid in water. The acidity constant for iodic acid is 0·157 mole litre^{-1} at 25°C (Leist, 1955; Pethybridge and Prue, 1967). The H_0 acidity functions are in accord with the observation that both iodic acid and periodic acid behave as weak acids in water. Furthermore comparison of the two H_0 scales reflects the conclusion, already drawn from knowledge of the ionization constants, that periodic acid is a weaker acid than iodic acid.

In dilute solution in water periodic acid exists predominantly as H_5IO_6. The relevant ionization equilibria are represented by equations (2.28)–(2.30)

$$H_5IO_6 + H_2O \rightleftharpoons H_3O^+ + H_4IO_6^- \tag{2.28}$$

$$H_4IO_6^- \rightleftharpoons 2H_2O + IO_4^- \tag{2.29}$$

$$H_4IO_6^- + H_2O \rightleftharpoons H_3O^+ + H_3IO_6^{2-} \tag{2.30}$$

(Crouthamel *et al.* 1951). The equilibrium constants at 25°C are $K(2.28) = 5{\cdot}1 \times 10^{-4}$; $K(2.29) = 40$; and $K(2.30) = 2{\cdot}0 \times 10^{-7}$. The ionization constant quoted by Näsänen (1954) for periodic acid is not a true equilibrium constant but is equal to $a_{H_3O^+}(a_{IO_4^-}+a_{H_4IO_6^-})/a_{H_5IO_6}$. The second ionization constant, $K(2.31)$, is so small that the formation of H_3O^+ ions in solution via equilibrium (2.31) will make a negligible contribution to the overall acidity of the solutions.

Pethybridge and Prue (1967) have studied aqueous iodic acid solutions using potentiometric, conductimetric, and kinetic techniques. Apart from confirming an earlier estimate (Leist, 1955) for the ionization constant of iodic acid they also reanalysed the cryoscopic data of Abel *et al.* (1934) and deduced a stability constant of 41 mole^{-1} for the $H(IO_3)_2^-$ ion. Cryoscopic data for $C_{HIO_3} > 0{\cdot}8$ mole litre^{-1} are consistent with the existence of polymeric species in the solutions. Dawber (1965) has suggested that the H_0 scale for aqueous iodic acid is also consistent with the proposal that polymeric species $[IO_3(HIO_3)_n]^-$ exist in solution. Equation (2.31) relates H_0 for a weak acid with α the degree of ionization of the weak acid and H_0 for a strong acid at the same water activity (Wyatt, 1957).

$$H_0(\text{weak}) - H_0(\text{strong}) = \log_{10}\left(\frac{1+\alpha}{2\alpha}\right) \tag{2.31}$$

Thus H_0(weak) should always be greater than H_0(strong). Comparison of H_0 for iodic acid (weak acid) with H_0 for sulphuric acid (strong acid) shows that if the water activity $a_w < 0{\cdot}984$ then $H_0(\text{strong}) > H_0(\text{weak})$. A possible explanation would be that in concentrated solutions $[IO_3(HIO_3)_n]^-$ is formed rather than $[IO_3(H_2O)_m]^-$. A water activity of 0·984 corresponds approximately to a concentration $C_{HIO_3} = 0{\cdot}9$ mole litre^{-1} which is approximately the same as the concentration for which the cryoscopic results were anomalous.

2.2.9. *Nitric Acid*

The H_0 acidity function for aqueous nitric acid at 20°C measured by Dawber and Wyatt (1960) is given in Table 2.15. For $C_{HNO_3} < 14{\cdot}2$ mole litre^{-1} the scale is based on the use of primary amines as indicators. At higher concentrations *N,N*-dimethyl-2,4,6-trinitroaniline and 4,4′-dinitrodiphenylamine were used and therefore the H_0 values are possibly less reliable. For $C_{HNO_3} < 7$ mole litre^{-1} Paul and Long (1957) have quoted an H_0 acidity function (Table 2.16) for aqueous nitric acid at 25°C based on the colorimetric measurements made by Hammett and Paul (1934). Comparison of the Dawber and Wyatt and the Paul and Long scales shows that they are approximately parallel functions of nitric acid concentration. Thus $H_0(25°) - H_0(20°) =$ ca. 0·1 units for $C_{HNO_3} < 5$ mole litre^{-1}. At $C_{HNO_3} = 7$ mole litre^{-1} the deviation is 0·2 units. The H_0 acidity function for aqueous nitric acid measured by Milyaeva (1958) refers to a temperature of 17°C. His results are consistent with the other two scales.

TABLE 2.15

The Hammett H_0 acidity function for aqueous nitric acid at 20°C (Dawber and Wyatt, 1960)

wt % HNO_3	C_{HNO_3} (mole litre^{-1})	$-H_0$	wt % HNO_3	C_{HNO_3} (mole litre^{-1})	$-H_0$
2	0·32	−0·33	45	9·13	2·62
4	0·65	0·08	50	10·4	2·88
6	0·97	0·29	55	11·7	3·13
8	1·32	0·46	60	13·0	3·42
10	1·68	0·64	65	14·3	3·72
15	2·58	0·97	70	15·7	3·99
20	3·55	1·28	75	17·1	4·30
25	4·56	1·57	80	18·5	4·62
30	5·62	1·85	85	19·8	4·96
35	6·75	2·10	90	21·2	5·31
40	7·92	2·36	95	22·5	5·75
			100	24·0	6·3

Indicators used were (pK_{BH^+} in brackets): 4-nitroaniline (1·03); 2-nitroaniline (−0·28); 4-chloro-2-nitroaniline (−1·11); 2,6-dichloro-4-nitroaniline (−2·55); 2,4-dichloro-6-nitroaniline (−3·30); *N,N*-dimethyl-2,4,6-trinitroaniline†; 4,4′-dinitrodiphenylamine (−6·2).

† H_0 values were deduced from measurements with this indicator as H_0 for the aqueous H_2SO_4 solution in which the ionization ratio of the indicator was identical to that in aqueous nitric acid. The relevant concentration range is 15 mole litre^{-1} < C_{HNO_3} < 19 mole litre^{-1}.

The degree of dissociation α of nitric acid in water over the whole concentration range has been measured by a study of the intensity of the 1049 cm^{-1} Raman absorption of the nitrate anion (Krawetz, 1955, quoted by Young *et al.* 1959). McKay (1956) has calculated the ionization constant to be ca. 24 mole litre^{-1} at 25°C. Comparison of H_0 for aqueous nitric acid with H_0 for aqueous sulphuric acid solutions of the same water activities (Vandoni and Laudy, 1952; Glueckauf and Kitt, 1956) shows that $H_0(H_2SO_4) < H_0(HNO_3)$. Dawber and Wyatt (1960) have shown that this is a result of the incomplete dissociation of nitric acid in water. Thus for 5·6 mole litre^{-1} < C_{HNO_3} < 17·1 mole litre^{-1} equation (2.31) is applicable. The sulphuric acid H_0 scale used in the correlation was that of Paul and Long (1957). The deviation between the calculated and experimental H_0 scales for aqueous nitric acid, was apart from one case always less than 0·09 H_0 units. In accord with the validity of equation (2.31) this good agreement suggests that the relative values (compared at the

TABLE 2.16

The Hammett H_0 acidity function for aqueous nitric acid at 25°C (Paul and Long, 1957; Hammett and Paul, 1934)

wt % HNO_3	C_{HNO_3} (mole litre^{-1})	$-H_0$	wt % HNO_3	C_{HNO_3} (mole litre^{-1})	$-H_0$
0·7	0·1	−0·98	19·8	3·5	1·17
1·6	0·25	−0·55	22·3	4·0	1·32
3·1	0·5	−0·21	24·7	4·5	1·46
4·6	0·75	0·02	27·1	5·0	1·57
6·1	1·0	0·18	29·4	5·5	1·69
9·0	1·5	0·45	31·7	6·0	1·79
11·8	2·0	0·67	33·9	6·5	1·89
14·6	2·5	0·85	33·9	7·0	1·99
17·2	3·0	1·02			

Indicators used were (pK_{BH^+} in brackets): 4-nitroaniline (0·99); 2-nitroaniline (−0·29); 4-chloro-2-nitroaniline (−1·03).

same water activity) of H_0 for aqueous solutions of weak acids are probably governed primarily by the relative strengths (pK_a) of the acids being compared.

100% nitric acid ($H_0 = -6{\cdot}3$) is much less acidic than anhydrous sulphuric acid or hydrogen fluoride. The self-ionization of nitric acid does not lead to the expected autoprotolysis products $H_2NO_3^+$ and NO_3^- but instead may be represented by the self-dehydration reaction (2.32) (Gillespie *et al.* 1950; Gillespie, 1967). The equilibrium constant at −40°C is 0·020 mole^3 kg.$^{-3}$

$$3HNO_3 \rightleftharpoons NO_2^+ + NO_3^- + HNO_3.H_2O \quad (2.32)$$

Dawber and Wyatt (1960) suggested that water would be protonated to a considerable extent when added in very small concentrations to 100% nitric acid. However vapour pressure (Lloyd and Wyatt, 1955) and cryoscopic (Gillespie *et al.* 1950) measurements show that water is not appreciably ionized in nitric acid. The water presumably exists predominantly as the solvated species $H_2O.HNO_3$ (Bethell and Sheppard, 1953) and also possibly to a lesser extent as $H_2O.2HNO_3$ (Young *et al.* 1959).

2.2.10. *Phosphorous Acid and Phosphorus Pentoxide*

Bascombe and Bell (1959) have measured an H_0 acidity function for aqueous phosphorous acid up to $C_{H_3PO_3} = 13{\cdot}8$ mole litre^{-1}. The highest concentration approximates to the limit of the solubility of H_3PO_3 in water. The H_0 scale given in Table 2.17 was interpolated from a smooth curve drawn through a plot of Bascombe and Bell's H_0 values against phosphorous acid concentration.

TABLE 2.17

The Hammett H_0 acidity function for aqueous phosphorous acid at 25°C (Bascombe and Bell, 1959)

$C_{H_3PO_3}$ (mole litre^{-1})	wt % H_3PO_3	H_0	$C_{H_3PO_3}$ (mole litre^{-1})	wt % H_3PO_3	H_0
0·096	0·436	1·26	6	22·3	−0·36
0·235	1·06	0·96	7	25·5	−0·48
0·5	2·2	0·69	8	28·6	−0·59
1·0	4·3	0·42	9	31·6	−0·70
1·5	6·4	0·31	10	34·6	−0·80
2	8·5	0·21	11	37·6	−0·90
3	12·2	0·05	12	40·6	−0·99
4	15·6	−0·09	13	43·6	−1·08
5	19·0	−0·24	14	46·6	−1·18

Indicators used were (pK_{BH^+} in brackets): 4-nitroaniline (1·02); 2-nitroaniline (−0·29); 4-chloro-2-nitroaniline (−1·02).

The first dissociation constant of phosphorous acid in water is 0·016 mole litre^{-1} at 25°C (Takahasi and Yui, 1941; Parsons, 1959). Thus H_3PO_3 is a weak acid in water and this explains the slow rate at which H_0 varies with phosphorous acid concentration. The second ionization of H_3PO_3, for which $K_a = 6{\cdot}9 \times 10^{-7}$ mole litre^{-1} (Parsons, 1959), is negligible and will not contribute to the measured acidity of the solutions.

The acidity of the P_2O_5–water system up to 80% H_3PO_4 (52·9% P_2O_5) at $19 \pm 2°$ has been studied by Heilbronner and Weber (1949) using four primary amines and one secondary amine (4-nitrodiphenylamine) as indicators. 4-Nitrodiphenylamine gave results which parallelled those for the primary amines in regions of overlap and therefore behaved as a Hammett base in these solutions. The results of Gel'bshtein *et al.* (1956a, 1956c) who measured H_0 for P_2O_5–water mixtures at four temperatures are given in Table 2.18. The scale for 20° agrees fairly closely with that determined by Heilbronner and Weber (1949). As for aqueous hydrochloric acid (Gel'bshtein *et al.* 1956a, 1956c) and aqueous sulphuric acid (Gel'bshtein *et al.* 1956a, 1956b; Tickle *et al.* 1970); the rate of change of H_0 with temperature (dH_0/dT) is a constant for a particular P_2O_5 concentration and temperatures in the range $20°C < t < 80°C$. The primary amine bases 4-nitroaniline, 2-nitroaniline, 4-chloro-2-nitroaniline and 2,4-dichloro-6-nitroaniline were used to establish the H_0 scales.

Gel'bshtein *et al.* (1956a, 1956c) extended their measurements up to 83·8 wt % P_2O_5 in water using benzalacetophenone as indicator. They found a maximum acidity at ca. 80% P_2O_5. However the existence of a maximum in a

TABLE 2.18

The Hammett H_0 acidity functions for aqueous phosphoric acid at four temperatures (Gel'bshtein *et al.* 1956a, 1956c)

wt % H_3PO_4	wt % P_2O_5	$-H_0$ 20°	40°	60°	80°	$10^2(dH_0/dT)$†
5	3·6	−1·06	−1·06	−1·06	−1·06	0
10	7·2	−0·75	−0·75	−0·75	−0·75	0
15	10·9	−0·52	−0·52	−0·52	−0·52	0
20	14·5	−0·30	−0·30	−0·30	−0·30	0
25	18·1	−0·08	−0·08	−0·08	−0·08	0
30	21·7	0·14	0·12	0·11	0·09	0·08
35	25·3	0·36	0·33	0·30	0·27	0·15
40	29·0	0·59	0·55	0·51	0·47	0·20
45	32·6	0·81	0·76	0·70	0·64	0·29
50	36·2	1·03	0·97	0·90	0·83	0·33
55	39·8	1·26	1·19	1·11	1·03	0·38
60	43·4	1·52	1·43	1·35	1·26	0·43
65	47·1	1·86	1·76	1·65	1·55	0·51
70	50·7	2·32	2·17	2·02	1·88	0·73
75	54·3	2·82	2·63	2·44	2·25	0·95
80	57·9	3·29	3·07	2·85	2·64	1·08
85	61·5	3·78	3·52	3·26	3·01	1·28
90	65·2	4·30	3·98	3·68	3·39	1·51
95	68·8	4·77	4·44	4·08	3·75	1·70
100	72·4	5·25‡	4·75‡	4·49‡	4·14‡	1·85

† $(dH_0/dT) = [H_0(80°) - H_0(20°)]/60$.
‡ Based on extrapolation of H_0 against wt % H_3PO_4 curve.

graph of $-H_0$ against concentration was later disproved by Downing and Pearson (1961). Gel'bshtein *et al.* (1964) repeated their measurements using the same indicators (2,4-dinitroaniline and 2-bromo-4,6-dinitroaniline) as Downing and Pearson and confirmed that H_0 decreased steadily with increasing P_2O_5 concentration. The H_0 acidity function for aqueous P_2O_5 at 25° is given in Table 2.19. For $C_{P_2O_5} < 60$ wt % the H_0 values are those of Gel'bshtein *et al.* (1964) corrected to $pK_{BH^+} = 0·99$ for the reference base, 4-nitroaniline. The H_0 values for $C_{P_2O_5} > 60$ wt % have been interpolated from a smooth curve drawn through a plot of the H_0 scales measured by Downing and Pearson (1961) and by Gel'bshtein *et al.* (1964).

The most unsatisfactory feature about the data in Table 2.19 is the value of $pK_{BH^+} = -5·2$ for 2,4-dinitroaniline. This is widely different from the figure of ca. −4·5 deduced from measurements using other concentrated acid solutions (Paul and Long, 1957; Vinnik *et al.* 1956; Ryabova *et al.* 1966). Plots of

TABLE 2.19

The Hammett H_0 acidity function for water/phosphorus pentoxide mixtures at 25°C (Downing and Pearson, 1961; Gel'bshtein *et al.* 1964)

wt % P_2O_5	$-H_0$	wt % P_2O_5	$-H_0$	wt % P_2O_5	$-H_0$
5	−0·80	35	1·09	65	4·16
10	−0·47	40	1·46	70	4·70
15	−0·16	45	1·87	75	5·24
20	0·15	50	2·32	80	5·92
25	0·46	55	2·84	85	6·96
30	0·78	60	3·46		

Indicators used were (pK_{BH^+} in brackets); 4-nitroaniline (0·99); 2-nitroaniline (−0·25); 4-chloro-2-nitroaniline (−1·04); 2,4-dichloro-6-nitroaniline (−3·10); 2,4-dinitroaniline (−5·2); 2-bromo-4,6-dinitroaniline (−6·9).

$\log_{10}(C_{BH^+}/C_B)$ against P_2O_5 concentration for the results of Gel'bshtein *et al.* (1964) show that their data are consistent with $pK_{BH^+} = -4{\cdot}63$ for 2,4-dinitroaniline rather than with the figure of −5·2 deduced by Downing and Pearson (1961). However if the H_0 scale is recalculated ($60\% < C_{P_2O_5} < 80\%$) using $pK_{BH^+} = -4{\cdot}63$, the acidity function is then not in satisfactory agreement with the earlier measurements of Gel'bshtein *et al.* (1956a, 1956c). Despite the apparent agreement between independently determined H_0 acidity functions for aqueous P_2O_5 it seems that certain anomalies between the results do exist when $C_{P_2O_5} > 60$ wt %. The validity of the scale at the higher P_2O_5 concentrations may therefore be in doubt.

The H_0 acidity function for dilute aqueous solutions of P_2O_5 reflects the fact that H_3PO_4 is a weak acid in water. Thus the first ionization constant K_1 is given by $pK_1 = 2{\cdot}148$ at 25°C (Bates, 1951). At P_2O_5 concentrations below 67·4 wt % the P_2O_5 exists predominantly as orthophosphoric acid (H_3PO_4) in aqueous solution. At the composition mole fraction ratio $(X_{H_2O}/X_{P_2O_5}) = 3$, which formally corresponds to 100% H_3PO_4, ca. 86% of the P_2O_5 exists as H_3PO_4, ca. 14% as pyrophosphoric acid, $H_4P_2O_7$, and a fraction of a percent as triphosphoric acid, $H_5P_3O_{10}$ (Jameson, 1959). As the concentration of P_2O_5 is further increased higher polymers with the general formula $H_{n+2}P_nO_{3n+1}$ are formed. No attempt has been made to correlate the acidity of the solutions containing appreciable concentrations of polymeric acids with their composition.

2.2.11. *Boron Trifluoride/Water Mixtures*

The Hammett acidity function for boron trifluoride/water mixtures has been measured by Vinnik *et al.* (1957b). For molar ratios $0{\cdot}07 < (X_{BF_3}/X_{H_2O}) < 0{\cdot}3$ the acidity function could not be measured because the irreversible hydrolysis of boron trifluoride to boric acid (precipitated out) takes place. As no indicator could be found which gave measurable ionization ratios when $(X_{BF_3}/X_{H_2O}) < 0{\cdot}07$ and when $(X_{BF_3}/X_{H_2O}) > 0{\cdot}3$ it was impossible to refer the results at higher BF_3 concentrations to standard state water by the usual stepwise comparison of successively weaker bases. Only solutions for which $(X_{BF_3}/X_{H_2O}) > 0{\cdot}3$ were therefore studied and the H_0 scale quoted was based on $pK_{BH^+} = -6{\cdot}6$ for 6-bromo-2,4-dinitroaniline (Vinnik *et al.* 1957a). The pK_{BH^+} for the other four indicators used were deduced by stepwise comparison with 6-bromo-2,4-dinitroaniline. The H_0 scale given in Table 2.20 was interpolated from a smooth curve drawn through the experimental points for all five indicators.

TABLE 2.20

The Hammett H_0 acidity function for boron trifluoride/water mixtures at 24 ± 2°C (Vinnik *et al.* 1957b)

(X_{BF_3}/X_{H_2O})	$-H_0$	(X_{BF_3}/X_{H_2O})	$-H_0$	(X_{BF_3}/X_{H_2O})	$-H_0$
0·33	4·42	0·60	7·66	0·85	9·54
0·35	4·65	0·65	8·00	0·90	9·95
0·40	5·35	0·70	8·34	0·95	10·41
0·45	6·20	0·75	8·71	0·975	10·80
0·50	6·85	0·80	9·12	0·996	11·4
0·55	7·31				

Indicators used were (pK_{BH^+} in brackets): 2,4-dinitroaniline (–4·44); 6-bromo-2,4-dinitroaniline (–6·6); anthraquinone (–8·6); 2,4,6-trinitroaniline (–9·28); 4-nitrotoluene (–10·89).

Boron trifluoride forms the stable hydrates $BF_3.H_2O$ and $BF_3.2H_2O$ with water (Greenwood and Martin, 1954). Both adducts are unionized as solids (Ford and Richards, 1956) but are ionized to some extent as liquids (Greenwood and Martin, 1951, 1953). Thus the dihydrate ($X_{BF_3}/X_{H_2O} = 0{\cdot}5$) exists in the liquid state predominantly in the ionic form $H_3O^+.BF_3OH^-$ (Wamser, 1951). The Hammett acidity of the liquid dihydrate, $H_0 = -6{\cdot}85$, is of the same order as the acidity of 100% nitric acid. The monohydrate $BF_3.H_2O$ in water also behaves as a strong acid and is largely ionized to the ionic form of the dihydrate. The acidity of the pure liquid monohydrate ($X_{BF_3}/X_{H_2O} = 1$), $H_0 < -11{\cdot}4$, is similar to that of anhydrous sulphuric acid and liquid hydrogen fluoride. The composition of BF_3/H_2O mixtures up to a BF_3 content corresponding to the dihydrate has been investigated by Wamser (1951). At equilibrium the solutions probably contain the species HBF_4, HBF_3OH, $HBF_2(OH)_2$, $HBF(OH)_3$, and $B(OH)_3$. The acid strength of HBF_4 is decreased as F is progressively substituted by OH. Thus HBF_4 is about as strong as HCl whereas for $B(OH)_3$ $pK_a = 9{\cdot}23$ at 25°C (Manov *et al.* 1944). The ionization of these compounds contributes to the overall acidity of the solutions by an amount dependent on their equilibrium concentrations and their acid ionization constants.

2.2.12. *Potassium Hydrogen Sulphate*

The anomalous (Paul and Long, 1957) H_0 scale determined by Bell and Brown (1954) for aqueous potassium hydrogen sulphate has been corrected by Satchell (1958). The results of Satchell's measurements are given in Table 2.21. The acidity function is in accord with expectation for solutions of a weak

TABLE 2.21

The Hammett H_0 acidity function for aqueous potassium hydrogen sulphate at 23 ± 2°C (Satchell, 1958)

C_{KHSO_4} (mole litre^{-1})	H_0	C_{KHSO_4} (mole litre^{-1})	H_0	C_{KHSO_4} (mole litre^{-1})	H_0
0·25	0·96	1·00	0·36	2·00	−0·03
0·50	0·68	1·50	0·14	2·25	−0·15

Indicators used were (pK_{BH+} in brackets, as given by Paul and Long, 1957): 4-nitroaniline (0·99); diphenylamine (0·78); 2-nitroaniline (−0·29).

acid. Thus a solution of $KHSO_4$ is slightly more acidic than a solution of equal concentration of phosphoric acid. This is consistent with the relative values of the ionization constants of the HSO_4^- ion, $pK_a(25°) = 2{\cdot}0$ (Redlich, 1946; Marshall and Jones, 1966), and of H_3PO_4, $pK_a(25°) = 2{\cdot}15$ (Bates, 1951).

2.2.13. *Carboxylic Acids*

Bascombe and Bell (1959) were unable to determine the Hammett acidity function for aqueous formic acid because irreversible reactions took place in the solutions. However an H_0 scale has been measured by Milyaeva (1958) who used 2-nitroaniline as indicator and by Stewart and Mathews (1960) who used 4-nitroaniline, 2-nitroaniline and 4-nitrodiphenylamine as indicators. The two scales (Table 2.22) deviate widely and this confirms Bascombe and

TABLE 2.22

The Hammett H_0 acidity function for aqueous formic acid

wt % HCOOH	C_{HCOOH} (mole litre^{-1})	$H_0(17°)$†	$H_0(25°)$‡
1	0·215		1·97
2	0·437		1·72
5	1·104	0·45	1·61
10	2·22	0·34	1·50
20	4·56	0·19	1·34
30	7·00	0·03	1·15
40	9·52	−0·09	0·96
50	12·20	−0·21	0·76
60	14·92	−0·35	0·55
70	17·74	−0·44	0·23
80	20·61	−0·55	−0·21
90	23·51		−0·72
95	25·1		−1·12
98	26·0		−1·51
100	26·6		−2·22

† Milyaeva (1958). 2-Nitroaniline, $pK_{BH^+}(17°) = -0·17$, indicator. This scale is not consistent with the ionization constant of formic acid in water (see text).

‡ Stewart and Mathews (1960). Indicators, $pK_{BH^+}(25°)$ in brackets, were: 4-nitroaniline (0·99); 2-nitroaniline (−0·29); 4-nitrodiphenylamine (−2·48).

Bell's conclusion that studies of amine protonation in formic acid may be subject to error. The deviation cannot be entirely due to the temperature difference for the two sets of data although this may contribute a small amount. For solutions containing moderate concentrations of a weak acid in water the approximate equation (2.33) should be applicable (Randles and Tedder,

$$H_0 \approx \tfrac{1}{2}pK_a - \tfrac{1}{2}\log_{10} C_{acid} \quad (2.33)$$

1955). K_a is the acid ionization constant of the weak acid. The H_0 acidity function deduced by Stewart and Mathews (1960) for aqueous formic acid, $pK_a(25°) = 3·752$ (Harned and Embree, 1934), is consistent with this equation. Their scale therefore appears to be reliable whereas that deduced by Milyaeva (1958) does not.

The experimental H_0 acidity functions for acetic acid, monochloracetic acid, dichloracetic acid and trichloracetic acid are given in Table 2.23. Bascombe

TABLE 2.23

The Hammett H_0 acidity functions for aqueous solutions of four aliphatic carboxylic acids

C_{acid} (mole litre^{-1})	$H_0(17°)$† CH_3COOH	$H_0(25°)$‡ $CH_2ClCOOH$	$H_0(25°)$§ $CHCl_2COOH$	$H_0(25°)$∥ CCl_3COOH
0·1			1·30	
0·2			1·02	0·81
0·5		1·63	0·76	0·40
1	0·90	1·54	0·66	0·10
2	0·52	1·47	0·63	0·00
3	0·42	1·47	0·63	0·00
4	0·36	1·47	0·63	0·00
5	0·33	1·43	0·63	−0·01
6	0·33	1·35	0·60	−0·10
7	0·32	1·25	0·50	−0·37
8	0·30	1·08	0·34	−0·78
9	0·29	0·91	0·08	−1·18
10	0·27	0·76	−0·20	
11	0·24	0·64	−0·61	
12	0·20		−0·75	
13	0·15			
14	0·11			
15	0·06			
16	0·00			

† Milyaeva (1958).
‡ Bascombe and Bell (1959).
§ Bell and Brown (1954); Bascombe and Bell (1959); Satchell (1958).
∥ Randles and Tedder (1955); Burkett *et al.* (1956); Bascombe and Bell (1959).

and Bell (1959) were unable to obtain satisfactory results from a study of the ionization of amine indicators in aqueous acetic acid. However Milyaeva (1958) measured an H_0 scale for aqueous acetic acid using 2-nitroaniline, $pK_{BH^+}(17°) = -0{\cdot}17$, as indicator. The scale is included in Table 2.23 but as for Milyaeva's formic acid results equation (2.33) is not obeyed and the data are probably unreliable. The acidity function for aqueous monochloracetic acid was determined by Bascombe and Bell (1959). The scale for dichloracetic acid has been interpolated from a smooth curve drawn through a plot of the H_0 results of Bascombe and Bell (1959), Bell and Brown (1954) and Satchell (1958). It is interesting to note that Satchell (1958) found that the secondary amines diphenylamine and 4-nitrodiphenylamine showed a quite different acidity function behaviour in dichloracetic acid solutions from that of primary amines. Hammett and Paul (1934) and Bell and Brown (1954) have studied trichloracetic acid solutions. More recent measurements, all in satisfactory agreement with each other, have been made by Randles and Tedder (1955), Burkett *et al.* (1956) and Bascombe and Bell (1959). The scale given in Table 2.23 for aqueous trichloracetic acid solutions has been deduced from these three independent studies.

The relative values of the H_0 acidity functions for monochlor-, dichlor- and trichloracetic acids at a particular acid concentration are in the order expected from the acid ionization constants of the three acids. At 25°C for monochloracetic acid $pK_a = 2{\cdot}87$ (Ives and Pryor, 1955), for dichloracetic acid $pK_a \approx 1{\cdot}4$ (Ciapetta and Kilpatrick, 1948; Harned and Hawkins, 1928) and for trichloracetic acid $pK_a = 0{\cdot}63$ (Halban and Brüll, 1944). Thus consideration of equation (2.33) for $C_{\text{acid}} = 1$ mole litre^{-1} shows that the three H_0 values are of the same order of magnitude as $0{\cdot}5\ pK_a$. The H_0 scale for acetic acid, for which $pK_a(17°) = 4{\cdot}76$ (Harned and Ehlers, 1932), is anomalous. Formic acid, $pK_a = 3{\cdot}75$ (Harned and Embree, 1934), gives as expected, solutions in water which are less acidic than those of monochloracetic acid at the same concentration.

Randles and Tedder (1955) measured the Hammett acidity function for aqueous solutions of trifluoracetic acid (Table 2.24) and their results were

TABLE 2.24

The Hammett H_0 acidity function for aqueous trifluoracetic acid (Randles and Tedder, 1955)†

X_{CF_3COOH} (mole fraction)	H_0	X_{CF_3COOH} (mole fraction)	H_0	X_{CF_3COOH} (mole fraction)	H_0
0·000 75	1·40	0·009	0·34	0·08	−0·37
0·001	1·22	0·010	0·29	0·09	−0·40
0·002	0·85	0·02	−0·02	0·10	−0·43
0·003	0·71	0·03	−0·14	0·2	−0·68
0·004	0·60	0·04	−0·20	0·3	−0·98
0·005	0·52	0·05	−0·26	0·4	−1·44
0·006	0·46	0·06	−0·30	0·5	−1·97
0·007	0·40	0·07	−0·34	0·6	−2·39
0·008	0·36				

Indicators used were (pK_{BH^+} in brackets): 4-nitroaniline (0·99); 2-nitroaniline (−0·25); 4-chloro-2-nitroaniline (−1·06).

† See also: Satchell (1958).

confirmed by Satchell (1958). Trifluoracetic acid is a fairly strong acid with $pK_a(25°) = -0{\cdot}26$ (Hood *et al.* 1955; but see also Henne and Fox, 1951) and the solutions are therefore more acidic than those of the other aliphatic carboxylic acids for which acidity functions have been measured. At high concentrations the acidity of the solutions is influenced by dielectric constant. Thus more ion association occurs in concentrated trichloracetic acid solutions than in concentrated trifluoracetic acid solutions and this partly accounts for the lower acidities of the former. Randles and Tedder (1955) noted for trifluoracetic acid ($0{\cdot}15 > X_{CF_3COOH} > 0{\cdot}03$) that a graph of H_0 against the logarithm of the acid mole fraction was approximately linear with slope 0·5. This is in accord with equation (2.33), and hence K_a = ca. 0·7 was deduced. This is in reasonable agreement with other values of 0·59 (Henne and Fox, 1951) and 1·8 (Hood *et al.* 1955).

2.2.14. *Sulphonic Acids*

The Hammett acidity function for aqueous methanesulphonic acid (Table 2.25) has been determined by Bascombe and Bell (1959). The solutions are similar in acidity to those of aqueous nitric acid. This is consistent with the

TABLE 2.25

The Hammett H_0 acidity function for aqueous methanesulphonic acid at 25°C (Bascombe and Bell, 1959)

C_{acid} (mole litre^{-1})	wt %	$-H_0$	C_{acid} (mole litre^{-1})	wt %	$-H_0$
0·302	3·0	−0·43	9·38	67·3	3·34
0·749	6·8	0·03	10·46	73·3	3·91
1·557	13·9	0·37	10·78	74·9	4·40
1·925	17·0	0·54	10·85	75·3	4·07
2·51	21·9	0·65	11·85	80·6	4·59
2·98	25·5	0·79	12·12	82·1	4·84
3·30	28·0	0·88	12·74	85·2	5·39
4·82	39·5	1·37	13·45	88·9	6·12
6·16	48·3	1·89	13·95	91·4	6·61
6·44	50·2	2·05	14·77	95·7	7·43
7·15	54·5	2·30	15·07	100·0	7·86
8·32	61·5	2·82			

Indicators were (pK_{BH^+} in brackets): 4-nitroaniline (1·02); 2-nitroaniline (−0·29); 4-chloro-2-nitroaniline (−1·02); 4,6-dichloro-2-nitroaniline (−3·61); 4-methyl-2,6-dinitroaniline (−3·96); 6-bromo-2,4-dinitroaniline (−6·64).

high ionization constant K_a of methanesulphonic acid in water. From Raman spectra Clarke and Woodward (1966) deduced that $K_a(25°) = 73$ mole litre^{-1}, whereas a proton magnetic resonance study gave $K_a = 16$ mole litre^{-1} (Covington and Lilley, 1967). Dawber (1968) has calculated degrees of dissociation of methanesulphonic acid in water from the acidity function data. The values are consistent with the above figures for pK_a but unfortunately do not give an indication of which is more correct. The degree of dissociation of methanesulphonic acid calculated by Bascombe and Bell (1959) from the H_0 scale are not consistent with the results of the spectroscopic studies of this acid in water.

The H_0 scale for 4-toluenesulphonic acid in water (Arnett and Mach, 1966) shows that this acid is not very effective for the protonation of weak bases. This arises because the 4-toluenesulphonate anion has a large salting-in effect on neutral solute species. Thus y_B (equation 1.29) is markedly decreased and H_0 decreases less rapidly with increasing acid concentration than would be expected for such a strong acid.

2.3. Comparisons between H_0 for Different Acids

In the above sections specific comments on the Hammett acidity functions for individual acids have been made where appropriate. We shall now consider more general theoretical treatments which have been found to give a satisfactory account of the acidity of solutions of both strong and weak acids in water.

2.3.1. *Strong Acids*

Why does the acidity (as defined by H_0) of aqueous solutions of strong acids increase much more rapidly than the stoichiometric acid concentration? Consideration of equation (1.29) would suggest that the difference between H_0 and C_{acid} (assumed equal to C_{H^+} for strong acids) results from variations of the activity coefficients of the species involved in equilibrium (1.25) with changing acid concentration. Thus the identity $(y_{H^+}y_B/y_{BH^+}) = 1$ may confidently be taken as applicable in dilute solutions but as the acid concentration increases it ceases to be valid. This interpretation of H_0 would require that $(y_{H^+}y_B/y_{BH^+})$ should become increasingly greater than unity as the acid concentration is raised. However Bascombe and Bell (1957; Bell, 1959) noted that the H_0 values for solutions of different strong acids (HCl, HBr, $HClO_4$, H_2SO_4) at the same molality were, within experimental error, the same up to $C_{acid} =$ ca. 8 mole litre^{-1}. This confirms the expectation that the acidity of the solutions is independent of the anions in solution and suggests that the ability of the acid solution to protonate a weakly basic neutral solute depends in some way upon the ration (C_{H^+}/C_{H_2O}). The related observation that the Hammett acidity function is a unique function of water activity for the acids H_2SO_4, HCl, $HClO_4$ and HNO_3 was made by Wyatt (1957). The agreement between the plots of $-H_0$ and a_w is illustrated in Fig. 2.3 (Yates and Wai, 1964) for aqueous $HClO_4$ and H_2SO_4. Acidity functions are those of Yates and Wai (1964) for $HClO_4$ and Jorgenson and Hartter (1963) for H_2SO_4, and water activities are given by Robinson and Baker (1946) and Robinson and Stokes (1959) respectively. Both Bascombe and Bell (1957) and Wyatt (1957) concluded from these correlations that solvation effects were playing a significant

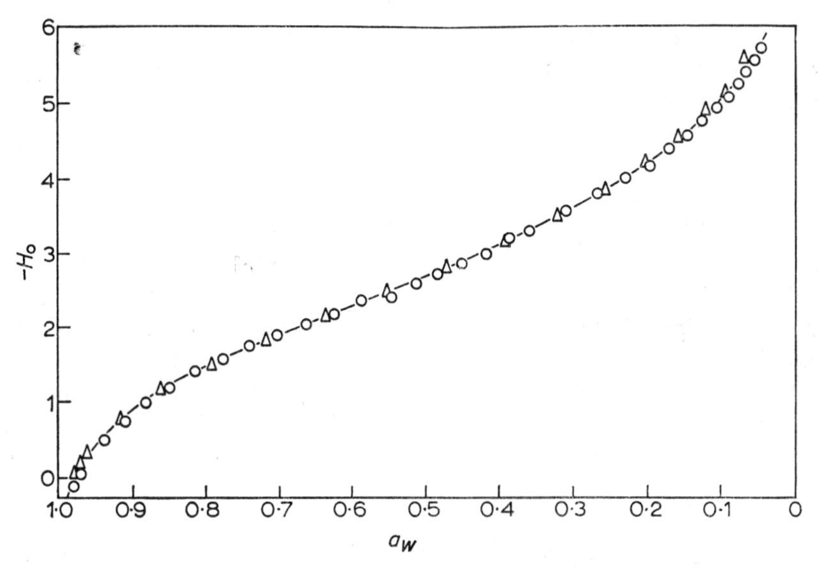

FIG. 2.3. Correlation of the Hammett H_0 acidity function and water activity. $\triangle = HClO_4$; $\circ = H_2SO_4$. (Reproduced by permission from K. Yates and H. Wai, 1964.)

role in determining the Hammett acidity of concentrated acid solutions. There emerged from this idea two different approaches which have since been extensively developed. A further related theory has also been suggested more recently by Högfeldt (1960b, 1961).

Bascombe and Bell (1957) proposed that the predominant factor contributing to the variation of $(y_{H^+}y_B/y_{BH^+})$ with acid concentration was the effect of solvation of the proton on y_{H^+}. The dependence of ionic activity coefficients in concentrated electrolyte solutions on the solvation requirements of the ions concerned has been treated theoretically by Stokes and Robinson (1948; Robinson and Stokes, 1959) and Glueckauf (1955, 1959). Bascombe and Bell used a simplified model based on the well-established (Azzam, 1962) hydration number of 4, for the proton in water. They considered that the protonation equilibrium of a neutral base in aqueous acid is correctly written as equation

(2.34) in which hH_2O is the number of water molecules which solvate the

$$B + H^+(hH_2O) \rightleftharpoons BH^+ + hH_2O \qquad (2.34)$$

proton. The solvation requirements of B and BH^+ were assumed negligible. Inclusion of a term a_w^h in equation (1.29) leads to equation (2.35). If C_{H^+} is taken as the stoichiometric acid concentration C then $\log_{10} y_B = AC$ where A is

$$-H_0 = \log_{10} C_{H^+} - h \log_{10} a_w + \log_{10}(y_{H^+} y_B / y_{BH^+}) \qquad (2.35)$$

the salting out coefficient for species B (Long and McDevit, 1952). Values of A are usually in the region of 0·1 for aqueous solutions. Equation (2.35) therefore rearranges to

$$-H_0 - \log_{10} C + h \log_{10} a_w = \log_{10} y_B = AC \qquad (2.36)$$

in which the identity $y_{H^+} = y_{BH^+}$ has been assumed. Substitution of known water activities and $h = 4$ in this equation gave linear plots for $C < 8$ mole litre^{-1} with slopes 0·07 for HCl, 0·1 for $HClO_4$ and 0·1 for H_2SO_4. These are typical salting-out coefficients and it may be concluded that it is realistic to explain the high acidity of concentrated solutions of strong acids in terms of the loss of hydration of the proton when a neutral base is protonated.

An alternative correlation which requires more questionable assumptions but eliminates the need to know water activities is based on substitution in equation (2.37) which follows from equation (2.35). The term C_w was taken

$$-H_0 = \log_{10} C - h \log_{10} C_w + \log_{10}(y_{H^+} y_B / y_{BH^+} y_w^h) \qquad (2.37)$$

by Bascombe and Bell as the concentration of water molecules in the solution other than those which were bound as hydration water to the proton. Also the approximation $(y_{H^+} y_B / y_{BH^+} y_w^h) = 1$ was made. Thus equation (2.38), in which

$$-H_0 = \log_{10} C - h \log_{10}(C/m)(1 - 0{\cdot}018\, hm) \qquad (2.38)$$

m is the molality of acid, becomes applicable. The absolute water concentration has been divided b y55·5 so that $C_w \rightarrow 1$ as $C \rightarrow 0$. H_0 values calculated using this equation with $h = 4$ are compared in Table 2.26 with the experimental H_0 scales for aqueous HCl, HBr, $HClO_4$ and H_2SO_4. The measured scales were taken from the Tables of Paul and Long (1957). For the concentrations at which the correlation is valid more recent measurements of H_0 have not appreciably altered the H_0 values.

Despite the many approximations made in this treatment the agreement between the calculated and experimental H_0 scales is very convincing for $C <$ ca. 7 mole litre^{-1}. A solvation number of four for the proton in water is found to give the best fit. This is in accord with other evidence that the proton is hydrated by four water molecules (Wicke *et al.* 1954; Glueckauf, 1955; Glueckauf and Kitt, 1955; Ackermann, 1957; van Panthaleon van Eck *et al.*

TABLE 2.26

Comparison between calculated (equation 2.38) and experimental H_0 acidity functions for four strong acids in water (Bascombe and Bell, 1957)

HBr							
C(mole litre^{-1})	1·00	2·00	3·00	4·00	5·00	6·00	7·00
$-H_0$ calc.	0·18	0·68	1·08	1·46	1·89	2·34	2·86
$-H_0$ exp.	0·20	0·71	1·11	1·50	1·93	2·38	2·85
HCl							
C(mole litre^{-1})	0·95	1·72	3·42	4·35	5·37	6·05	7·00
$-H_0$ calc.	0·14	0·53	1·18	1·52	1·91	2·19	2·60
$-H_0$ exp.	0·17	0·56	1·20	1·52	1·89	2·12	2·48
$HClO_4$							
C(mole litre^{-1})	0·98	1·82	3·05	4·39	5·96	7·79	
$-H_0$ calc.	0·19	0·66	1·23	1·86	2·77	4·38	
$-H_0$ exp.	0·20	0·69	1·26	1·91	2·81	4·19	
H_2SO_4							
C(mole litre^{-1})	1·03	2·00	3·01	4·00	5·32	6·00	6·93
$-H_0$ calc.	0·19	0·69	1·18	1·66	2·25	2·77	3·67
$-H_0$ exp.	0·31	0·84	1·28	1·72	2·42	2·76	3·18

1957; Azzam, 1962). When $0{\cdot}018hm > 1$ the calculation breaks down because this implies that there are then insufficient water molecules to solvate all the protons in the system. For $h = 4$ this occurs at $m =$ ca. 14. The number of water molecules liberated when a neutral base is protonated will almost certainly decrease as the total concentration of water in solution decreases. Bascombe and Bell (1957) considered the solvated proton to exist entirely as $H_9O_4^+$ at all acid concentrations. Other calculations have suggested that all the possible alternatives H_3O^+, $H_5O_2^+$, $H_7O_3^+$, $H_9O_4^+$, ... $H(H_2O)_n^+$ for the solvated proton should be considered (Wyatt, 1957; Högfeldt, 1960b; Perrin, 1964; Robertson and Dunford, 1964). For example Perrin (1964) considered equation (2.39) which follows from equation (2.35). Here the term a_w^h is replaced by $\sum_n q_n a_w^n$ which is summed over all the different solvation numbers n in the solution. The coefficients q_n are the fraction of protons which are solvated as $[H(H_2O)_n]^+$. From the tangents to the graphs of $(-H_0 - \log_{10} C_{H^+})$ against

$$-H_0 = \log_{10} C_{H^+} - \log_{10}\left(\sum_n q_n a_w^n\right) \tag{2.39}$$

($-\log_{10} a_w$) slopes $\bar{n}$ where measured. These are weighted average values of n. Perrin showed that deduction of n as a function of water activity for aqueous perchloric acid enabled calculation of H_0 for perchloric acid–sodium perchlorate solutions via equation (2.40). The calculated values agreed well with

$$-H_0 = \log C_{H^+} + \int_0^{-\log_{10} a_w} \bar{n}\, d(-\log_{10} a_w) \qquad (2.40)$$

the experimental measurements for mixtures with total ionic strengths up to ca. 7 mole litre^{-1}. Robertson and Dunford (1964) have taken a model for sulphuric acid solutions in the range 0·1 mole kg^{-1} $< C_{H_2SO_4} <$ 1000 mole kg^{-1} in which the series of ten proton hydrate species from one solvating water molecule H_3O^+ to ten water molecules $H_{21}O_{10}{}^+$ have been considered. Concentrations of each of the species $[H(H_2O)_n]^+$ have been tabulated as a function of sulphuric acid concentration.

Treatments which consider series of different degrees of solvation for the proton are effectively fitting the acidity function to a power series in water activity. Whether the solvation numbers which emerge from the correlations are realistic remains to be proven. As acid concentrations and ionic strengths increase the usual approximations inherent in the calculations concerning activity coefficients become more and more incorrect. Thus although for the more dilute solutions the results may be meaningful it is unlikely that those at higher concentrations are very significant. However the conclusion that the solvation number of the proton decreases as the water concentration falls is logical. As will be seen later it is also probably better to consider h in equation (2.34) as the difference in solvation between $(B + H^+)$ and BH^+ rather than to neglect the hydration of B and (more particularly) the BH^+ ion.

Wyatt (1957) was the first to correlate H_0 acidity functions with successive hydration equilibria. For dilute ($C_{acid} < 60\%$) solutions of strong acids in water a model was considered in which up to four water molecules solvated the H_3O^+ ion. The activity of water a_w was taken as a measure of the mole fraction X_w of "free" water in solution. "Free" water implies all that water which is not bound to ions in solvation. Wyatt wrote the Deno and Taft (1954) equation (2.5) as $(H_0 + k) = \log_{10}(X_w/X_{H_3O^+})$ and deduced equation (2.41) for the relationship between H_0 and water activity. This equation gave the most

$$H_0 + k + \log_{10}\left(\frac{1 - a_w}{2a_w}\right) = \log_{10}(1 + 20a_w + 150a_w{}^2 + 500a_w{}^3 + 625a_w{}^4) \quad (2.41)$$

satisfactory fit between calculated and experimental H_0 scales for aqueous H_2SO_4, $HClO_4$ and HCl. For 2·5 mole kg^{-1} $< C_{acid} < 60\%$ the agreement was better than 0·1 units if the value 5·02 was taken for the constant k. Despite the many assumptions the concept that in aqueous acid solutions the acidity is largely determined by hydration effects is consistent with these satisfactory correlations. As in equation (2.36) Wyatt's equation (2.41) can be improved by inclusion on the left-hand side of the term $\log_{10} y_B = AC$ (Ojeda and Wyatt, 1964). More recent calculations along these lines by Dawber (1966b) have shown perchloric acid ceases to be completely ionized in water at concentrations greater than 3·5 mole litre^{-1}. Wyatt (1957) has considered the possible

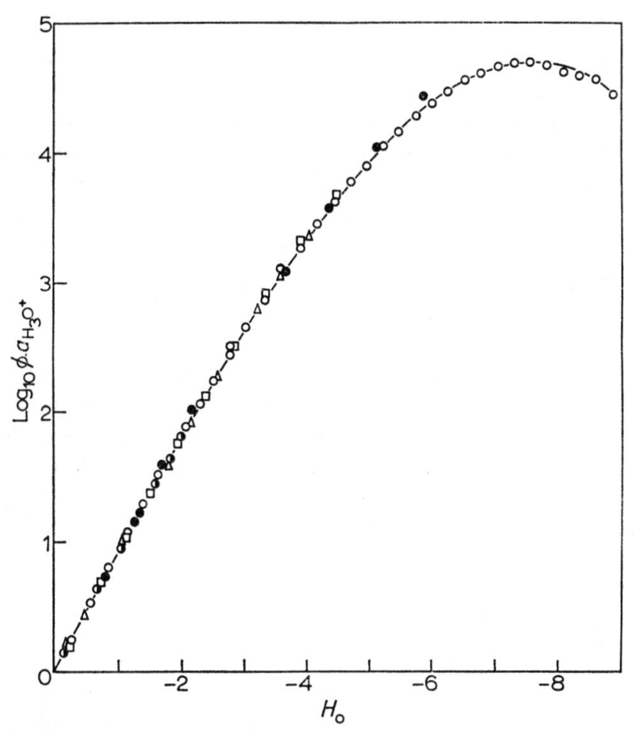

FIG. 2.4. The cation activity functions $\log_{10}\phi.a_{H_3O^+}$ for the acids: ○ H_2SO_4; ● $HClO_4$; ◑ HNO_3; △ HCl; □ HBr. (Reproduced by permission from E. Högfeldt, 1960.)

effect on acidity of the hydration of the anions derived from the strong acids.

The hydration of the anions has also been investigated by Högfeldt (1960a, 1960b). He deduced equation (2.42) which relates the activity a_n of the solvated

$$\log_{10}\phi.a_n = -H_0 + n\log_{10} a_w \tag{2.42}$$

proton $[H(H_2O)_n]^+$ with the Hammett acidity function. The function ϕ equals (y_B/y_{BH^+}) where B is the indicator used to measure H_0. Equation (2.43) is the corresponding "anionic activity function" for the solvated anion $[A(H_2O)_{n'}]^-$ of the strong acid. Figure (2.4) shows the results of Högfeldt's calculations of the cation activity functions (with $n = 1$) for five acids. All the points lie on a common curve and this is consistent with Wyatt's observation (1957) that the

$$\log_{10}\phi^{-1}a_{A^-} = H_0 + n'\log_{10}a_w + \log_{10}(a_{H^+}a_{A^-}) \tag{2.43}$$

H_0 scales for strong acids are a unique function of water activity. For the anions ($n' = 0$) no such agreement was obtained. Thus each anionic activity function showed its own individualistic dependence on H_0. This presumably arises because of the variation in structure and solvation requirements of the different anions. This contrasts with the behaviour of the cationic activity functions for which a particular water activity leads to the same value of ϕ and the same extent of solvation of the proton for all strong acids. For a particular water activity ϕ must also be the same for all Hammett bases (Vinnik and Librovich, 1966). It is interesting that nitric acid, for which $K_a =$ ca. 24 mole litre^{-1} at 25° (McKay, 1956), at low concentrations gives the same correlation between H_0 and a_w as the strong acids $HClO_4$, HCl, HBr and H_2SO_4 (Wyatt, 1957; Högfeldt, 1960b). At high concentrations nitric acid is incompletely dissociated ($\alpha < 1$) and equation (2.31) becomes applicable (Dawber and Wyatt, 1960).

Högfeldt (1960a, 1960b) also deduced equations (2.44) and (2.45) for the average hydration numbers of the protons and anions respectively in solutions of strong acids. The activity coefficients y_{H^+} and y_{A^-} are given by equation (2.46). For sulphuric acid $\bar{n}_{H^+}$ was about 4 in dilute solutions (< ca. 6 mole

$$\bar{n}_{H^+} = -d\log_{10}\phi . y_{H^+}/d\log_{10}a_w \tag{2.44}$$

$$\bar{n}_{A^-} = d\text{–}\log_{10}\phi^{-1}y_{A^-}/d\log_{10}a_w \tag{2.45}$$

$$\phi . y_{H^+} = (\phi . a_{H^+}/C_{H^+}) \qquad \phi^{-1} . y_{A^-} = (\phi^{-1} . a_{A^-}/C_{A^-}) \tag{2.46}$$

litre^{-1}). This agrees with the findings of the Bascombe and Bell treatment (1957). As the H_2SO_4 concentrations increased $\bar{n}_{H^+}$ fell to 1 in concentrated solutions (ca. 16 mole litre^{-1}). The solvation of the HSO_4^- anion fell from 3 to 0 in the same range of concentrations. This treatment therefore presents the realistic picture that as the water concentration is reduced the extent of solvation of the ions decreases and this is reflected in the measured acidity of the solutions. This is in accord with Wyatt's approach (1957). In moderately concentrated solutions the ClO_4^- and HSO_4^- ions are solvated to about the same extent, whereas above ca. 10 mole litre^{-1} ClO_4^- seems to be more solvated than HSO_4^-. The solvation number of the NO_3^- anion is less by ca. 3 than those for ClO_4^- and HSO_4^- in dilute solutions.

Zarakhani and Vinnik (1962) noted that the proton magnetic resonance chemical shift δ_H for aqueous solutions of $HClO_4$, HCl, H_2SO_4 and HNO_3 varied linearly with the Hammett acidity function. Also $(-H_0 + \log_{10}a_w)$, which equals Högfeldt's $\log_{10}\phi . a_{H_3O^+}$ activity function, gave straight lines through the origin when plotted against δ_H. Thus the chemical shift is apparently directly proportional to the acidity of the solutions as measured by H_0. The four acids give plots with different slopes presumably because of the differing effects of the anions which are present. Campbell (1963) has correlated p.m.r. chemical shifts with H_0 for aqueous hydrogen fluoride solutions.

2.3.2. *Weak Acids*

For dilute (< ca. 1 mole litre^{-1}) solutions of weak acids equation (2.33) should be applicable (Randles and Tedder, 1955). In Table 2.27 H_0 when $C_{acid} = 1$ mole litre^{-1} is compared with $\frac{1}{2}pK_a$ for a series of acids. The agreement

TABLE 2.27

Test of equation (2.33) for weak acids in water. H_0 when $C_{acid} = 1$ mole litre^{-1} compared with pK_a. Data taken from references given in Section 2.2

Acid	pK_a	$\frac{1}{2}pK_a$	H_0	$(\frac{1}{2}pK_a - H_0)$
HCOOH	3·75	1·88	1·62	0·26
HF	3·17	1·58	1·20	0·38
CH_2ClOOH	2·87	1·44	1·54	−0·10
H_3PO_4	2·15	1·08	0·63	0·45
HSO_4^-	2·0	1·00	0·36	0·64
H_3PO_3	1·80	0·90	0·42	0·48
HIO_4	1·55	0·78	0·58	0·20
$CHCl_2COOH$	1·4	0·70	0·66	0·04
HIO_3	0·80	0·40	0·18	0·32
CCl_3COOH	0·63	0·32	0·10	0·22
CF_3COOH	−0·26	−0·13†	0·04	0·17
$MeSO_3H$	−1·20	−0·60†	−0·15	0·45

† These acids are too strong for equation (2.33) to be applicable. Thus for a strong ($\alpha = 1$) acid when $C = 1$ mole litre^{-1} $H_0 = -\log_{10} C = 0$. Experimental values of H_0 for strong acids at $C = 1$ mole litre^{-1} are $H_0 =$ ca. −0·2.

is not particularly good but the increase in acidity (decrease in H_0) as pK_a decreases is as expected. In general when $C_{acid} = 1$ mole litre^{-1} the solutions are more acidic than equation (2.33) would predict. As $C_{acid} \rightarrow 0$ the equation should become less approximate. At higher concentrations equation (2.31) may be used to correlate H_0 with the degree of ionization α of weak acids (Wyatt, 1957). This equation has been proved satisfactory for aqueous nitric acid (Dawber and Wyatt, 1960) but is not applicable for iodic acid (Dawber, 1965).

Extensions of the Bascombe and Bell (1957) equations (2.36) and (2.38) have been used to correlate H_0 with water activities or concentrations for weak acids.

Thus equation (2.47) is found to give a reasonable fit of the H_0 acidity function

$$-H_0 = \log_{10} \alpha_m - h \log_{10} (C/m)(1 - 0{\cdot}018\, \alpha m) \tag{2.47}$$

for nitric acid (Bascombe and Bell, 1959; Bell, 1959). For iodic acid the fit is not so good. However Dawber (1965) has noted that values of α calculated

$$-H_0 = \log_{10} \alpha C + 4 \log_{10} a_w \tag{2.48}$$

for HIO_3 using equation (2.48) are consistent with the corresponding values deduced from conductivity measurements (Abel *et al.* 1934).

Degrees of dissociation of methanesulphonic acid calculated by Bascombe and Bell's method are in poor agreement with Raman and proton magnetic resonance studies of that acid (Bascombe and Bell, 1959; Bell, 1959; Dawber, 1968a). Dawber (1968a) has concluded that the simple treatment which is based on $H_9O_4^+$ for the hydrated proton is unsatisfactory. Calculations for aqueous perchloric acid support this view (Dawber, 1966a, 1966b). It is probably better to incorporate the concept of successive hydration equilibria in the calculations. Applying Wyatt's (1957) approach to weak acids equation (2.42) becomes (Dawber, 1966b, 1968a, 1968b):

$$-\left\{\log_{10}\left[\frac{1 - a_w}{a_w F(1 + 1/\alpha)}\right] + H_0\right\} = A\alpha C + k \tag{2.49}$$

where

$$F = 1 + 20a_w + 150a_w^2 + 500a_w^3 + 625a_w^4 \tag{2.50}$$

If $\alpha = 1$ equation (2.49) reduces to equation (2.42) with the inclusion of the term AC for $\log_{10} y_B$ (Ojeda and Wyatt, 1964). Degrees of dissociation for methanesulphonic acid calculated via equation (2.49) are consistent with the Raman (Clarke and Woodward, 1966) and p.m.r. (Covington and Lilley, 1967) results. The H_0 scales for aqueous HNO_3 and HIO_3 may also be calculated via equation (2.49) (Dawber, 1968b). Agreement between the calculated and experimental scales is slightly improved if the effect on H_0 of the formation of complex anions of the type $A^-.HA$ is considered. Dawber (1966b) plotted the left-hand side of equation (2.49) with $\alpha = 1$ against C_{HClO_4} for perchloric acid. The graph was linear up to ca. 3·7 mole litre^{-1} and this shows that $HClO_4$ is completely ionized in water up to that concentration. The slope of the line was 0·053 which is a reasonable figure for the salting-out coefficient. At higher concentrations the function being plotted decreased as C_{HClO_4} increased. This strongly suggests that the $HClO_4$ is then incompletely ionized. A similar conclusion was also reached from a plot of the left-hand side of equation (2.36), with $h = 4$, against perchloric acid concentration (Dawber, 1966a). This also showed a linear section up to ca. 3·8 mole litre^{-1} and then passed through a maximum indicating that $\alpha < 1$. Högfeldt's (1960b) approach is also applicable for weak acids (Högfeldt, 1961; Dawber, 1966b). In accord with the Bascombe and Bell and the Wyatt treatments it leads to the conclusion that perchloric acid becomes incompletely ionized in the region of C_{HClO_4} = ca. 3–4 mole litre^{-1} (Dawber 1966b).

2.4. Ionization Constants for Hammett Bases

Paul and Long (1957) tabulated pK_{BH+} for indicators which had been used to establish Hammett acidity functions for aqueous acids between 1932 and 1957. The data was summarized as a set of "best values" of pK_{BH+} for the bases. Since then further results have been obtained which suggest that some of these values are incorrect. This is particularly so for some of the weaker bases whose quoted pK_{BH+} was based on a stepwise extrapolation via results for several indicators which have since been proved to be unsatisfactory for the

TABLE 2.28

"Best values" of pK_{BH+} for primary amine Hammett bases in water at 25°

Base	pK_{BH+}	References†
3-Nitroaniline	2·50	a, b, e
2-4-Dichloroaniline	2·02	a, e
4-Nitroaniline	0·99	a, b, d, e, h
2-Nitroaniline	−0·29	a, b, c, d, e, f, h
4-Chloro-2-nitroaniline	−1·03	a, b, c, e, g, h
5-Chloro-2-nitroaniline	−1·54	d, f
2,5-Dichloro-4-nitroaniline	−1·82	c, d, g
2-Chloro-6-nitroaniline	−2·46	c, d, g
2,6-Dichloro-4-nitroaniline	−3·24	c, g
2,4-Dichloro-6-nitroaniline	−3·29	a, b, d, e, f, h
2,6-Dinitro-4-methylaniline	−4·28	a, b, e
2,4-Dinitroaniline	−4·48	a, d, e, f, g, h
2,6-Dinitroaniline	−5·48	c, d, g
4-Chloro-2,6-dinitroaniline	−6·17	c, d, g
6-Bromo-2,4-dinitroaniline	−6·71	a, b, c, d, g, h
3-Methyl-2,4,6-trinitroaniline	−8·37	c, d, g
3-Bromo-2,4,6-trinitroaniline	−9·62	c, g
3-Chloro-2,4,6-trinitroaniline	−9·71	c
2,4,6-Trinitroaniline	−10·04	c, d

† a. Paul and Long (1957); b. Bascombe and Bell (1959); c. Jorgenson and Hartter (1963); d. Ryabova *et al.* (1966); e. Högfeldt and Bigeleisen (1960); f. Vinnik *et al.* (1956); g. Yates and Wai (1964); h. Gel'bshtein *et al.* (1964).

strict definition of H_0. As Paul and Long's summary proved very useful in many later acidity function studies the Table 2.28, containing revised best pK_{BH^+} values, is included here. Data for several bases which had not been studied before 1957 are also given. All the indicators listed are primary amines and are therefore suitable for the measurement of Hammett H_0 acidity functions. As recommended by Paul and Long (1957) many published acidity function scales are referred to the reference state pure water via $pK_{BH^+} = 0{\cdot}99$ for 4-nitroaniline. This value has deliberately been kept the same in Table 2.28. In any case more recent determinations (refs. b, d, e, h in Table 2.28) suggest that this figure is correct to within $\pm$ 0·03 units.

REFERENCES

Abel, E., Redlich, O., and Hersch, P. (1934). *Z. phys. Chem. A*, **170**, 112.
Ackermann, Th. (1957). *Discuss. Faraday Soc.* **24**, 180.
Akitt, J. W., Covington, A. K., Freeman, J. G., and Lilley, T. H. (1965). *Chem. Comm.* 349.
Arnett, E. M., and Mach, G. W. (1964). *J. Amer. Chem. Soc.* **86**, 2671.
Arnett, E. M., and Mach, G. W. (1966). *J. Amer. Chem. Soc.* **88**, 1177.
Azzam, A. M. (1962). *Z. phys. Chem.* **32**, 309.
Bascombe, K. N., and Bell, R. P. (1957). *Discuss. Faraday Soc.* **24**, 158.
Bascombe, K. N., and Bell, R. P. (1959). *J. Chem. Soc.* 1096.
Bass, S. J., and Gillespie, R. J. (1960). *J. Chem. Soc.* 814.
Bass, S. J., Gillespie, R. J., and Robinson, E. A. (1960). *J. Chem. Soc.* 821.
Bates, R. G. (1951). *J. Res. Natl. Bur. Std.* **47**, 127.
Bell, R. P. (1959). "The Proton in Chemistry", Ch. 6. Methuen, London.
Bell, R. P., and Brown, A. H. (1954). *J. Chem. Soc.* 774.
Bell, R. P., Dowding, A. L., and Noble, J. A. (1955). *J. Chem. Soc.* 3106.
Bell, R. P., Bascombe, K. N., and McCoubrey, J. C. (1956). *J. Chem. Soc.* 1286.
Bethell, D. E., and Sheppard, N. (1953). *J. Chem. Phys.* **21**, 1421.
Bonner, T. G., and Lockhart, J. C. (1957). *J. Chem. Soc.* 2840.
Bonner, T. G., and Phillips, J. (1966). *J. Chem. Soc. B.* 650.
Boyd, R. H. and Wang, C-H. (1965). *J. Amer. Chem. Soc.* **87**, 430.
Brand, J. C. D. (1950). *J. Chem. Soc.* 997.
Brand, J. C. D., and Rutherford, A. (1952). *J. Chem. Soc.* 3916.
Brand, J. C. D., Horning, W. C., and Thornley, M. B. (1952a). *J. Chem. Soc.* 1374.
Brand, J. C. D., James, J. C., and Rutherford, A. (1952b). *J. Chem. Phys.* **20**, 530.
Brand, J. C. D., James, J. C., and Rutherford, A. (1953). *J. Chem. Soc.* 2447.
Braude, E. A. (1948). *J. Chem. Soc.* 1971.
Broene, H. H., and De Vries, T. (1947). *J. Amer. Chem. Soc.* **69**, 1644.
Burkett, H., Murphy, R., and Yarian, D. (1956). *Proc. Indiana Acad. Sci.* **66**, 86.
Campbell, M. D. (1963). Thesis. Purdue University, Indiana.
Ciapetta, F. G., and Kilpatrick, M. (1948). *J. Amer. Chem. Soc.* **70**, 633.
Clarke, J. H. R., and Woodward, L. A. (1966). *Trans. Faraday Soc.* **62**, 2226.
Coryell, C. D., and Fix, R. C. (1955). *J. Inorg. Nuclear Chem.* **1**, 119.
Covington, A. K., and Lilley, T. H. (1967). *Trans. Faraday Soc.* **63**, 1749.
Covington, A. K., Tait, M. J., and Wynne-Jones, W. F. K. (1965). *Proc. Roy. Soc. A.* **286**, 235.

Crouthamel, C. E., Hayes, A. M., and Martin, D. S. (1951). *J. Amer. Chem. Soc.* **73**, 82.
Dawber, J. G. (1965). *J. Chem. Soc.* 4111.
Dawber, J. G. (1966a). *Chem. Comm.* 3.
Dawber, J. G. (1966b). *J. Chem. Soc. A*, 1056.
Dawber, J. G. (1968a). *Chem. Comm.* 58.
Dawber, J. G. (1968b). *J. Chem. Soc. A.* 1532.
Dawber, J. G., and Wyatt, P. A. H. (1960). *J. Chem. Soc.* 3589.
de Fabrizio, E. (1966). *Ricerca Sci.* **36**, 1321.
Deno, N. C., and Taft, R. W. (1954). *J. Amer. Chem. Soc.* **76**, 244.
Downing, R. G., and Pearson, D. E. (1961). *J. Amer. Chem. Soc.* **83**, 1718.
Drucker, C. (1937). *Trans. Faraday Soc.* **33**, 660.
Ford, P. T., and Richards, R. E. (1956). *J. Chem. Soc.* 3870.
Gel'bshtein, A. I., Shcheglova, G. G., and Temkin, M. I. (1956a). *Doklady Akad. Nauk S.S.S.R.* **107**, 108.
Gel'bshtein, A. I., Shcheglova, G. G., and Temkin, M. I. (1956b). *Russ. J. Inorg. Chem.* **1**, 167.
Gel'bshtein, A. I., Shcheglova, G. G., and Temkin, M. I. (1956c). *Zhur. neorg. Khim.* **1**, 282.
Gel'bshtein, A. I., Airapetova, R. P., Shcheglova, G. G., and Temkin, M. I. (1964). *Russ. J. Inorg. Chem.* **9**, 816.
Giaque, W. F., Hornung, E. W., Kunzler, J. E., and Rubin, T. R. (1960). *J. Amer. Chem. Soc.* **82**, 62.
Gillespie, R. J. (1966). "Chemical Physics of Ionic Solutions" (B. E. Conway and R. G. Barradas, eds), p. 599. Wiley, New York.
Gillespie, R. J. (1967). "Physico-chemical Processes in Mixed Aqueous Solvents" (F. Franks, ed.), p. 129. Heinemann, London.
Gillespie, R. J., and Cole, R. H. (1956). *Trans. Faraday Soc.* **52**, 1325.
Gillespie, R. J., and Malhotra, K. C. (1967). *J. Chem. Soc. A*, 1994.
Gillespie, R. J., and Oubridge, J. V. (1956). *J. Chem. Soc.* 80.
Gillespie, R. J., and Robinson, E. A. (1962). *Canad. J. Chem.* **40**, 658.
Gillespie, R. J., and White, R. F. M. (1958). *Trans. Faraday Soc.* **54**, 1846.
Gillespie, R. J., and White, R. F. M. (1960). *Canad. J. Chem.* **38**, 1371.
Gillespie, R. J., Hughes, E. D., and Ingold, C. K. (1950). *J. Chem. Soc.* 2552.
Glueckauf, E. (1955). *Trans. Faraday Soc.* **51**, 1235.
Glueckauf, E. (1959). "The Structure of Electrolytic Solutions" (W. J. Hamer, ed.), p. 97. Wiley, New York.
Glueckauf, E., and Kitt, G. P. (1955). *Proc. Roy. Soc. A*, **228**, 322.
Glueckauf, E., and Kitt, G. P. (1956). *Trans. Faraday Soc.* **52**, 1074.
Greenwood, N. N., and Martin, R. L. (1951). *J. Chem. Soc.* 1915.
Greenwood, N. N., and Martin, R. L. (1953). *J. Chem. Soc.* 1427.
Greenwood, N. N., and Martin, R. L. (1954). *Quart. Rev.* **8**, 1.
Halban, H. v., and Brüll, J. (1944). *Helv. Chim. Acta*, **27**, 1719.
Hammett, L. P., and Deyrup, A. J. (1932). *J. Amer. Chem. Soc.* **54**, 2721.
Hammett, L. P., and Paul, M. A. (1934). *J. Amer. Chem. Soc.* **56**, 527.
Harned, H. S., and Ehlers, R. W. (1932). *J. Amer. Chem. Soc.* **54**, 1350.
Harned, H. S., and Embree, N. D. (1934). *J. Amer. Chem. Soc.* **56**, 1042.
Harned, H. S., and Hawkins, J. E. (1928). *J. Amer. Chem. Soc.* **50**, 85.
Heilbronner, E., and Weber, S. (1949). *Helv. Chim. Acta*, **32**, 1513.
Heinzinger, K., and Weston, R. E. (1965). *J. Chem. Phys.* **42**, 272.

Henne, A. L., and Fox, C. J. (1951). *J. Amer. Chem. Soc.* **78**, 2323.
Högfeldt, E. (1960a). *Acta Chem. Scand.* **14**, 1597.
Högfeldt, E. (1960b). *Acta Chem. Scand.* **14**, 1627.
Högfeldt, E. (1961). *J. Inorg. Nuclear Chem.* **17**, 302.
Högfeldt, E., and Bigeleisen, J. (1960). *J. Amer. Chem. Soc.* **82**, 15.
Hood, G. C., and Reilly, C. A. (1957). *J. Chem. Phys.* **27**, 1126.
Hood, G. C., and Reilly, C. A. (1960). *J. Chem. Phys.* **32**, 127.
Hood, G. C., Redlich, O., and Reilly, C. A. (1954). *J. Chem. Phys.* **22**, 2067.
Hood, G. C., Redlich, O., and Reilly, C. A. (1955). *J. Chem. Phys.* **23**, 2229.
Hyman, H. H., Kilpatrick, M., and Katz, J. J. (1957). *J. Amer. Chem. Soc.* **79**, 3668.
Ives, D. J. G., and Pryor, J. H. (1955). *J. Chem. Soc.* 2104.
Jameson, R. F. (1959). *J. Chem. Soc.* 752.
Jones, L. H., and Penneman, R. A. (1954). *J. Chem. Phys.* **22**, 781.
Jorgenson, M. J., and Hartter, D. R. (1963). *J. Amer. Chem. Soc.* **85**, 878.
Kaandorp, A. W., Cerfontain, H., and Sixma, F. L. (1962). *Rec. Trav. Chim.* **81**, 969.
Kachurin, O. I. (1967). *Russ. J. Phys. Chem.* **41**, 1209.
Katritzky, A. R., Waring, A. J., and Yates, K. (1963). *Tetrahedron*, **19**, 465.
Kunzler, J. E., and Giaque, W. F. (1952). *J. Amer. Chem. Soc.* **74**, 3472.
Leist, M. (1955). *Z. phys. Chem.* (*Leipzig*), **205**, 16.
Lewis, G. N., and Bigeleisen, J. (1943). *J. Amer. Chem. Soc.* **65**, 1144.
Lloyd, L., and Wyatt, P. A. H. (1955). *J. Chem. Soc.* 2248.
Long, F. A., and McDevit, W. F. (1952). *Chem. Rev.* **51**, 119.
Malmberg, C. G. (1958). *J. Res. Natl. Bur. Std.* **60**, 609.
Manov, G. G., Delollis, N. J., and Acree, S. F. (1944). *J. Res. Natl. Bur. Std.* **33**, 287.
Marshall, W. L., and Jones, E. V. (1966). *J. Phys. Chem.* **70**, 4028.
McKay, H. A. C. (1956). *Trans. Faraday Soc.* **52**, 1568.
Milyaeva, N. M. (1958). *Russ. J. Inorg. Chem.* **3**, 2011.
Näsänen, R. (1954). *Acta Chem. Scand.* **8**, 1587.
Ojeda, M., and Wyatt, P. A. H. (1964). *J. Phys. Chem.* **68**, 1857.
Parsons, R. (1959). "Handbook of Electrochemical Constants." Butterworths, London.
Paul, M. A. (1954). *J. Amer. Chem. Soc.* **76**, 3236.
Paul, M. A., and Long, F. A. (1957). *Chem. Rev.* **57**, 1.
Perrin, C. (1964). *J. Amer. Chem. Soc.* **86**, 256.
Pethybridge, A. D., and Prue, J. E. (1967). *Trans. Faraday Soc.* **63**, 2019.
Randles, J. E. B., and Tedder, J. M. (1955). *J. Chem. Soc.* 1218.
Redlich, O. (1946). *Chem. Rev.* **39**, 333.
Robertson, E. B., and Dunford, H. B. (1964). *J. Amer. Chem. Soc.* **86**, 5080.
Robinson, R. A. (1948). *J. Amer. Chem. Soc.* **70**, 1870.
Robinson, R. A., and Baker, O. J. (1946). *Trans. Roy. Soc. N.Z.* **76**, 256.
Robinson, R. A., and Stokes, R. H. (1959). "Electrolyte Solutions." Butterworths, London.
Ryabova, R. S., Medvetskaya, I. M., and Vinnik, M. I. (1966). *Russ. J. Phys. Chem.* **40**, 182.
Salomaa, P. (1957). *Acta Chem. Scand.* **11**, 125.
Satchell, D. P. N. (1958). *J. Chem. Soc.* 3904.
Schubert, W. M., and Burkett, H. (1956). *J. Amer. Chem. Soc.* **78**, 64.
Schubert, W. M., and Myhre, P. C. (1958). *J. Amer. Chem. Soc.* **80**, 1755.
Stewart, R., and Mathews, T. (1960). *Canad. J. Chem.* **38**, 602.
Takahasi, K., and Yui, N. (1941). *Bull. Inst. Phys. Chem. Res.* **20**, 521.

Tickle, P., Briggs, A. G., and Wilson, J. M. (1970). *J. Chem. Soc. B.*, 65.
Vandoni, M. R., and Laudy, M. (1952). *J. Chim. Phys.* **49**, 99.
van Panthaleon van Eck, C. L., Mendel, H., and Boog, W. (1957). *Discuss. Faraday Soc.* **24**, 200.
Vinnik, M. I., and Librovich, N. B. (1966). *Tetrahedron*, **22**, 2945.
Vinnik, M. I., and Ryabova, R. S. (1964). *Russ. J. Phys. Chem.* **38**, 320.
Vinnik, M. I., Kruglov, R. N., and Chirkov, N. M. (1956). *Zhur. fiz. Khim.* **30**, 827.
Vinnik, M. I., Kruglov, R. N., and Chirkov, N. M. (1957a). *Zhur. fiz. Khim.* **31**, 832.
Vinnik, M. I., Manelis, G. B., and Chirkov, N. M. (1957b). *Russ. J. Inorg. Chem.* **2**, No. 7, 306.
Wamser, C. A. (1951). *J. Amer. Chem. Soc.* **73**, 409.
Wicke, E., Eigen, M., and Ackermann, Th. (1954). *Z. phys. Chem.* **1**, 340.
Wyatt, P. A. H. (1957). *Discuss. Faraday Soc.* **24**, 162.
Wyatt, P. A. H. (1960). *Trans. Faraday Soc.* **56**, 490.
Wyatt, P. A. H. (1969). *Trans. Faraday Soc.* **65**, 585.
Yates, K., and Wai, H. (1964). *J. Amer. Chem. Soc.* **86**, 5408.
Yates, K., and Wai, H. (1965). *Canad. J. Chem.* **43**, 2131.
Young, T. F., and Walrafen, G. E. (1961). *Trans. Faraday Soc.* **57**, 34.
Young, T. F., Maranville, L. F., and Smith, H. M. (1959). "The Structure of Electrolytic Solutions" (W. J. Hamer, ed.), p. 35. Wiley, New York.
Zarakhani, N. G., and Vinnik, M. I. (1962). *Russ. J. Phys. Chem.* **36**, 483.

CHAPTER 3

The Behaviour of Solutes other than Neutral Hammett Bases in Concentrated Aqueous Acid Solutions

3.1. The J_0 (or H_R) Acidity Function

A van't Hoff *i* factor of 4 for solutions of many triarylcarbinols in sulphuric acid has been found from cryoscopic measurements (Hantzsch, 1909; Hammett and Deyrup, 1933; Newman and Deno, 1951a). The relevant ionization equilibrium is given by equation (3.1).

$$ROH + 2H_2SO_4 \rightleftharpoons R^+ + H_3O^+ + 2HSO_4^- \quad (3.1)$$

This mode of ionization could be regarded as a protonation of the carbinol followed by dehydration of the conjugate acid (Gold and Hawes, 1951). It clearly differs from the simple behaviour of primary amines as Brønsted bases in concentrated acid solutions. The ionization ratios C_{R^+}/C_{ROH} for 4,4′,4″-trinitrotriphenylcarbinol in 80–95% aqueous sulphuric acid were measured colorimetrically by Westheimer and Kharasch (1946). The variation of C_{R^+}/C_{ROH} with sulphuric acid concentration differed significantly from that for the ionization ratio C_{BH^+}/C_B for the conventional (albeit non-Hammett) base anthraquinone. Lowen *et al.* (1950) and Murray and Williams (1950) confirmed this result and further showed that different triarylcarbinols gave ionization ratios which were parallel functions of acid concentration. They therefore suggested the definition of a new acidity function based on measurement of ionization ratios C_{R^+}/C_{ROH} for indicators which ionize in concentrated sulphuric acid solutions according to equation (3.1). The acidity function was originally given the symbol H_R (Lowen *et al.* 1950). However C_0 (Deno *et al.* 1955) and J_0 (Gold and Hawes, 1951; Gold, 1955a, 1955b; Paul and Long, 1957) have also been suggested. At present H_R and J_0 share popularity while C_0 is rarely used (Deno *et al.* now use H_R). J_0 will be used here.

The J_0 acidity function is defined by equation (3.2) in which K_{ROH} is the

$$J_0 = -pK_{ROH} - \log_{10}(C_{R^+}/C_{ROH}) \quad (3.2)$$

equilibrium constant for the reaction (3.3). Rearrangement of equation (3.2)

$$ROH + H^+ \rightleftharpoons R^+ + H_2O \quad (3.3)$$

leads to equation (3.4). It follows that J_0, like H_0, tends to pcH at low ionic

$$J_0 = -\log_{10} \frac{a_{H^+} y_{ROH}}{a_w y_{R^+}} \tag{3.4}$$

strengths and becomes identical with pH at infinite dilution. The measurement of the J_0 function is analogous to the determination of H_0. Indicators used must ionize according to equation (3.3). A useful criterion for this is that in accord with equation (3.1) the indicators should give a depression of the freezing point of sulphuric acid consistent with a van't Hoff factor of 4. Providing R^+ and ROH have measurably different electronic absorption spectra, (C_{R^+}/C_{ROH}) can be evaluated from spectrophotometric results via a similar equation to (1.34). For indicators which ionize appreciably in low acid concentrations pK_{ROH} is determined as the intercept at $C_{acid} = 0$ of a plot of $\log_{10}(C_{ROH} C_{acid}/C_{R^+})$ against C_{acid}. Other pK_{ROH} are deduced from the stepwise comparison of $\log_{10}(C_{R^+}/C_{ROH})$ for indicators which ionize in overlapping ranges of acid concentration.

Deno *et al.* (1955) measured J_0 for aqueous sulphuric acid solutions (0·5–98% H_2SO_4) at 25° using 18 triarylcarbinols as indicators. For $C_{H_2SO_4} > 93\%$ the J_0 values were calculated by an indirect method. This has been criticized by Vinnik *et al.* (1959b) who deduced J_0 for $85{\cdot}46\% < C_{H_2SO_4} < 98{\cdot}97\%$ from measurements of the ionization of 2-benzoylbenzoic acid in these solutions (Vinnik *et al.* 1959a). 2-Benzoylbenzoic acid gives a van't Hoff factor $i = 4$ in 100% H_2SO_4 (Newman *et al.* 1945) and is therefore apparently a satisfactory indicator for J_0 determination. The J_0 scale is given in Table 3.1. It should be noted that the function denoted J_0 by Deno *et al.* (1955) was calculated as $(H_0 + \log_{10} a_w)$ and is not the same as the J_0 function of the present notation which is defined by equation (3.2). Deno *et al.* denoted the present J_0 by C_0. The variation of J_0 with temperature was investigated for a small range of sulphuric acid concentrations by Entelis *et al.* (1960) who used triphenylcarbinol as indicator. More recently Arnett and Bushick (1964) using a series of triarylcarbinols have established J_0 scales for $0{\cdot}75\% < C_{H_2SO_4} < 98\%$ at 0°, 15°, 30° and 45°. Their results are included in Table 3.1 and are compatible with the scale at 25° up to $C_{H_2SO_4}$ = ca. 70%. Above this concentration rather large deviations occur. This is reflected in the discrepancy between pK_{ROH} at 25° for 4,4′,4″-trinitrophenylcarbinol deduced as 16·27 by Deno *et al.* (1955) and as 18·08 by Arnett and Bushick (1964). The latter workers employed only 4 indicators to study $C_{H_2SO_4} > 55\%$ and therefore the stepwise comparison of the ionization equilibria involved unsatisfactorily large differences between pK_{ROH} for successive indicators. Deno *et al.* (1955) used eight indicators for studying this range of sulphuric acid concentrations and it therefore follows that their results are probably the more reliable. De Fabrizio (1966) has found

TABLE 3.1

The J_0 (H_R) acidity function for aqueous sulphuric acid solutions at five temperatures

	$-J_0$†				
% H_2SO_4	25°	0°	15°	30°	45°
0·5	−1·25				
1	−0·92	−0·94	−0·99	−1·05	−1·10
2	−0·62	−0·69	−0·70	−0·84	−0·72
5	0·07	−0·14	−0·12	−0·37	−0·25
10	0·72	0·49	0·62	0·20	0·29
20	1·92	1·69	1·70	1·33	1·49
30	3·22	3·09	2·84	2·78	2·80
40	4·80	4·71	4·45	4·50	4·35
50	6·60	6·51	6·47	6·40	6·23
60	8·92	8·56	8·80	8·61	8·88
70	11·52	11·30	11·55	11·67	12·01
80	14·12	14·28	14·81	15·28	15·80
90	16·72	16·85	18·04	18·42	19·43
94	17·24	17·80	19·08	19·53	20·47
96	18·49	18·20	19·53	20·07	20·96
98	19·27	18·63	20·00	20·57	21·43
98·5	19·63				
98·8	19·88				
98·97	20·03				

† References: J_0 at 25°, Deno *et al.* (1955); Vinnik *et al.* (1959a); J_0 at other temperatures, Arnett and Bushick (1965).

that the J_0 acidity function for solutions of D_2SO_4 in D_2O is identical to that for H_2SO_4 in H_2O up to a sulphuric acid concentration of 10 molar.

The experimental J_0 scales at 25° for aqueous solutions of $HClO_4$ and HNO_3 (Deno *et al.* 1959a), H_3PO_4, HCl and toluene-4-sulphonic acid (Arnett and Mach, 1966) and HCOOH (Stewart and Mathews, 1960) are given in Table 3.2. They are all entirely based on measurements of the ionization of triarylcarbinols in the solutions. Epple *et al.* (1962) have also measured J_0 for 0–22% HCl in water using 4,4′-dimethyldiphenylmethylcarbinol and 4,4′-dimethoxytriphenylcarbinol as indicators. However their scale differs somewhat from that determined by Arnett and Mach (1966).

Comparison of the J_0 acidity function with the corresponding H_0 scale for any given acid shows that J_0 decreases more rapidly than does H_0 with increasing acid concentration. The deviation between the two scales for

TABLE 3.2

The $J_0(H_R)$ acidity function for aqueous solutions of six acids at 25°

	$-J_0$					
% acid	HCl†	$HClO_4$‡	HNO_3‡	H_3PO_4†	HCOOH§	HOTs†‖
0·1	−1·70					
0·2	−1·36		−1·50			
0·5	−0·95	−1·30	−1·10	−1·85		−1·60
1	−0·56	−0·97	−0·74	−1·61		−1·20
2	−0·11	−0·54	−0·27	−1·42		−0·82
5	0·75	0·24	0·49	−1·00	−1·62	−0·26
10	1·74	1·04	1·30	−0·55	−1·37	0·20
20	3·80	2·43	2·63	0·30	−0·80	0·72
30	6·23	3·79	3·83	1·19	−0·15	1·12
40		5·54	5·07	2·12	0·57	1·50
50		7·86	6·40	3·30	1·30	1·88
60		11·14		4·65	2·00	
70				6·51	2·75	
80				8·20	3·60	
90					4·72	
95					5·57	
99·5					ca. 7·91	

† Arnett and Mach (1966).
‡ Deno *et al.* (1959a).
§ Stewart and Mathews (1960).
‖ Toluene-4-sulphonic acid.

aqueous sulphuric acid is shown in Fig. 3.1. For the range $10\% < C_{H_2SO_4} < 95\%$ H_2SO_4 the J_0 function is approximately given by $2H_0$. Combination of equations (1.29) and (3.4) leads to equation (3.5) for the relationship between J_0 and H_0.

$$J_0 = H_0 + \log_{10} a_w + \log_{10}(y_B\, y_{R^+}/y_{BH^+}\, y_{ROH}) \tag{3.5}$$

Despite the important studies of indicator solubility by Boyd (1963a) little is known about activity coefficient behaviour of primary amines and their conjugate acids and triarylcarbinols and their carbonium ions in concentrated solutions of acids. Following Hammett's postulate that (y_B/y_{BH^+}) is the same for different bases in a particular solution it might be possible to equate the activity coefficient term in equation (3.5) to zero (Gold and Hawes, 1951; Gold, 1955b). This leads to the approximate equation

$$J_0' = H_0 + \log_{10} a_w \tag{3.6}$$

However calculation of J_0' for sulphuric acid shows that this function is quite different from the experimental J_0 scale (Fig. 3.1). A similar discrepancy exists for all the other acids which have been studied (Table 3.2). Arnett and Mach (1966) showed that $(J_0' - H_0 - \log_{10} a_w)$ was a linear function of molarity of

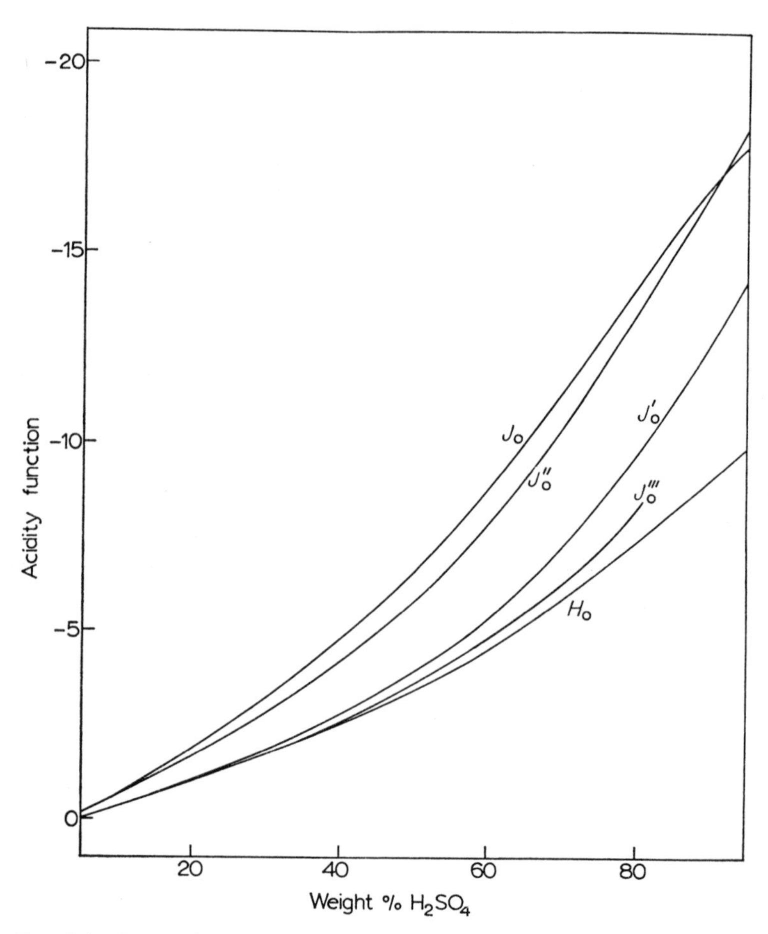

FIG. 3.1. Comparison of H_0, J_0, J_0', J_0'' and J_0''' for aqueous sulphuric acid.

acid for H_2SO_4, HCl, H_3PO_4 and $HClO_4$ up to concentrations of ca. 10 mole litre^{-1}. Thus equation (3.7) is obeyed, where A is a combination of the

$$J_0 - H_0 - \log_{10} a_w = \log_{10}(y_B\, y_{R^+}/y_{BH^+}\, y_{ROH}) = AC_{\text{acid}} \qquad (3.7)$$

Setschenow coefficients (Long and McDevit, 1952) for the medium effects of the activity coefficients y_B, y_{R^+}, y_{BH^+} and y_{ROH}. The values of A were different for each acid in the order $HClO_4 > H_2SO_4 > H_3PO_4 > HCl$. For sulphuric acid concentrations in the range 13 mole litre^{-1} < $C_{H_2SO_4}$ < 18 mole litre^{-1} the activity coefficient term in equation (3.5) appears to be approximately constant at ca. −4·5 (Deno *et al.* 1955; Högfeldt, 1962; Arnett and Mach, 1966). There is support for the conclusion that activity coefficients of individual species may be fairly constant in this concentration range (Deno and Taft, 1954) although this has been questioned by Gold (1955b). Unfortunately Boyd's (1963a) measurements do not extend to these high concentrations.

Gold (1955b) investigated the effect of setting different activity coefficient terms equal to zero in theoretical equations for the J_0 acidity function. Thus the two equations (3.8) and (3.9) were proposed as being possible relationships between the J_0 and H_0 acidity functions.

$$J_0'' = 2H_0 + \log_{10} X_{H_3O^+} + \log_{10}(55{\cdot}5) \tag{3.8}$$

$$J_0''' = H_0 + \log_{10} X_{H_2O} \tag{3.9}$$

Correction of Gold's calculations in accord with the more recent determinations of H_0 for aqueous H_2SO_4 (Jorgenson and Hartter, 1963; Ryabova *et al.* 1966) show that J_0'' is a much better approximation to J_0 than either J_0' or J_0'' (Fig. 3.1). A similar result was obtained by Stewart and Mathews (1960) for formic acid. The definition of J_0'' was based on the assumption that

$$(y_B{}^2\, y_{R^+}\, y_{H_3O^+}/y_{BH^+}^2\, y_{ROH}) = 1 \tag{3.10}$$

As this expression contains activity coefficients for five different species it is difficult to envisage why its validity should be approximately maintained for large changes in acid concentration.

Clearly in order to help in the understanding of deviations between J_0 and H_0 it is desirable to have information about the variation with acid concentration of the activity coefficients of individual species. In this way the individual components of equations (1.29) and (3.4) may be considered and their separate contributions to the acidity functions recognized. Boyd (1963a) has measured the solubilities of several primary amines and their conjugate acids and also one J_0 indicator (triphenylcarbinol) and one carbonium ion (4,4′,4″-trimethoxytriphenyl carbonium cation) in aqueous sulphuric acid solutions up to 70% H_2SO_4. In accord with earlier conclusions (Deno *et al.* 1955; Deno and Perizzolo, 1957) Boyd's measurements showed that (y_B/y_{ROH}) does not vary too much with changing acid concentration but that (y_{R^+}/y_{BH^+}) decreases appreciably with increasing acid concentration. Thus in addition to the formal difference of $\log_{10} a_w$, the deviation between J_0 and H_0 may be attributed mainly to differences between the carbonium ion and ammonium ion activity coefficients in concentrated acid solutions.

The deviation between y_{R^+} and y_{BH^+} may be explained by contrasting the solvation requirements and effect on solvent structure of ammonium ions and carbonium ions in water. Thus Taft (1960) noted that if $(J_0 - H_0)$ was plotted against $-\log_{10} a_w$ a linear graph of slope 4·0 resulted. Furthermore a plot against $-\log_{10} a_w$ of J_0 minus the acidity function defined by 4-nitrodiphenylamine in aqueous H_2SO_4 was linear with slope 3·0. These results are explicable in terms of the equilibria (3.11) and (3.12) in which the solvent molecules

$$\left[\begin{array}{l} \mathrm{H\cdots OH_2} \\ | \\ \mathrm{ArN{-}H\cdots OH_2} \\ | \\ \mathrm{H\cdots OH_2} \end{array}\right]^+ + \mathrm{ROH} \rightleftharpoons \mathrm{ArNH_2} + \mathrm{R^+} + 4\mathrm{H_2O} \qquad (3.11)$$

$$\left[\begin{array}{l} \mathrm{H\cdots OH_2} \\ | \\ \mathrm{Ar_2N{-}H\cdots OH_2} \end{array}\right]^+ + \mathrm{ROH} \rightleftharpoons \mathrm{Ar_2NH} + \mathrm{R^+} + 3\mathrm{H_2O} \qquad (3.12)$$

hydrogen bonding to the amine cations are formally included in the equilibrium expressions. This is a simplified model. Arnett and Bushick (1964) have extended these ideas and conclude that in aqueous solution the carbonium ions derived from triarylcarbinols undergo generalized dielectric solvation whereas ammonium ions undergo specific solvation effects through hydrogen bonding. In the light of these views the Bascombe and Bell (1957) and related treatments for predicting the Hammett acidity function must be modified. Thus Bascombe and Bell ignored the solvation requirements of the neutral indicators and the ammonium ions and only considered the solvation number (4) of the proton to be significant. The fact that the expected solvation number of 4 for the proton in water gave a satisfactory correlation with H_0 is probably fortuitous (Arnett and Bushick, 1964). The solvation requirements of all the species (both neutral and charged) in the ionization equilibrium for a solute in concentrated acid solution must be taken into account. The concept that solvation effects are important in deciding acidity function behaviour is now generally accepted.

In Table 3.3 pK_{ROH} values are given for triarylcarbinols which have been used to establish J_0 acidity functions for several aqueous acids. Apart from the results for 4-MeO, 4-Me and unsubstituted triphenylmethanol the values show reasonable agreement from one acid to another. As for the Hammett acidity function, which is generally anchored to standard state pure water via $pK_{BH^+}(25°) = 0·99$ for 4-nitroaniline (Paul and Long, 1957), it would seem reasonable that $pK_{ROH}(25°) = -0·82$ for 4,4′4″-trimethoxytriphenylcarbinol should be a common basis for all J_0 acidity functions. It is relevant to note that amino substituted triarycarbinols should be avoided as J_0 indicators because the amine groups are susceptible to simple protonation. Thus Belotserkovskaya and Ginsburg (1964) have studied the ionization of seven aminotriphenyl-

TABLE 3.3

Indicators for J_0 determination. pK_{ROH} values for triarylcarinols at 25° as given by ionization measurements in aqueous solutions of five acids

Substituted triphenylcarbinol	pK_{ROH}				
	H_2SO_4†	HCl‡	$HClO_4$§	HNO_3§	H_3PO_4‡
4,4′,4″-Trimethoxy	−0·82	−0·82	−0·82	−0·80	−0·82
4,4′-Dimethoxy	1·24	1·06	1·14	1·11	1·23
4-Methoxy	3·40	3·08	3·59	3·41	3·42
2,2′,2″-Trimethyl	3·40				
4,4′,4″-Trimethyl	3·56				
4-Methyl	5·24	4·93	5·67		5·34
3,3′,3″-Trimethyl	6·35		5·95		
4,4′,4″-Triisopropyl	6·54				
Unsubstituted	6·63	6·12	6·89	6·60	6·50
4,4′,4″-Trichloro	7·74		8·01		
4-Nitro	9·15		9·76		
3,3′,3″-Trichloro	11·03				
4,4′,4″-Trinitro	16·27				

† Deno *et al.* (1955).
‡ Arnett and Mach (1966).
§ Deno *et al.* (1959a).

carbinols in aqueous sulphuric acid. The relevant competing ionization equilibria are depicted for one case in equation (3.13). A is the neutral indicator,

$$\underset{\text{(B)}}{Ph-\overset{\overset{+}{PhNHMe_2}}{\underset{Ph}{C}}-OH} \rightleftharpoons \underset{\text{(A)}}{Ph-\overset{Ph.NMe_2}{\underset{Ph}{C}}-OH} + H^+ \rightleftharpoons \underset{\text{(C)}}{\left[\overset{Ph.NMe_2}{\underset{Ph\quad Ph}{C}}\right]^+} + H_2O \quad (3.13)$$

B the ammonium cation formed by simple protonation and C is the carbonium ion formed by a "protonation–dehydration" ionization. The formation of C should follow the J_0 acidity function whereas the formation of B might be expected to follow H_0. In accord with this it was found that $\log_{10}(C_C/C_B)$ was a linear function of $(H_0 - J_0)$ for all seven indicators. The slopes were 0·70–0·75 and not 1·0 as might be expected. However this is in accord with the different acidity function behaviour of primary and tertiary amines in aqueous acids (Arnett and Mach, 1964, 1966). Other solutes whose ionization in concentrated acids have been studied includes nitrous acid (Turney and Wright, 1958; Deno *et al.* 1959a; Bayliss *et al.* 1963) and the pseudo bases of *N*-cyano-quinolinium and -isoquinolinium ions (Huckings and Johnson, 1964).

3.2. The Acidity Functions defined by the Ionization of Neutral Bases which do not follow the Hammett H_0 Acidity Function

3.2.1. *Secondary and Tertiary Amines*

The acidity functions defined by the ionization of secondary or tertiary amines in concentrated acid solutions differ from the Hammett H_0 acidity function defined from the ionization of primary amines. Taft (1960) has compared the ionization of secondary and primary amines and has attributed the differences in acidity function behaviour to the different (H-bonded) solvation requirements of the conjugate acid cations derived from the amines (equations 3.11 and 3.12). Boyd (1963a) has shown that the anilinium ion and the *N*-ethylanilinium ion have widely deviating activity coefficients in concentrated sulphuric acid solutions. Tertiary amines have been investigated in much more detail by Arnett and Mach (1964, 1966). Using standard methods (Section 1.4) they have measured the acidity functions defined by the ionization of a series of tertiary amines in aqueous solutions of sulphuric, hydrochloric, phosphoric and 4-toluenesulphonic acids. Similar measurements for aqueous perchloric acid have been made by Yates and Wai (1968). The symbol H_0''' has been given for these scales in order to maintain a clear distinction from H_0 (sometimes called H_0' particularly by E. M. Arnett) defined by primary amines. For all four acids which have been studied H_0''' deviates significantly from H_0 and lies somewhere between H_0 and J_0.

The formal difference between H_0''' and H_0 is given by equation (3.14) in which B is a primary (Hammett base) and B‴ a tertiary amine (Arnett and

$$H_0''' - H_0 = \log_{10}(y_{\mathrm{B}}\, y_{\mathrm{B'''H^+}}/y_{\mathrm{B'''}}\, y_{\mathrm{BH^+}}) \qquad (3.14)$$

Mach, 1966). Boyd (1963a) has shown that $(y_{\mathrm{BH^+}}/y_{\mathrm{B'''H^+}})$, where BH^+ is the anilinium ion and $B'''H^+$ is the *N*,*N*-dimethylanilinium ion, is ca. 10 in 50% aqueous H_2SO_4 for which $(H_0''' - H_0) = $ ca. $-1{\cdot}2$. This suggests that $(y_{\mathrm{BH^+}}/y_{\mathrm{B'''H^+}})$ is probably the predominant variant which decides the difference between H_0 and H_0'''. Explanation of the $J_0 - H_0$ deviation is based on a similar conclusion (Section 3.1). Inclusion of the solvation requirements of the cations in the formal equilibria for the ionization of primary and tertiary amines leads to equation (3.15) and hence equation (3.16) for the relationship between H_0 and H_0''' (Arnett and Mach, 1964).

$$[\mathrm{BH}.n\mathrm{H_2O}]^+ + \mathrm{B'''} \rightleftharpoons [\mathrm{B'''H}.n'''\mathrm{H_2O}]^+ + \mathrm{B} + (n - n''')\mathrm{H_2O} \qquad (3.15)$$

$$H_0''' - H_0 = \log_{10}(y_{\mathrm{B}}\, y_{[\mathrm{B'''H}.n'''\mathrm{H_2O}]^+}/y_{\mathrm{B'''}}\, y_{[\mathrm{BH}.n\mathrm{H_2O}]^+}) + (n - n''')\log_{10} a_w \qquad (3.16)$$

By an argument similar to that of Taft (1960) it would be expected that $n > n'''$ and therefore $H_0''' < H_0$ (remembering that $\log_{10} a_w$ is negative). Although H_0 is a unique function of water activity for several strong acids (Wyatt, 1957) the differences $(H_0''' - H_0)$ for H_2SO_4 and HCl are not the same function of $\log_{10} a_w$ (Arnett and Mach, 1966). Thus the formal explanation of acidity function and activity coefficient differences in terms of hydration effects is probably not simple. Arnett and Burke (1966) have determined entropies of transfer of anilinium and *N,N*-dimethylanilinium ions from aqueous 10% H_2SO_4 to higher concentrations of H_2SO_4 up to 70%. The entropies of solvation of the two anilinium ions varied with acidity in the same way. There were no significant differences which could be clearly identified with the different hydration requirements of the primary and tertiary anilinium ions. However in view of the small change in the partial molal entropy of water between 10% and 70% H_2SO_4 this result is reasonable.

The ionization of nitroquanidines in aqueous H_2SO_4 parallels the H_0 acidity function (Bonner and Lockhart, 1958).

3.2.2. *Amides and Pyridine-1-oxides*

Cryoscopy in 100% H_2SO_4 has shown that amides ionize as Brønsted bases in concentrated acid (Hantzsch, 1931; O'Brien and Niemann, 1950; Fraenkel and Franconi, 1960). A p.m.r. study of many amides showed that all of them were predominantly protonated on oxygen (Fraenkel and Franconi, 1960). The acidity function behaviour of amides in aqueous acids differs significantly from that of amines. Thus Edward and Wang (1962) noted that propionamide was not a Hammett base and suggested that this could be interpreted by assuming that the protonated amide cation is more hydrated than primary anilinium cations. Katritzky and Waring (1962) found that for several cyclic amides (uracil derivatives) although $\log_{10}(C_{BH^+}/C_B)$ were linear functions of $-H_0$ the slopes were often appreciably less than one. This result was confirmed for many primary, secondary and tertiary benzamide and naphthamide derivatives (Katritzky *et al.* 1963; Moodie *et al.* 1963; Homer and Moodie, 1963; Yates and Wai, 1965). The appreciable difference between the ionization behaviour of amines and amides led Yates *et al.* (1964) to define a new acidity function H_A based on measurement of the extent of protonation of primary amides in concentrated acid solutions.

Eight primary amides were used to establish the H_A scale for aqueous sulphuric acid. Unfortunately no amide has yet been found which is sufficiently basic to link the measurements for the weakly basic amides in concentrated solutions with standard state pure water. However 2-pyrrolecarboxamide, which is measurably ionized in 10–30% H_2SO_4, gives a curve of $\log_{10}(C_{BH^+}/C_B)$ against % H_2SO_4 which overlaps and is reasonably parallel to the corresponding curve for the ionization of 4-nitroaniline. Yates *et al.* (1964) therefore used

4-nitroaniline as the primary indicator for establishing the H_A scale and deduced pK_a for 2-pyrrolecarboxamide and the other more basic amides by the usual stepwise comparison. The H_A scales for aqueous HCl (Yates and Riordan, 1965) and $HClO_4$ (Yates and Wai, 1968) have also been measured.

In general plots of $\log_{10}(C_{BH^+}/C_B)$ against $-H_0$ for amides have slopes in the range 0·5–0·8 although slopes as low as 0·39 and as high as 1·00 (for 3-methylbenzamide which therefore behaves as a Hammett base) have been recorded (Homer and Moodie, 1963; Katritzky *et al.* 1963; Armstrong and Moodie, 1968). This contrasts with the observation (Edward and Stollar, 1963) that C_{BH^+}/C_B for three thiolactams in aqueous H_2SO_4 increases more rapidly than H_0 with increasing H_2SO_4 concentration. The difference between the H_0 and H_A acidity functions is greater for aqueous sulphuric acid than for hydrochloric acid. Thus for $C_{H_2SO_4} = 66{\cdot}5\%$, $H_A = -3{\cdot}50$ and $(H_A - H_0) = 1{\cdot}77$ whereas for $C_{HCl} =$ ca. 31%, $H_A = -3{\cdot}50$ and $(H_A - H_0) = 0{\cdot}40$. H_A is not a unique function of water activity for the two acids. Yates and Riordan (1965) have suggested that these differences arise because competition for water molecules between the acid anions (Cl^- and HSO_4^-) and the amide cations is not so important in HCl as in H_2SO_4. This is consistent with the relative hydration requirements ($h_{Cl^-} < h_{HSO_4^-}$) for the chloride and bisulphate anions (Wyatt, 1957).

More detailed discussions of the deviation between H_0 and H_A have approached the problem by consideration of either the variation of individual activity coefficients with acid concentration or the possible different hydration requirements of the protonated amides and primary amines. The latter approach has led to the conclusion that the amide cations are H-bonded to more (solvating) water molecules than are primary anilinium ions. Thus representation of a protonated amide like (I) has been visualized (Homer and Moodie, 1963; Yates *et al.* 1964). However this is a considerable oversimplification of

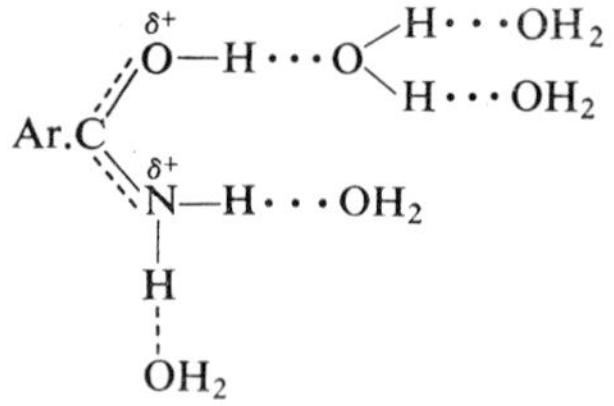

```
            δ+              H···OH2
          ,O—H···O<
         //                 H···OH2
    Ar.C
         \\ δ+
          `N—H···OH2
           |
           H
           :
           OH2
```

(I)

what must be the true state of affairs. Sweeting and Yates (1966) have deduced activity coefficients for some benzamide derivatives and their conjugate acids in 0–70% H_2SO_4 from solubility measurements. Comparison of the activity coefficient behaviour of the benzamide cation AH^+ with that for the anilinium ion BH^+ (Boyd, 1963a) shows that y_{AH^+} increases faster than does y_{BH^+} with increasing H_2SO_4 concentration. This is consistent with $H_A > H_0$ and extends Boyd's finding that in accord with $H_0 > H_0'''$ y_{BH^+} increased faster than the activity coefficient of the *N,N*-dimethylanilinium ion with increasing acid concentration. Equation (3.17) follows from the definition of H_A. Sweeting and Yates (1966) concluded that for 5–40% H_2SO_4 the relative importance of

$$H_A = -\log_{10}(C_{H^+} y_{H^+} y_A / y_{AH^+}) \tag{3.17}$$

the various parameters in this equation in deciding the variation of H_A is

$$y_{H^+} > y_{AH^+} \geqslant C_{H^+} > y_A$$

whereas for $C_{H_2SO_4} = 40\%$ the order becomes

$$y_{H^+} \gg y_{AH^+} \gg y_A > C_{H^+}$$

Of course studies involving amides in concentrated acid solutions should be correlated with the H_A rather than H_0 acidity function. However, there are some cases, notably propionamide, 4-nitrobenzamide, 2-toluamide and *N,N*-dimethyl-2-naphthamide, in which the ionization of amides does not appear to parallel H_A (Yates and Stevens, 1965). The ionization equilibria of some carbamic acid esters correlate much closer with the amide acidity function than with H_0 (Armstrong and Moodie, 1968). The ionization of several pyridine-1-oxides in sulphuric acid solutions also parallels the amide acidity function (Johnson *et al.* 1967). Thus the acidity function defined from a study of the protonation of these pyridine-1-oxides is, within experimental error, identical to the amide scale. Furthermore the more basic pyridine-1-oxides studied enable a rigorous reference of the data to standard state water. This has yet to be achieved for the amide acidity function. Johnson *et al.* (1967) also found that the second protonation of phenazine 5,10-dioxide overlapped and paralleled the protonation of 2,4,6-trinitrobenzamide at high sulphuric acid concentrations. They have therefore extended the H_A scale to 93% H_2SO_4 using measurements with phenazine 5,10 -dioxide as indicator.

The parallellism between H_A and the protonation equilibria of pyridine-1-oxides is not general. Many pyridine-1-oxides behave as Hammett bases and their ionization closely follows H_0 (Johnson *et al.* 1965). Thus even within a set of molecules which retain the same functional (basic) site other changes within the molecules can lead to significantly different acidity function behaviour. This is exemplified by the data in Table 3.4. The ionization ratio for a series of pyridine-1-oxides is correlated with the Hammett H_0 acidity function according to equation (3.18). The resulting values of n are a function

$$H_0 = pK_{BH^+} - n\log_{10}(C_{BH^+}/C_B) \tag{3.18}$$

of the substituent groups and although n = ca. 1 for many of the bases there are some notable exceptions. This result is not unique to pyridine-1-oxides since

TABLE 3.4

The ionization of pyridine-1-oxides in aqueous sulphuric acid. Parameters of equation (3.18). (Johnson *et al.* 1965)

Pyridine-1-oxide	pK_a	n
4-Chloro	0·33	1·10
3-Chloro	−0·13	1·10
3-Bromo	−0·16	1·08
2-Acetyl	−0·45	1·01
4-Acetyl	−0·47	1·02
2-Chloro	−0·77	1·47
3,5-Dibromo	−0·85	1·06
3-Nitro	−1·07	1·00
4-Nitro	−1·73	1·05
2-Cyano	−2·08	1·23
2-Nitro	−2·71	1·44

many other series of related bases contain members whose acidity function behaviour differs significantly from the other members of the series. The data in Table 1.5 also demonstrate this point. In fact if ionization ratios could be measured much more accurately than is possible at present it may well be that the variation of (C_{BH^+}/C_B) with acid concentration would be different for every base in a given series. Existing data show that the differences would be small and bases with related structure would in general show similar behaviour. However it is important to realize that deviations, of varying magnitude, from standard behaviour can and do exist.

3.2.3. *Pyridines, Indoles and Pyrroles*

The parameters of equation (3.18) have been deduced for twelve halogen substituted pyridine derivatives which were protonated in aqueous H_2SO_4 (Johnson *et al.* 1965). The values of n range from 0·80 for pentachloropyridine to 1·14 for 2,3,4,5-tetrabromopyridine. This latter value is exceptional, the next highest value being 1·03 for 2,3,6-tribromopyridine. It may be concluded that the ionization behaviour of five of the pyridines ($0{\cdot}95 < n < 1{\cdot}03$) approximately parallel the H_0 acidity function. This is perhaps surprising as pyridines might be expected to protonate in accord with the tertiary amine acidity function H_0''' (Arnett and Mach, 1946, 1966). A plot of H_0 against H_0''' has a slope of 0·74 for aqueous H_2SO_4. Thus none of the pyridines studied ionize according to H_0''' although some show acidity function behaviour between H_0 and H_0''' and two (pentachloropyridine $n = 0{\cdot}80$ and 2,3,5,6-

tetrachloropyridine $n = 0{\cdot}82$) give much closer agreement with H_0''' than with H_0. Again it appears that throughout a series of related bases certain appreciable differences in acidity function behaviour occur.

The phenomenon is also apparent from a study of the protonation of indole and 24 of its derivatives in aqueous sulphuric acid (Hinman and Lang, 1964). Plots of $\log_{10}(C_{BH^+}/C_B)$ against $C_{H_2SO_4}$ showed that the ionization behaviour fell into three groups. Indole and derivatives which were unsubstituted in the hetero ring formed one group and 1,3-dialkyl indoles another. The largest group was formed by 18 compounds which were substituted by alkyl groups in some other way. Similar results were also obtained for some of the bases in aqueous perchloric acid. Although there were minor deviations in acidity function behaviour for the 18 indoles which formed the largest group there was generally reasonable agreement throughout the set. This led Hinman and Lang (1964) to define a new acidity function H_I based on measurement of the protonation of 9 indoles for aqueous H_2SO_4 and 4 indoles for aqueous $HClO_4$.

The H_I acidity function for aqueous sulphuric acid is nearly identical to the H_0''' scale measured by Arnett and Mach (1964) using tertiary anilines as indicators. Thus although the indoles are predominantly carbon bases (Hinman and Whipple, 1962) their indicator behaviour agrees closely with that of the tertiary nitrogen bases studied by Arnett and Mach (1964, 1966). However, the other two groups of indoles identified by Hinman and Lang do not conform with this agreement. Thus the ionization of indole itself parallels an acidity function similar to the amide H_A scale (Yates *et al.* 1964). It is difficult to explain the large changes in acidity function behaviour which results when alkyl groups are substituted in the hetero ring of indole. This is even more so since no such effect is observed for pyrrole and some of its mono- and di-substituted methyl derivatives all of which show similar protonation behaviour in aqueous sulphuric acid (Chiang and Whipple, 1963). The ionization of pyrrole and its methyl derivatives approximately parallels the H_I and H_0''' acidity functions.

3.2.4. *Ketones*

In 100% anhydrous sulphuric acid many ketones are fully protonated and cryoscopic measurements give a van't Hoff factor $i = 2$ (Hantzsch, 1907; Hammett and Deyrup, 1933; Newman and Deno, 1951b). In accord with this Hammett and Deyrup (1932) included some ketones among the indicators which they used in their pioneering acidity function study. However it is now known that ketones and primary amines show differing acidity function behaviour (Yates and Wai, 1965; Bonner and Phillips, 1966). Thus Bonner and Phillips (1966) have measured an acidity function for aqueous sulphuric acid in the concentration range $40\% < C_{H_2SO_4} < 90\%$ using 13 benzophenone derivatives as indicators. Up to $C_{H_2SO_4} = 60\%$ H_0 and the benzophenone acidity scale are identical. Bonner and Phillips called their scale H_0 but as this is unsatisfactory in the nomenclature being used here it would be better to designate it H_B (B for benzophenone). This brings the nomenclature in line with H_I for the indole scale and H_A for the amide acidity function. At sulphuric acid concentrations greater than 60% H_0 and H_B deviate until in 90% H_2SO_4 $(H_B - H_0) = 1{\cdot}71$ units. In fact for $40\% < C_{H_2SO_4} < 90\%$ H_B is a linear function of % H_2SO_4 with slope $-0{\cdot}097$ over this entire range of H_2SO_4 concentration.

The indicators used to define H_B gave individual differences in $d\log_{10}(C_{BH^+}/C_B)/dC_{H_2SO_4}$ between 0·082 and 0·116. It is interesting that the lowest slope, which corresponds to an acidity function which decreases the least steeply with increasing acid concentration, was for 2,4-dihydroxybenzophenone. The conjugate acid of this ketone would be expected to be more solvated via hydrogen bonding than the conjugate acids of the other ketones, none of which contained two hydroxy substituents. As discussed above in Sections 3.2.1 and 3.2.2 the greater the solvation of the protonated form of a base the less steep is the variation of the relevant acidity function with changing acid concentration. The case of 2,4-dihydroxybenzophenone compares with the contrasting ionization behaviour of phloroglucinol and its methyl ethers. This will be discussed in the following section.

Stewart *et al.* (1963) have evaluated pK_{BH^+} for many ketones by correlating the ionization curves for the ketones in aqueous sulphuric acid with the H_0 acidity function. As $H_B \neq H_0$ when $C_{H_2SO_4} > 60\%$ it is necessary to adjust some of the pK_{BH^+} values in accord with the H_B scale. Thus, for example, pK_{BH^+} for 4-chlorobenzophenone becomes $-6{\cdot}15$ (previously $-6{\cdot}64$) and for 4,4′-dichlorobenzophenone becomes $-6{\cdot}40$ (previously $-6{\cdot}96$). This clearly shows the errors which may be incurred if ionization equilibria are correlated with an acidity function based on measurements with indicators which are structurally dissimilar from the bases under investigation. The determination of pK_{BH^+} of very weak bases via correlation with acidity functions has been reviewed by Arnett (1963).

Wasif (1967) has measured the H_B acidity function for selenic acid using 8 substituted benzophenones as indicators. The scale is given in Table 3.5 and is the only acidity function for selenic acid which has been measured to date. Comparison of H_B for sulphuric acid and selenic acid shows that the two acidity scales are similar functions of acid concentration. This is consistent with the nearly identical ionization constants of H_2SO_4 and H_2SeO_4 and of HSO_4^- and $HSeO_4^-$ (Sherrill and Lyons, 1932; Gelbach and King, 1942; Marshall and Jones, 1966).

The protonation of several aliphatic α,β-unsaturated carbonyl compounds in aqueous H_2SO_4 gives a better correlation with the amide H_A (Section 3.2.2) acidity function than with H_0 (Zalewski and Dunn, 1968). Zalewski and Dunn (1968) obtained a similar result for benzophenone and deduced $pK_{BH^+} =$

TABLE 3.5

The H_B (benzophenone) acidity function for aqueous selenic acid at ca. 25° (Wasif, 1967)

% H_2SeO_4	$-H_B$	% H_2SeO_4	$-H_B$	% H_2SeO_4	$-H_B$
53·9	2·61	71·9	4·43	85·0	5·92
56·5	2·85	75·0	4·67	86·9	6·25
59·0	3·05	77·1	4·86	88·3	6·49
60·1	3·22	79·9	5·37	91·0	6·90
62·6	3·46	80·4	5·40	92·0	7·08
65·1	3·69	81·1	5·46	93·0	7·24
67·7	3·95	82·0	5·57	94·0	7·41
70·4	4·13	84·0	5·91	95·0	7·80

Substituted benzophenones used as indicator bases were (pK_{BH^+} in brackets): 2,4,2′,4′-tetramethoxy (–3·34); 2,4,4′-trimethoxy (–3·60); 4,4′-dimethoxy (–4·41); 4-methoxy (–4·93); 4-hydroxy (–5·03); 4-chloro-4′-methoxy (–5·25); 4-chloro (–6·16); 4-nitro (–6·88).

–4·01 for this base. However this value differs appreciably from that of –5·70 quoted by Bonner and Phillips (1966). Greig and Johnson (1968) found that $\log_{10}(C_{BH^+}/C_B)$ for benzophenone increased more rapidly than $-H_0$ with increasing H_2SO_4 concentration and evaluated a figure of ca. –8·2 for pK_{BH^+}. This contrasts with Bonner and Phillips's (1966) and Zalewski and Dunn's (1968) results for which a plot of $\log_{10}(C_{BH^+}/C_B)$ against $-H_0$ had a slope less than unity for benzophenone. The discrepancies between these data emphasize the additional difficulty inherent in the interpretation of spectra for which there are appreciable medium effects. The protonation behaviour of carbonyl compounds is also very sensitive to the structure of the particular molecule being protonated (Greig and Johnson, 1968).

3.2.5. *Alkenes, Azulenes, Phloroglucinol and its Methyl Ethers*

From equations (3.5) it follows that the relationship between J_0 and an acidity function H defined by the protonation of weak neutral bases should be given by equation (3.19) providing the identity (3.20) is obeyed. None of the

$$H = J_0 - \log_{10} a_w \tag{3.19}$$

$$\log_{10}(y_B\, y_{R^+}/y_{BH^+}\, y_{ROH}) = 0 \tag{3.20}$$

acidity functions H_0, H_A, H_I, H_B or H_0''' satisfy this equation. Thus in these cases equation (3.20) is not valid. This is partly due to differences between y_B and y_{ROH} but is predominantly caused by differences between y_{R^+} and y_{BH^+}. These arise mainly because of the markedly different solvation requirements of the carbonium ions R^+ and the protonated bases BH^+. However Deno *et al.* (1959) showed that the protonation equilibria between certain diarylolefins and the corresponding diarylalkyl cations were in accord with equation (3.21) in which C_{ol} is the alkene concentration, C_{R+} is the concentration of the conjugate acid

$$J_0 - \log_{10} a_w = pK_{R+} - \log_{10}(C_{R+}/C_{ol}) \qquad (3.21)$$

cation (carbonium ion) and pK_{R+} is its acid ionization constant. The right-hand side of this equation corresponds to equation (1.27) and it therefore follows that equations (3.19) and (3.20) are applicable if B is a diarylolefin and H (given the symbol H_R') is the acidity function which parallels the protonation behaviour of diarylolefins. This agreement of the experimental and formal relationship between H_R' and J_0 arises because the triaryl carbonium ions derived from the ionization of J_0 indicators have very similar structures to the diarylalkyl carbonium ions derived from protonation of diaryl alkenes. The solvation requirements and activity coefficient variation in concentrated acid solution will be similar for the two sets of ions. Thus $y_{R+} \approx y_{BH+}$ and equation (3.20) becomes a valid approximation. In general determination of an H_R' scale is made via measurement of the J_0 scale which is combined with water activities according to equation (3.21) (for example see Arnett and Mach, 1966). It is relevant to note that the protonation of a series of aliphatic dienes in aqueous H_2SO_4 paralleled H_0 and not the H_R' acidity scale (Deno *et al.* 1963).

Phloroglucinol and its methyl and ethyl ethers form a series of structurally similar weak carbon bases. However the acidity function behaviour of the various members of the series show distinct differences which are a function of the number of hydroxy groups present in the molecules (Kresge and Chiang, 1961; Kresge *et al.* 1962; Schubert and Quacchia, 1962, 1963). This is exemplified by comparing the values of $-d\log_{10}(C_{BH+}/C_B)/dH_0$ for 1,3,5-trimethoxybenzene (1·26), 1-hydroxy-3,5-dimethoxybenzene (ca. 1·02 is lower limit), 1,3-dihydroxy-5-methoxybenzene (0·98) and phloroglucinol (0·85) in aqueous perchloric acid (Schubert and Quacchia, 1963). The ionization equilibria of 1-hydroxy-3,5-diethoxybenzene and 1,3,5-triethoxybenzene show similar behaviour to that of the appropriate methoxy analogues (Kresge *et al.* 1962). These results are explicable in terms of the ability of the conjugate acids of the molecules concerned to hydrogen bond to the solvent water. The more hydroxy groups in the ion the greater will be its ability to hydrogen bond and therefore the less steep will be the variation of $\log_{10}(C_{BH+}/C_B)$ with changing acid concentration.

The protonation equilibria of none of these bases follow the H_R' acidity function. Thus even for 1,3,5-trimethoxybenzene a plot of $\log_{10}(C_{BH+}/C_B)$ against $-H_R'$ has a slope of 0·78, and the acidity function appropriate to the ionization of the base therefore lies between H_0 and H_R' (Schubert and

Quacchia, 1963). It should be noted that there is some doubt about this result as Kresge *et al.* (1962) and Kresge and Chiang (1961) found $d\log_{10}(C_{BH^+}/C_B)/dH_0$ was ca. 2 for 1,3,5-trimethoxy and 1,3,5-triethoxybenzene. They concluded that the protonation of these bases does in fact parallel H'_R. However the accumulation of evidence shows that carbon bases do not all obey the same acidity function. This is further exemplified by the ionization of indoles and pyrroles which also protonate on carbon but follow neither H_0 nor H'_R (Hinman and Whipple, 1962; Whipple *et al.* 1963; Hinman and Lang, 1960, 1964; Chiang and Whipple, 1963; Abraham *et al.* 1959).

Long and Schulze (1961, 1964; Schulze and Long, 1964) have made a detailed study of the ionization of several azulenes in aqueous perchloric acid. They found that in general azulenes either protonate on the carbon at the 3-position or in some cases on the oxygen atoms of 1-substituents. It is interesting that the two types of ionization give clearly divergent acidity function behaviour. Thus for 1-formylazulene and 1-nitroazulene, which protonate on oxygen, $-d\log_{10}(C_{BH^+}/C_B)/dH_0$ is 1·1 and 1·0 respectively. Both of these compounds behave as Hammett bases and their ionization equilibria follow H_0. However five other 1-substituted azulenes all of which are protonated on carbon gave values in the range 1·6–1·9 for $-d\log_{10}(C_{BH^+}/C_B)dH_0$. Thus these compounds protonate in approximate agreement with H'_R.

Long and Schulze (1964) stressed that the J_0 and H'_R acidity functions are based on measurements using carbinols as indicators. These are susceptible to hydrogen bonding with solvent water and in accord with this the neutral carbinols show rather unusual activity coefficient behaviour (Boyd, 1963a). It has been suggested that a series of hydrocarbon bases are much more likely to provide an acidity function which would be more generally applicable. The protonation equilibria of the azulenes leads to the definition of an acidity function which differs from both J_0 and H'_R and which could fulfil this purpose.

3.2.6. *Conclusions and Comments on the Measurement of* pK_{BH^+} *using the Acidity Function Method*

The original concept that the activity coefficient ratio (y_{BH^+}/y_B) might be the same for all bases in a particular acid solution has been proved incorrect. The concept is often true for structurally similar bases but sometimes even quite trivial variations in structure can produce appreciable differences in (y_{BH^+}/y_B). Particularly important is the introduction of groups which have the ability to be solvated via hydrogen bonding. Such groups can lead to dramatic changes in the way y_{BH^+} and/or y_B vary with changing electrolyte concentration. However several sets of bases whose activity coefficient behaviour is similar within the set have been studied and appropriate acidity function scales have been defined. Some are plotted in Fig. 3.2 which clearly demonstrates the magnitude of the deviations in the behaviour of the different classes of base.

The spectrophotometric method for the determination of acidity functions also gives valuable information about the strengths of the weakly basic indicator molecules. By the stepwise comparison of a series of progressively

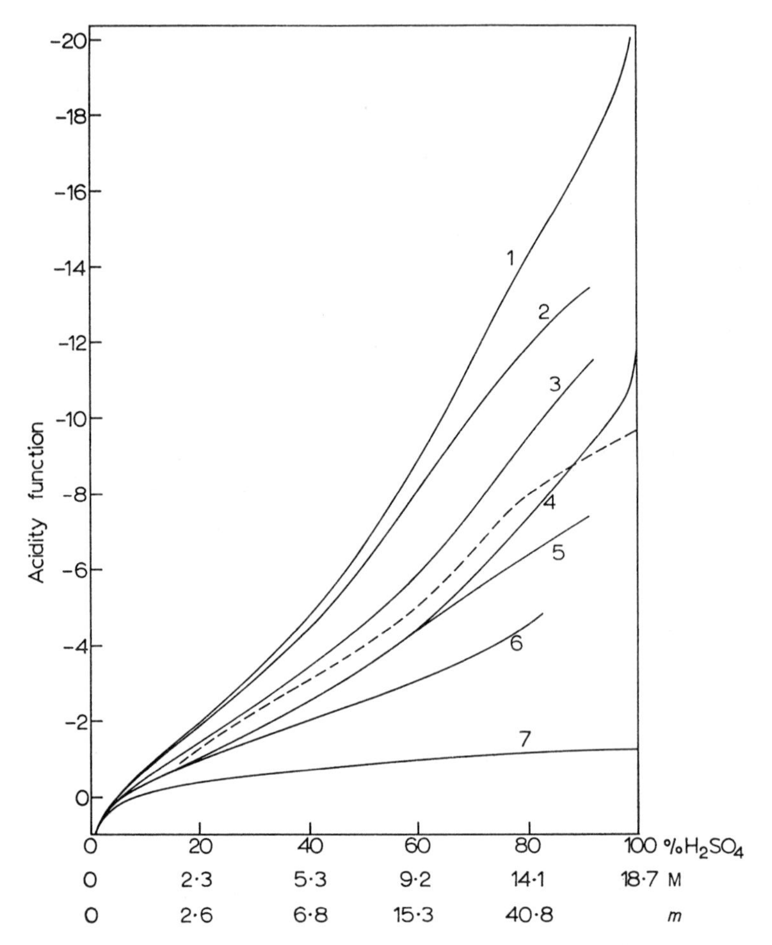

FIG. 3.2. Comparison of acidity functions for aqueous sulphuric acid. Scales are (class of indicator in brackets): (1) J_0 (triarylcarbinols), (2) H'_R (diarylalkenes), (3) H'''_0 (tertiary amines) = ca. H_I (indoles), (4) H_0 (primary amines), (5) H_B (benzophenones), (6) H_A (amides), (7) $-\log_{10} C_{H_2SO_4}$. The dashed line is the redox function R_0(H).

weaker structurally similar bases not only an acidity function but also a value for pK_{BH^+} of each base is deduced. If the basic postulates of acidity function theory are applicable these are thermodynamic pK_{BH^+} values referred to standard state water. Alternatively ionization ratios (C_{BH^+}/C_B) may be measured for a base of unknown pK_{BH^+} in a series of acid solutions. $Log_{10}(C_{BH^+}/C_B)$ is then correlated via equation (1.27) with a known acidity function for the strong acid. Providing the protonation behaviour of the base in question parallels the acidity function with the which ionization ratios are being correlated then a consistent set of pK_{BH^+} values should be obtained. The criterion which *must* be obeyed if the pK_{BH^+} is to stand any chance of being the correct thermodynamic value is that $\log_{10}(C_{BH^+}/C_B)$ should be a linear function *with unit slope* of the acidity function being used to deduce the results.

If there is no acidity function available which satisfactorily correlates with $\log_{10}(C_{BH^+}/C_B)$ in this way then it is difficult to make a reliable estimate of pK_{BH^+}. This situation often arose in the days when only H_0 acidity scales had been measured and the behaviour of classes of indicator other than primary amines had not been investigated in any detail. It follows that early measurements of pK_{BH^+} were usually only reliable providing B was a Hammett base whose protonation followed H_0. If $\log_{10}(C_{BH^+}/C_B)$ does follow H_0 then when $C_{BH^+} = C_B$, $H_0 = pK_{BH^+}$. Davis and Geissman (1954) have suggested a graphical method for the determination of pK_{BH^+} based on the identity of pK_{BH^+} and H_0 at half protonation. This method has often been applied to the determination of pK_{BH^+} for weak bases whose protonation does not parallel H_0. Thus the H_0 value corresponding to half protonation of a base may be taken as a measure of the strength of that base. It is, however, not the strictly defined thermodynamic pK_{BH^+}. This is exemplified by the study of the protonation of ketones by Stewart *et al.* (1963). The pK_{BH^+} they quote give a useful comparison of the relative base strengths of the ketones investigated. However the recent determination of the ketone H_B acidity function for aqueous H_2SO_4 (Bonner and Phillips, 1966) enables a more rigorous determination of what are probably reliable thermodynamic ionization constants. The values are significantly altered (see also however Greig and Johnson, 1968).

It is recommended that the tabulation of the results of studies of the ionization of weak bases in strong acid solutions should include an indication of the acid concentrations at which half protonation of the bases occurs. This enables the data to be easily held under review in the light of later discoveries or perhaps the determination of more appropriate acidity functions with which to correlate the results (Arnett and Anderson, 1963). The strengths of weak bases has been the subject of a comprehensive review by Arnett (1963). This included a detailed section on the determination of base ionization constants via acidity function measurements. These need not be confined to ultraviolet or visible spectral studies. Thus p.m.r. spectra sometimes enable accurate evaluation of ionization ratios (Taft and Levins, 1962; O'Reilly and Leftin, 1960). Jaques and Leisten (1964) have determined the base strength of ether by correlating H_0 with the vapour pressures of ether in aqueous $HClO_4$ and aqueous H_2SO_4

solutions. It also is interesting to note the investigation by Hopkinson and Wyatt (1967) of the effect on the pK_{BH^+} of weak bases of excitation to the first excited singlet state. In some cases the excited state pK_{BH^+} estimated directly from the fluorescence spectra agreed fairly well with pK_{BH^+} calculated from the ground state values using the Förster cycle method (Förster, 1950; Weller, 1952). Noyce and Jorgenson (1962) have compared different methods for the evaluation of pK_{BH^+} from acidity functions.

Bunnett and Olsen (1965, 1966) have suggested a method for the determination of pK_{BH^+} using equation (3.22) via which the ionization ratio (C_{BH^+}/C_B) for any weak base is related to the Hammett H_0 acidity function. Plots of

$$\log_{10}(C_{BH^+}/C_B) + H_0 = \phi(H_0 + \log_{10} C_{H^+}) + pK_{BH^+} \qquad (3.22)$$

$\log_{10}(C_{BH^+}/C_B) + H_0$ against $H_0 + \log_{10} C_{H^+}$ are generally linear and extrapolation to the intercept at $(H_0 + \log_{10} C_{H^+}) = 0$ gives pK_{BH^+}. This approach has been tested for the ionization of amides (Yates *et al.* 1964), tertiary amines (Arnett and Mach, 1964), indoles (Hinman and Lang, 1964) and triarylcarbinols (Deno *et al.* 1955). The deduced pK_{BH^+} were often in good agreement but sometimes up to ca. 2 pK units different from the figures obtained from the conventional stepwise comparison method. The estimation of the intercept requires an extrapolation of the graph which is longer the weaker the base being studied. For long extrapolations the plots must be accurately linear if the deduced value of pK_{BH^+} is to be at all reliable. Furthermore it is an assumption inherent in the method that the slope ϕ of the line remains constant down to $H_0 + \log_{10} C_{H^+} = 0$. It can be shown that ϕ is given by equation (3.23) in which B is the base under investigation and B′ is a primary amine base used in the determination of H_0.

$$\log_{10}(y_{B'H^+} y_B / y_{BH^+} y_{B'}) = \phi \log_{10}(y_{B'H^+} / y_{B'} y_{H^+}) \qquad (3.23)$$

For bases showing similar acidity function behaviour the values of ϕ should be identical. Thus $0{\cdot}42 < \phi < 0{\cdot}55$ for the amides (apart from 2-pyrrolecarboxamide) studied by Yates *et al.* (1964). This range is sufficiently narrow for the appropriate amides to be acceptable as a consistent set of bases suitable for the definition of an acidity function. For the indoles three groups of bases were distinguished. However, apart from indoles with no substituent in the hetero ring, the groups were not the same as those given by the more conventional acidity function treatment used by Hinman and Lang (1964).

3.3. The Protonation of Charged Bases in Aqueous Acid Solutions

All the acidity functions which have been considered above are pertinent to the protonation of neutral bases in solution. The protonation of ionic base species leads to acidity functions which are not necessarily identical to their neutral base counterparts. Thus the definition of the H_- and H_+ acidity functions are given by equations (3.24) and (3.25) respectively. The H_-

$$H_- = -\log_{10}(a_{H^+}y_{S^-}/y_{SH}) = pK_{SH} - \log_{10}(C_{SH}/C_{S^-}) \quad (3.24)$$

$$H_+ = -\log_{10}(a_{H^+}y_{R^+}/y_{RH^{2+}}) = pK_{RH^{2+}} - \log_{10}(C_{RH^{2+}}/C_{R^+}) \quad (3.25)$$

function is appropriate to weak bases (S^-) which are mononegatively charged and H_+ is appropriate to singly positively charged bases (R^+). The subscript placed after H indicates the charge of the base form of the indicator (thus H_0 refers to protonation of bases for which $z = 0$). In accord with this nomenclature acidity functions J_+, J_-, H_{2+} (or H_{++}), etc. may be similarly defined. In dilute acid solutions there will be a difference between H_-, H_0, H_+, etc. as a result of the different electrostatic (Debye–Hückel) contributions to the activity coefficient ratios in equations (3.24), (1.29), (3.25), etc.

Measurements of the second protonation of a series of substituted 1,2-phenylenediamines and 2,3-dimethylquinoxalines in aqueous H_2SO_4 (0·02–14·0 mole litre^{-1}) have led to the definition of an H_+ acidity function which closely follows the corresponding H_0 scale (Vetešník *et al.* 1968). The second protonations of 4-aminoacetophenone (Bonner and Lockart, 1957) and a series of aminopyridines (Brignell *et al.* 1967) in aqueous sulphuric acid also follow H_0. This parallel between H_0 and H_+ is apparently extended into oleum solutions (Brand *et al.* 1952). Staples (1966) has shown that the protonation of the complex cations of a series of isothiocyanatobisethylenediamine-cobalt(III) salts follows the H_0 acidity function for 78–91% H_2SO_4 in water. However the second protonations of $[Fe(o\text{-phen})_2(CN)_2]^+$ and its 2,2-′ dipyridyl analogue (appropriate acidity function H_{2+}) do not parallel H_0. H_+ is a function of the molecular structure of the cationic bases being protonated. There is not a unique H_+ acidity scale. This follows since although the second protonation of the aminopyridines correlates with H_0, the second protonations of pyrazine and pyrimidine follow the amide H_A scale (Brignell *et al.* 1967). The protonations of the conjugate acids of quinazoline and quinoxaline do not correlate satisfactorily with any existing acidity function. In their study of the J_0 acidity function Lowen *et al.* (1950; Murray and Williams, 1950) used several 2+ or 3+ charged cations as indicators. The observed acidity function behaviour was similar to that for neutral carbinols and this indicates that J_{2+} and J_{3+} may approximately parallel J_0.

The H_- acidity functions for aqueous perchloric and sulphuric acid have been investigated by Boyd (1961, 1963b) using a series of cyanocarbon acids as indicators. The H_- and H_0 scales for the two acids are compared in Fig. 3.3. Also plotted is $(H_0 - H_-)$ which from combination of equations (1.29) and (3.24) is given by equation (3.26).

$$H_0 - H_- = \log_{10}(y_{BH^+}y_{S^-}/y_B y_{SH}) \quad (3.26)$$

At low acid concentrations $(H_0 - H_-)$ decreases, then passes through a minimum, and finally starts to increase rather slowly with increasing acid concentration. The initial decrease is consistent with the fact that at low ionic strengths the acitivity coefficients in equation (3.26) will be primary governed by electrostatic contributions. Boyd (1963a) has shown that at high acid concentrations

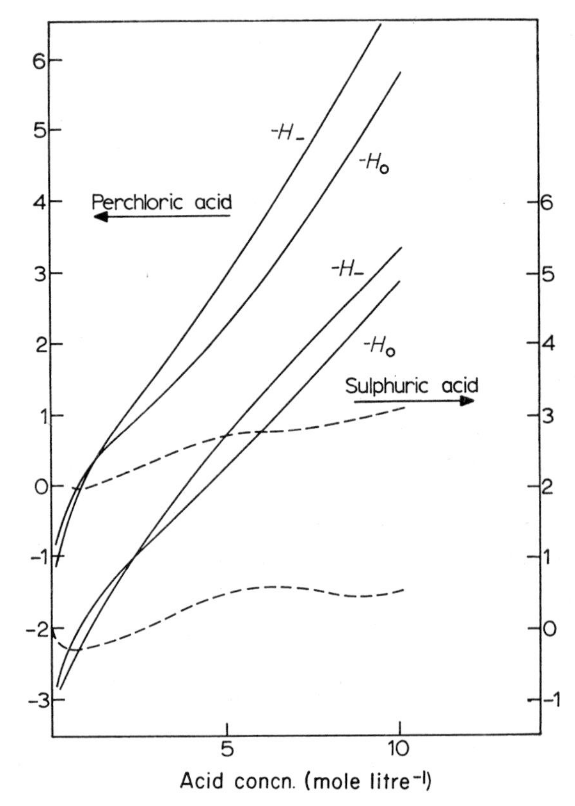

FIG. 3.3. Comparison of H_0 and H_- for aqueous perchloric acid (upper curves) and for aqueous sulphuric acid (lower curves). Dashed curves are $(H_0 - H_-)$. Reproduced with permission from Boyd (1963b).

y_{BH^+} and y_{S^-} vary with increasing ionic strength in such a way that the product $y_{BH^+}y_{S^-}$ is relatively insensitive to changing acid concentration. If it is assumed that $y_B y_{SH}$ also does not change very much then the magnitude of the trends in $(H_0 - H_-)$ are reasonable.

The H_- acidity functions for aqueous sulphuric and perchloric acid solutions deduced from kinetics measurements by Cox and McTigue (1964) do not agree with Boyd's (1961) scales. This suggests that (as for H_+ and H_0) there is no unique H_- scale for a given aqueous acid. Added evidence for this conclusion is gained from a study of the protonation of several anionic azobenzene derivatives in aqueous sulphuric acid (Reeves, 1966). In some cases the protonation of the mononegatively charged bases approximately paralleled the H_0 acidity function (Jorgenson and Hartter, 1963) and in some cases it paralleled the Boyd (1961) H_- scale. There were some bases for which no existent acidity function scale for aqueous sulphuric acid could be satisfactorily correlated with the protonation equilibria. Reeves (1966) studied the protonation of one anionic base with charge 2– (appropriate acidity function H_{2-}) and found that $\log_{10}(C_{S^{2-}}/C_{SH^-})$ decreased less steeply with increasing H_2SO_4 concentration (ca. 50% H_2SO_4) than either H_0 or H_-.

An H_- scale for aqueous hydrochloric acid has been measured using a series of phosphorus-containing indicators (Phillips, 1961). The scale was shown to be consistent with equation (3.27) the derivation of which required the assumption that (y_{BH^+}/y_B) for the Hammett bases used to define H_0 is

$$\tfrac{1}{2}(H_0 + H_-) = \text{ca.} -\log_{10} y_{\pm} C_{HCl} \tag{3.27}$$

identical to (y_{S^-}/y_{SH}) for the bases used to define H_-. This relationship is not obeyed by the H_- scales for perchloric and sulphuric acids defined using cyanocarbon acid indicators. H_- acidity functions have also been quoted for nitric acid (Cox and McTigue, 1964) and for trichloracetic and trifluoracetic acids (Randles and Tedder, 1955). The latter two were measured using the picrate anion as weak base. Compared at equal mole fractions of carboxylic acid $H_-(CF_3COOH) < H_-(CCl_3COOH)$ and this is consistent with the related observation that $H_0(CF_3COOH) < H_0(CCl_3COOH)$. Thus comparison of either H_0 or H_- indicates that CF_3COOH is more acidic than CCl_3COOH. As the carboxylic acid content of the solutions increases the dielectric constant rapidly decreases. Although this leads to a decrease in the rate of change of H_0 with acid concentration it has the opposite effect on H_-. Thus the protonation of an anionic base is, on electrostatic considerations, favoured by a decrease in dielectric constant.

3.4. Acidity Functions from E.m.f. Measurements

The pH scale, which is meaningful for dilute electrolyte solutions only, is defined in terms of the e.m.f. values of cells in which one electrode is reversible to hydrogen ions in solution. Michaelis and Granick (1942) have shown that the electrode potentials of certain redox electrodes give a response which is a function of acidity for concentrated aqueous acid solutions. The reversible redox equilibria between the reduced, semi-oxidized, and fully oxidized forms of each of the three solutes 3-hydroxythiazine, 3-aminothiazine, and 3,9-diaminothiazine enabled the setting up of appropriate electrodes. The use of these allowed a "pH" scale to be measured for aqueous sulphuric acid up to ca. 12 mole litre^{-1}. The pH scale deduced is remarkably similar to the H_0 acidity function for aqueous sulphuric acid. Thus even for $C_{H_2SO_4} = 10$ mole litre^{-1} the deviation between H_0 and pH is only ca. 0·5 units.

Similar measurements were made by Strehlow and Wendt (1961) using a ferrocene, ferricinium redox electrode. By combination of polarographic and e.m.f. data the e.m.f. $E(x)$ of the cell (3.28) was evaluated for a series of aqueous

$$\text{Pt, } H_2(\text{g, 1 atm}) \,|\, \text{aq. } H_2SO_4\text{, Ferrocene, Ferricinium(1:1)} \,|\, \text{Pt} \tag{3.28}$$

sulphuric acid solutions up to 100% H_2SO_4. Hence a "redox function" R_0(H) was defined by equation (3.29) in which $E(1)$ is the e.m.f. of cell (3.28) when $a_{H+} = 1$.

$$R_0(\text{H}) = F[E(x) - E(1)]/2{\cdot}303RT \tag{3.29}$$

The R_0(H) scale agreed very closely with the pH values deduced by Michaelis and Granick (1942). R_0(H) is plotted in Fig. 3.2 where it is compared with the spectrophotometrically determined acidity functions.

The possible use of glass electrodes to measure acidity functions of aqueous sulphuric acid has been investigated by Clerc *et al.* (1965). Providing the electrodes were made of suitable glass they were capable of reproducibly responding to changes of acidity in sulphuric acid solutions for which $-1 < H_0 < -9{\cdot}6$. The response correlated with the H_0 scale.

3.5. Phase Equilibria Relevant to Acidity Function Studies

3.5.1. *Cryoscopy*

The determination of the van't Hoff *i* factor for solutes in 100% sulphuric acid is a useful and simple method for gaining information about the modes of ionization of the solute molecules in acid solutions. The relevance of this to the determination of acidity functions has already been discussed in Section 1.4 above. The cryoscopic technique can sometimes be improved by using as solvent sulphuric acid to which is added small concentrations of another substance such as water or sulphur trioxide. Several suitable cryoscopic mixtures have been described (Leisten, 1961; Leisten and Wright, 1964; Leisten and Walton, 1964). Robinson and Quadri (1967a, 1967b) have coupled cryoscopic measurements in aqueous sulphuric acid with conductimetric measurements in dilute oleum solutions in order to find out how a wide variety of organic species ionize in concentrated acids. Combination of cryoscopic and conductimetric data is sometimes much more informative than results from only one of these techniques (for example, Gillespie and Malhotra, 1967). Other physical methods available for the study of organic ions in solution have recently been reviewed with special reference to carbonium ions (Bethell and Gold, 1967). In some cases methanesulphonic acid is a more suitable solvent than sulphuric acid for cryoscopic studies (Craig *et al.* 1950).

3.5.2. *Partition Equilibria*

The partition equilibria of weak bases between an organic solvent and an aqueous acid solution can be correlated with an appropriate acidity function for the latter. Equation (3.30) relates the acidity function H with the concentration C_{BH^+} of protonated base in the aqueous phase, the concentration C'_B of unprotonated base in the organic phase, and P the partition coefficient $(= C_B/C'_B)$ for the neutral base between the two phases.

$$H = pK_{BH^+} + \log_{10} P - \log_{10}(C_{BH^+}/C'_B) \tag{3.30}$$

The applicability of this equation has been tested for the partition of some azulenes between petrol ether or toluene and aqueous H_2SO_4 or H_3PO_4 (Plattner *et al.* 1949), three conjugate hydrocarbons between cyclohexane and aqueous H_2SO_4 (Gold and Tye, 1952), and dithizone between carbon tetrachloride and aqueous H_2SO_4 (Akaiwa and Kawamoto, 1967). In all these cases $\log_{10}(C_{BH^+}/C'_B)$ plotted against H_0 gave straight lines. However the slopes were approximately −2 and not −1 as required by the equation. This makes the reliable estimation of $(pK_{BH^+} + \log_{10} P)$ at infinite dilution difficult. Akaiwa and Kawamoto (1967) avoided this problem for dithizone by using J_0 for H in equation (3.30). The slope of the appropriate plot was then ca. −1 as predicted and hence a more or less constant figure for $(pK_{BH^+} + \log_{10} P)$ was obtained from the set of experimental points. The calculation of pK_{BH^+} requires knowledge of $\log_{10} P$. Gold and Tye (1952) pointed out that considerable errors may be inherent in values of P for hydrocarbons distributed between water and an organic solvent. A corresponding error would appear in pK_{BH^+}. Arnett and Wu (1960, 1962) have devised a graphical method for the determination of P and pK_{BH^+} via a rearranged form of equation (3.30). The pK_{BH^+} deduced for a series of ethers agreed with other literature values in some cases but not in others (Edward, 1963; Edward *et al.* 1962).

It has been suggested that the values of H_0 for which $C_{BH^+} = C'_B$ may be taken as a comparative measure of the acidities of the conjugate acids of a series of hydrocarbon bases (Plattner *et al.* 1949). However this is rather an empirical approach and could be subject to serious error.

The activity coefficients of 4-nitroaniline, 2-nitroaniline and diphenylamine in aqueous sulphuric acid have been evaluated from partition measurements (Vinnik and Librovich, 1966).

3.5.3. *Solubility Measurements*

Activity coefficients of species in concentrated electrolyte solutions may be deduced from appropriate solubility determinations (Long and McDevit,

1952). Thus Deno and Perizzolo (1957) have investigated the activity coefficients of several neutral molecules in aqueous sulphuric acid solutions. Ojeda and Wyatt (1964) have measured y_B for 4-nitroaniline in aqueous hydrochloric acid solutions, containing added salts. The most important study from the acidity function point of view is that of Boyd (1963a) who determined the variation of activity coefficients with H_2SO_4 concentration for several typical H_0, J_0 and H_- indicators. By choosing suitable solutes y_B, y_{BH^+} (equation 1.29), y_{SH}, y_{S^-} (equation 3.24), y_{R^+} and y_{ROH} (equation 3.4) were studied up to 70% sulphuric acid. The observed activity coefficient behaviour of each group of solutes was in accord with the observed differences between the H_0, J_0 and H_- acidity function scales. Vinnik and Librovich (1966) have shown that 4-nitroaniline, 2-nitroaniline and diphenylamine give similar variations of (y_B/y_{BH^+}) with sulphuric acid concentration over a wide range of acidity. That diphenylamine should give results paralleling those for the two primary amines is surprising in view of the different acidity function behaviour of primary and secondary amines (Taft, 1960).

Hammett and Chapman (1934) were the first to combine solubility determinations with acidity function data in order to estimate ionization constants for very weak bases. The evaluation of pK_{BH^+} required that the activity coefficient y_B of the weak bases should be assumed constant for the range of concentrations of strong acid over which the measurements were made. This is now known not to be true (Boyd, 1963a; Deno and Perizzolo, 1957). Thus for example the steep rise in the solubility of benzoic acid which occurs at ca. 70% H_2SO_4 is caused primarily by the decrease in y_B as the sulphuric acid concentration is increased. The value of pK_{BH^+} (benzoic acid) $= -5{\cdot}9$ deduced by Hammett and Chapman is therefore spurious, the correct value probably being about 2 units lower (Yates and Wai, 1965).

The ionization constants of several aromatic sulphonic acids have been deduced from measurements of the solubility of the acids as a function of sulphuric acid concentration (Horyna, 1959a, 1959b, 1962). The sulphonic acids studied could exist in three forms $ArSO_3^-$, $ArSO_3H$ and $ArSO_3H_2^+$ in sulphuric acid solutions. The concentrations of each form at equilibrium will be related to the H_- and H_0 acidity functions for aqueous sulphuric acid according to equations (3.31) and (3.32). Let S be the total solubility ($ArSO_3^- +$ $ArSO_3H$ in dilute H_2SO_4 and $ArSO_3H + ArSO_3H_2^+$ in more concentrated

$$H_0 = pK_{ArSO_3H_2^+} - \log_{10}(C_{ArSO_3H_2^+}/C_{ArSO_3H}) \tag{3.31}$$

$$H_- = pK_{ArSO_3H} - \log_{10}(C_{ArSO_3H}/C_{ArSO_3^-}) \tag{3.32}$$

H_2SO_4) of sulphonic acid, and S_0 be the solubility of unionized $ArSO_3H$ in sulphuric acid solutions in which only the unionized form exists. Then for H_2SO_4 solutions in which $ArSO_3^-$ and $ArSO_3H$ are in equilibrium ($C_{ArSO_3H_2^+} =$ ca. 0) equation (3.33) is applicable. Similarly when only $ArSO_3H$ and $ArSO_3H_2^+$

exist to any appreciable extent ($C_{ArSO_3^-}$ = ca. 0) equation (3.34) should be obeyed (Horyna, 1959a).

$$\log_{10}[(S - S_0)/S_0] = H_- - pK_{ArSO_3H} \quad (3.33)$$

$$\log_{10}[(S - S_0)/S_0] = pK_{ArSO_3H_2^+} - H_0 \quad (3.34)$$

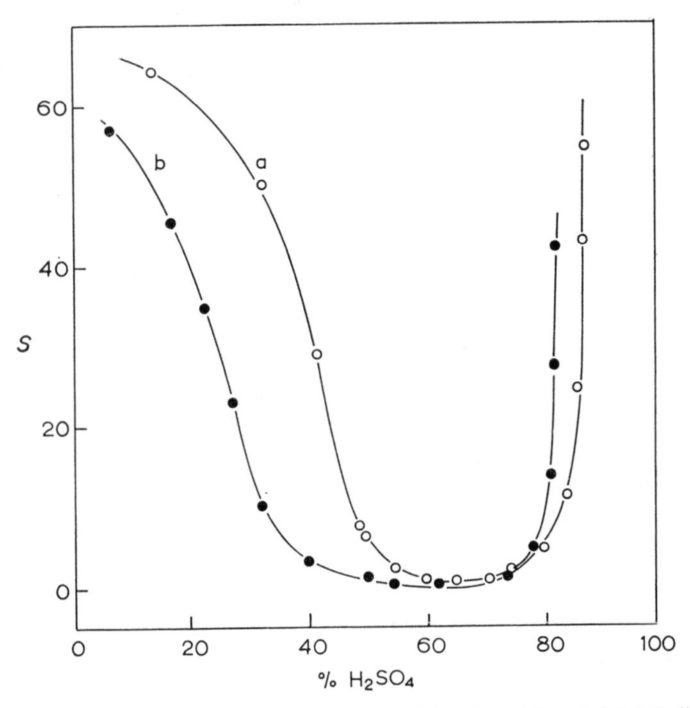

FIG. 3.4. Solubility of (a) 4-methylbenzenesulphonic acid and (b) 2,5-dimethylbenzenesulphonic acid in aqueous sulphuric acid solutions. Reproduced with permission from Horyna (1959a).

Figure 3.4 shows the variation of solubility with sulphuric acid concentration for 4-methylbenzenesulphonic acid and 2,5-dimethylbenzene sulphonic acid. The corresponding plots (Fig. 3.5) of $\log_{10}[(S - S_0)/S_0]$ against the Hammett acidity function H_0 have two linear sections from which pK_{ArSO_3H} and $pK_{ArSO_3H_2^+}$ may be deduced. This method relies on the assumption that S_0 is independent of sulphuric acid concentration. As discussed above this may not be true. However the plots in Fig. 3.5 have the correct slopes of +1 in accord with equation (3.33) and −1 in accord with equation (3.34). The substitution of H_0 for H_- in equation (3.33) is probably not serious in view of the similarity between the H_0 and H_- scales for aqueous sulphuric acid (Boyd 1961, 1963b).

Using the above approach Horyna (1960) has also investigated the solubilities of a series of carboxylic acids in concentrated HCl, H_2SO_4 or HNO_3 solutions. The slopes of the appropriate graphs were found to differ significantly from 1 in some cases. The treatment was adapted such that this was

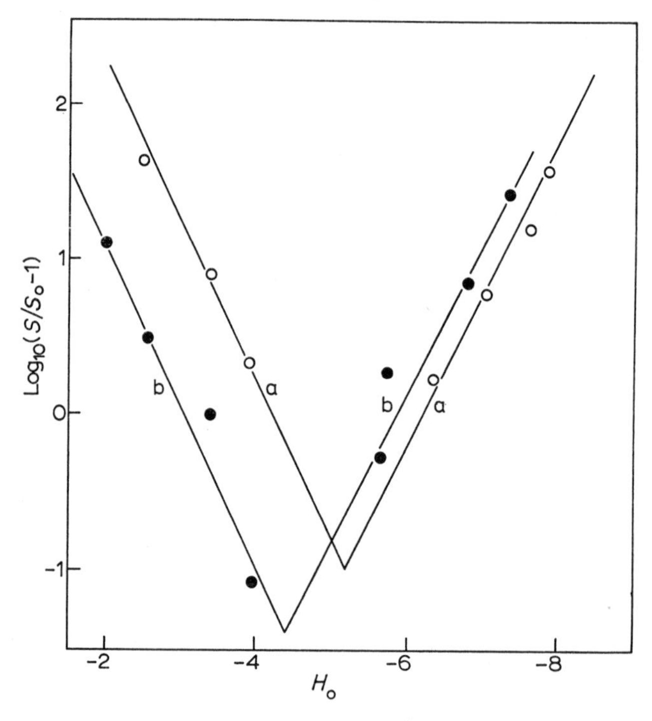

FIG. 3.5. Test of equations (3.33) and (3.34) for (a) 4-methylbenzene sulphonic acid and (b) 2,5-dimethylbenzenesulphonic acid. Reproduced with permission from Horyna (1959a).

equated with the variations of the activity coefficients of the neutral unionized carboxylic acids at different concentrations of strong acid. Possible differences in the acidity function behaviour of different carboxylic acids should also be considered. For the protonation of some acids the solubility data correlated

$$RCOOH + H^+ \rightleftharpoons RCO^+ + H_2O \tag{3.35}$$

with J_0 and not with H_0. This is consistent with the mode of ionization (3.35) and not with a simple protonation to $RCOOH_2^+$.

3.6. The Effects of Neutral Salts on Acidity Functions

The addition of neutral salts to solutions of strong acids in water has a profound effect on the acidity of the solutions. The majority of quantitative studies of this effect has been concerned with the changes in H_0 produced by the addition of high concentrations of salts. This is exemplified by the results (Fig. 3.6) obtained by Paul (1954) and Long and McIntyre (1954) for the variation of H_0 for 0·1 mole litre^{-1} hydrochloric acid on the addition of several

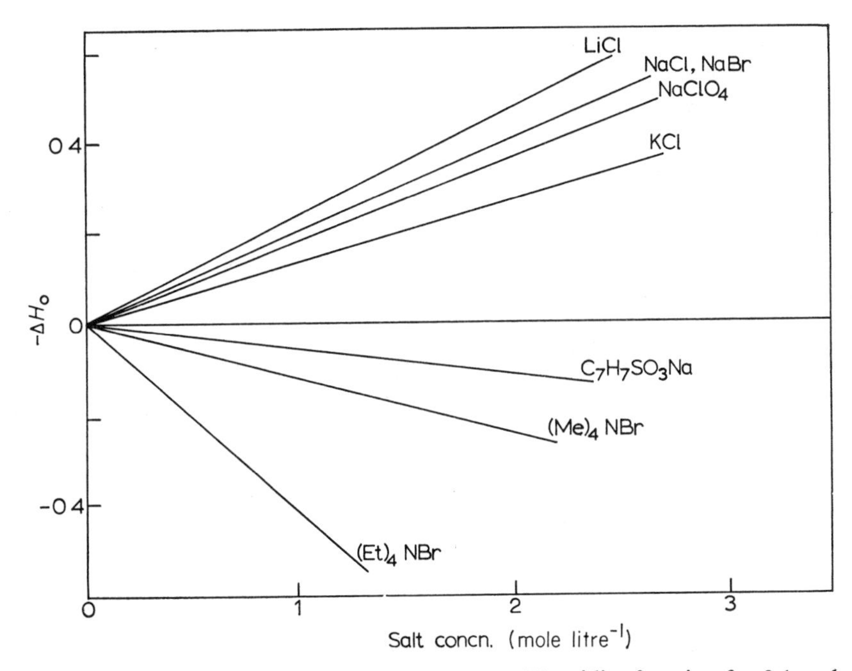

FIG. 3.6. Effect of added salts on the Hammett H_0 acidity function for 0·1 mole litre^{-1} aqueous hydrochloric acid at 25°. 4-Nitroaniline indicator. Reproduced with permission from Long and McIntyre (1954).

salts. ΔH_0 is equal to H_0 for (0·1 M HCl + salt) minus H_0 for 0·1 M HCl alone. In all cases ΔH_0 is a linear function of the salt concentration. Equation (3.36) in which C_S is the salt concentration is therefore applicable (Long and McIntyre, 1954). However whereas KCl, $NaClO_4$, NaCl, NaBr and LiCl produce an

$$-\Delta H_0 = \alpha C_S \tag{3.36}$$

increase in acidity (α positive), the larger solutes Et_4NBr, Me_4NBr and $C_7H_7SO_3Na$ give a decrease (α negative). The results are based on use of 4-nitroaniline as indicator.

Several other independent studies have given results which are in accord with those figured here. Thus sodium chlorate increases the acidity of aqueous perchloric acid solutions (Harbottle, 1951; Challis and Ridd, 1962) and lithium perchlorate has a larger effect (Day and Wyatt, 1966). The relative abilities of lithium and sodium salts to increase the acidity of aqueous acid solutions is also reflected in data for $LiNO_3$–HNO_3 and $NaNO_3$–HNO_3 mixtures. Thus, for example, for 2·5 *m* $LiNO_3$ in 0·5 *m* HNO_3 $H_0 = -0{\cdot}71$ whereas for 2·5 *m* $NaNO_3$ in 0·5 *m* HNO_3 $H_0 = -0{\cdot}28$ (Day and Wyatt, 1966).

Challis *et al.* (1962) have noted that sodium perchlorate and perchloric acid have similar effects on the deviation between H_0 and $-\log_{10} C_{acid}$ for concentrations up to ca. 3 mole litre^{-1}. Similarly for aqueous hydrochloric acid solutions the deviation between H_0 and $-\log_{10} C_{H^+}$ is equivalent to the magnitude of the salt effect which would be produced by adding potassium chloride at the same concentration (Moiseev and Flid, 1954; Paul and Long, 1957). Moiseev and Flid found that for several acid–salt mixtures the ionization ratio (C_{BH^+}/C_B) of 4-nitroaniline fitted equation (3.37) in which μ is the total ionic

$$\log_{10}(C_{BH^+}/C_B) = \log_{10} C_{H^+} + pK_{BH^+} + L\mu \qquad (3.37)$$

strength of the solutions and L is a constant. For aqueous HCl or for aqueous KCl–HCl, $CaCl_2$–HCl or $MgCl_2$–HCl mixtures $L = 0{\cdot}1$, and $L = 0{\cdot}129$ for aqueous H_2SO_4 or KCl–H_2SO_4, KNO_3–H_2SO_4 or $NaNO_3$–H_2SO_4 mixtures. For a series of acid–salt mixtures of constant ionic strength equation (3.37) would predict that $(H_0 + \log_{10} C_{H^+})$ should be a constant. This is true for different small concentrations of acid in a fixed high concentration of salt (Rosenthal and Dwyer, 1962, 1963a). However Brubaker *et al.* (1962) measured H_0 for HCl–$AlCl_3$ and HCl–KCl mixtures at fixed total ionic strengths ($\mu = 1$ and $\mu = 3$) and deduced that $(H_0 + \log_{10} C_{H^+})$ was a linear function of HCl concentration. The H_0 acidity functions for equimolar mixtures of $NaClO_4$ and HNO_3 and of $NaClO_4$ and H_3PO_4 have been reported (Lee and Stewart, 1964). Sodium perchlorate is nearly identical to sodium nitrate in its ability to enhance the acidity of aqueous nitric acid.

From the definitions of H_0 in equation (1.29) it follows that ΔH_0 (Fig.3.6) will be given by equation (3.38) in which the activity coefficients are referred to

$$-\Delta H_0 = \log_{10}(y_{H^+}\, y_B/y_{BH^+}) \qquad (3.38)$$

0·1 mole litre^{-1} aqueous HCl standard state. It is interesting to consider the contribution which y_B makes to the variation of ΔH_0 with salt concentration. Long and McIntyre (1954) have measured the solubility of 4-nitroaniline in water and in 0·5 and 1 mole litre^{-1} solutions of the series of salts for which acidity function data are available, In all cases $\log_{10} y_B$ was a linear function

$$\log_{10} y_B = K_B\, C_S \qquad (3.39)$$

of C_S in accord with equation (3.39) in which K_B is the salting out parameter (salting in if K_B is negative). The value of K_B for each salt is compared in Table 3.6 with the corresponding value of α. In general the trends in K_B are approximately reflected in the figures for α. Furthermore the difference $(\alpha - K_B)$, which is the proportionality constant between $\log_{10}(y_{H^+}/y_{BH^+})$ and C_S, is influenced less than either α or K_B in going from one salt to another. The results show that variation of both y_B and (y_{H^+}/y_{BH^+}) contributes to the observed salt effects. However y_B is apparently influenced more than (y_{H^+}/y_{BH^+}) on changing the salt.

TABLE 3.6

Proportionality constants of equations (3.36) and (3.39) (Long and McIntyre, 1954; Paul and Long, 1957), and the hydration parameters h for some salts in water (Glueckauf, 1955, 1959)

Salt	α	K_B	$(\alpha - K_B)$	h
LiCl	0·245	0·082	0·16	5·31
NaCl	0·205	0·072	0·13	3·60
NaBr	0·205	0·040	0·17	3·94
$NaClO_4$	0·180	−0·082	0·26	2·60†
KCl	0·145	0·030	0·12	2·58
$C_7H_7SO_3Na$	−0·06	−0·51	0·45	
Me_4NBr	−0·12	−0·272	0·15	
Et_4NBr	−0·41	−0·44	0·03	

† Taken as $h_{NaClO_4} = h_{NaCl} + h_{HClO_4} - h_{HCl}$.

The activity coefficient behaviour of solutes in concentrated solutions is related to the solvation requirements and the effect on solvent structure of the species present. Thus in accord with the lead given by Bascombe and Bell (1957) and Wyatt (1957) consideration of solvent effects often greatly facilitates our understanding of acidity function results. Rosenthal and Dwyer (1963a) have tested this approach for 1–9 mole litre^{-1} lithium chloride solutions containing small concentrations of hydrochloric acid. Equation (3.40) was applicable where $B =$ ca. 0·15 and $n =$ ca. 2. The parameter n may be looked

$$H_0 = -\log_{10} C_{H^+} + n \log_{10} a_w - BC_{LiCl} \qquad (3.40)$$

upon as the difference in hydration number between $(B + H^+)$ and BH^+ in the lithium chloride solutions.

Comparison of the effects of a series of lithium and sodium salts on H_0 led Day and Wyatt (1966) to the conclusion that the relative magnitude $(Li^+ > Na^+)$ of the effects was connected with the solvation requirements $(Li^+ > Na^+)$

of the two cations. The greater the solvation requirements of an added ion the greater its effect on acidity function. This is also reflected in the thermodynamic properties of the salt solutions. Thus it is interesting that the values of α for LiCl, NaCl, NaBr, $NaClO_4$ and KCl are in the same order as the hydration numbers (Table 3.6) for these salts deduced by Glueckauf (1955, 1959) from activity coefficient data. This is consistent with McTigue's (1964) formulation for the relationship between acidity function and the hydration requirements of a swamping electrolyte.

The effect of lithium perchlorate and sodium bromide on J_0 for aqueous perchloric acid has been studied by Huckings and Johnson (1964). The salt effects ΔJ_0 are linear functions of the salt concentration. In accord with the deviation between J_0 and H_0 the magnitude of ΔJ_0 is ca. 2·3 times greater than ΔH_0 for a particular salt concentration. The acidity function J_0 for aqueous sulphuric acid is increased (becomes less negative) for the addition of Et_4NBr. This is analogous to the effect of Et_4NBr on H_0 (Long and McIntyre, 1954).

Potentiometric measurements using a hydrogen or a glass electrode have enabled the evaluation of "pH" values for a series of solutions containing high concentration of salts (Rosenthal and Dwyer, 1962, 1963b). As with pH scales for concentrated acids without added salt (Michaelis and Granick, 1942; Strehlow and Wendt, 1961) there is a close correlation between the pH values and the corresponding H_0 data based on the ionization of primary amines. This was confirmed by Schwabe (1967) who measured the effect of 34 salts on the potentiometrically defined pH of solutions of either $HClO_4$, HCl or HBr. In most, but not all cases the pH was a linear function of the salt concentration. The difference in pH between that (pH_0) for a dilute acid solution and that (pH_m) in the presence of a concentration m_S of salt is given by equation (3.41). The charges z_+ and z_- and the size parameters r_+ and r_- refer to the individual

$$pH_m - pH_0 = -0{\cdot}0555\, m_S |z_+ z_-| \left(\frac{h_+}{r_+} - \frac{h_-}{r_-}\right) \quad (3.41)$$

ions of the salt. The factors h_+ and h_- are a function of the hydration requirements of the ions. The derivation of equation (3.41) is based on a consideration of the electrostatic interactions occurring between the hydrogen ions and the ions derived from the salt. The good agreement between the calculated and observed salt effects confirms that the effect of added salt on acidity depends on the size, charge and solvation requirements of the ions.

REFERENCES

Abraham, R. J., Bullock, E., and Mitra, S. S. (1959). *Canad. J. Chem.* **37**, 1859.
Akaiwa, H., and Kawamoto, H. (1967). *J. Inorg. Nuclear Chem.* **29**, 541.
Armstrong, V. C., and Moodie, R. B. (1968). *J. Chem. Soc. B*, 275.

Arnett, E. M. (1963). *Progr. Phys. Org. Chem.* **1**, 223.
Arnett, E. M., and Anderson, J. N. (1963). *J. Amer. Chem. Soc.* **85**, 1542.
Arnett, E. M., and Burke, J. J. (1966). *J. Amer. Chem. Soc.* **88**, 2340.
Arnett, E. M., and Bushick, R. D. (1964). *J. Amer. Chem. Soc.* **86**, 1564.
Arnett, E. M., and Mach, G. W. (1964). *J. Amer. Chem. Soc.* **86**, 2671.
Arnett, E. M., and Mach, G. W. (1966). *J. Amer. Chem. Soc.* **88**, 1177.
Arnett, E. M., and Wu. C. Y. (1960). *J. Amer. Chem. Soc.* **82**, 4999.
Arnett, E. M., and Wu, C. Y. (1962). *J. Amer. Chem. Soc.* **84**, 1680.
Bascombe, K. N., and Bell, R. P. (1957). *Discuss. Faraday Soc.* **24**, 158.
Bayliss, N. S., Dingle, R., Watts, D. W., and Wilkie, R. J. (1963). *Aust. J. Chem.* **16**, 933.
Belotserkovskaya, N. G., and Ginsburg, O. F. (1964). *Doklady Akad. Nauk. S.S.S.R.* **155**, 1098.
Bethell, D., and Gold, V. (1967). "Carbonium Ions—An Introduction." Academic, London and New York.
Bonner, T. G., and Lockhart, J. C. (1957). *J. Chem. Soc.* 364.
Bonner, T. G., and Phillips, J. (1966). *J. Chem. Soc. B*, 650.
Boyd, R. H. (1961). *J. Amer. Chem. Soc.* **83**, 4288.
Boyd, R. H. (1963a). *J. Amer. Chem. Soc.* **85**, 1555.
Boyd, R. H. (1963b). *J. Phys. Chem.* **67**, 737.
Brand, J. C. D., Horning, W. C., and Thornley, M. B. (1952). *J. Chem. Soc.* 1374.
Brignell, P. J., Johnson, C. D., Katritzky, A. R., Shakir, N., Tarhan, H. O., and Walker, G. (1967). *J. Chem. Soc. B*, 1233.
Brubaker, C. H., Rasmussen, P. G., and Luehrs, D. C. (1962). *J. Chem. and Eng. Data.* **7**, Part 2, 563.
Bunnett, J. F., and Olsen, F. P. (1965). *Chem. Comm.* 601.
Bunnett, J. F., and Olsen, F. P. (1966). *Canad. J. Chem.* **44**, 1899.
Challis, B. C., and Ridd, J. H. (1962). *J. Chem. Soc.* 5208.
Challis, B. C., Larkworthy, L. F., and Ridd, J. H. (1962). *J. Chem. Soc.* 5203.
Chiang, Y., and Whipple, E. B. (1963). *J. Amer. Chem. Soc.* **85**, 2763.
Clerc, J. T., Stefanac, Z., and Simon, W. (1965). *Helv. Chim. Acta*, **48**, 1566.
Cox, B. G., and McTigue, P. T. (1964). *J. Chem. Soc.* 3893.
Craig, R. A., Garrett, A. B., and Newman, M. S. (1950). *J. Amer. Chem. Soc.* **72**, 163.
Davis, C. T., and Geissman, T. A. (1954). *J. Amer. Chem. Soc.* **76**, 3507.
Day, J. S., and Wyatt, P. A. H. (1966). *J. Chem. Soc. B*, 343.
De Fabrizio, E. (1966). *Ricerea Sci.* **36**, 1321.
Deno, N. C., and Perizzolo, C. (1957). *J. Amer. Chem. Soc.* **79**, 1345.
Deno, N. C., and Taft, R. W. (1954). *J. Amer. Chem. Soc.* **76**, 244.
Deno, N. C., Jaruzelski, J. J., and Schriesheim, A. (1955). *J. Amer. Chem. Soc.* **77**, 3044.
Deno, N. C., Berkheimer, H. E., Evans, W. L., and Peterson, H. J. (1959a). *J. Amer. Chem. Soc.* **81**, 2344.
Deno, N. C., Groves, P. T., and Saines, G. (1959b). *J. Amer. Chem. Soc.* **81**, 5790.
Deno, N. C., Bollinger, J., Friedman, M., Hafer, K., Hodge, J. D., and Houser, J. J. (1963). *J. Amer. Chem. Soc.* **85**, 2998.
Edward, J. T. (1963). *Chem. and Ind.* 489.
Edward, J. T., and Stollar, H. (1963). *Canad. J. Chem.* **41**, 721.
Edward, J. T., and Wang, I. C. (1962). *Canad. J. Chem.* **40**, 966.
Edward, J. T., Leane, J. B., and Wang, I. C. (1962). *Canad. J. Chem.* **40**, 1521.
Entelis, S. G., Eckel, G. V., and Chirkov, N. M. (1960). *Doklady Akad. Nauk. S.S.S.R.* **130**, 826.

Epple, G. V., Odintsova, V. P., and Entelis, S. G. (1962). *Izvest. Akad. Nauk. S.S.S.R. Otdel. Khim. Nauk.* 1365.
Förster, T. (1950). *Z. Elektrochem.* **54**, 42.
Fraenkel, G., and Franconi, C. (1960). *J. Amer. Chem. Soc.* **82**, 4478.
Gelbach, R. W., and King, G. B. (1942). *J. Amer. Chem. Soc.* **64**, 1054.
Gillespie, R. J., and Malhotra, K. C. (1967). *J. Chem. Soc. A*, 1994.
Glueckauf, E. (1955). *Trans. Faraday Soc.* **51**, 1235.
Glueckauf, E. (1959). "The Structure of Electrolytic Solutions" (W. J. Hamer, ed.), p. 97. Wiley ,New York.
Gold, V. (1955a). *Chem. and Ind.* 172.
Gold, V. (1955b). *J. Chem. Soc.* 1263.
Gold, V., and Hawes, B. W. V. (1951). *J. Chem. Soc.* 2102.
Gold, V., and Tye, F. L. (1952). *J. Chem. Soc.* 2181.
Greig, C. C., and Johnson, C. D. (1968). *J. Amer. Chem. Soc.* **90**, 6453.
Hammett, L. P., and Chapman, R. P. (1934). *J. Amer. Chem. Soc.* **56**, 1282.
Hammett, L. P., and Deyrup, A. J. (1932). *J. Amer. Chem. Soc.* **54**, 2721.
Hammett, L. P., and Deyrup, A. J. (1933). *J. Amer. Chem. Soc.* **55**, 1901.
Hantzsch, A. (1907). *Z. phys. Chem.* **61**, 257.
Hantzsch, A. (1909). *Z. phys. Chem.* **65**, 41.
Hantzsch, A. (1931). *Chem. Ber.* **64**, 667.
Harbottle, G. (1951). *J. Amer. Chem. Soc.* **73**, 4024.
Hinman, R. L., and Lang, J. (1960). *Tetrahedron Letters*, **21**, 12.
Hinman, R. L., and Lang, J. (1964). *J. Amer. Chem. Soc.* **86**, 3796.
Hinman, R. L., and Whipple, E. B. (1962). *J. Amer. Chem. Soc.* **84**, 2534.
Högfeldt, E. (1962). *Acta Chem. Scand.* **16**, 1054.
Homer, R. B., and Moodie, R. B. (1963). *J. Chem. Soc.* 4377.
Hopkinson, A. C., and Wyatt, P. A. H. (1967). *J. Chem. Soc. B*, 1333.
Horyna, J. (1959a). *Coll. Czech. Chem. Comm.* **24**, 1596.
Horyna, J. (1959b). *Coll. Czech. Chem. Comm.* **24**, 2637.
Horyna, J. (1960). *Coll. Czech. Chem. Comm.* **25**, 2720.
Horyna, J. (1962). *Coll. Czech. Chem. Comm.* **27**, 1324.
Huckings, B. J., and Johnson, M. D. (1964). *J. Chem. Soc.* 5371.
Jaques, D., and Leisten, J. A. (1964). *J. Chem. Soc.* 2683.
Johnson, C. D., Katritzky, A. R., Ridgewell, B. J., Shakir, N., and White, A. M. (1965). *Tetrahedron*, **21**, 1055.
Johnson, C. D., Katritzky, A. R., and Shakir, N. (1967). *J. Chem. Soc. B*, 1235.
Jorgenson, M. J., and Hartter, D. R. (1963). *J. Amer. Chem. Soc.* **85**, 878.
Katritzky, A. R., and Waring, A. J. (1962). *J. Chem. Soc.* 1540.
Katritzky, A. R., Waring, A. J., and Yates, K. (1963). *Tetrahedron*, **19**, 465.
Kresge, A. J., Barry, G. W., Charles, K. R., and Chiang, Y. (1962). *J. Amer. Chem. Soc.* **84**, 4343.
Kresge, A. J., and Chiang, Y. (1961). *Proc. Chem. Soc.* 81.
Lee, D. G., and Stewart, R. (1964). *Canad. J. Chem.* **42**, 486.
Leisten, J. A. (1961). *J. Chem. Soc.* 2191.
Leisten, J. A., and Walton, P. R. (1964). *J. Chem. Soc.* 3180.
Leisten, J. A., and Wright, K. L. (1964). *J. Chem. Soc.* 3173.
Long, F. A., and McDevit, W. F. (1952). *Chem. Rev.* **51**, 119.
Long, F. A., and McIntyre, D. (1954). *J. Amer. Chem. Soc.* **76**, 3243.
Long, F. A., and Schulze, J. (1961). *J. Amer. Chem. Soc.* **83**, 3340.
Long, F. A., and Schulze, J. (1964). *J. Amer. Chem. Soc.* **86**, 327.

Lowen, A. M., Murray, M. A., and Williams, G. (1950). *J. Chem. Soc.* 3318.
Marshall, W. L., and Jones, E. V. (1966). *J. Phys. Chem.* **70**, 4028.
McTigue, P. T. (1964). *Trans. Faraday Soc.* **60**, 127.
Michaelis, L., and Granick, S. (1942). *J. Amer. Chem. Soc.* **64**, 1861.
Moiseev, I. I., and Flid, R. M. (1954). *J. Appl. Chem. U.S.S.R.* **27**, 1047.
Moodie, R. B., Wale, P. D., and Whaite, T. J. (1963). *J. Chem. Soc.* 4273.
Murray, M. A., and Williams, G. (1950). *J. Chem. Soc.* 3322.
Newman, M. S., and Deno, N. C. (1951a). *J. Amer. Chem. Soc.* **73**, 3644.
Newman, M. S., and Deno, N. C. (1951b). *J. Amer. Chem. Soc.* **73**, 3651.
Newman, M. S., Kuivila, H. G., and Garrett, A. B. (1945). *J. Amer. Chem. Soc.* **67**, 704.
Noyce, D. S., and Jorgenson, M. J. (1962). *J. Amer. Chem. Soc.* **84**, 4312.
O'Brien, J. L., and Niemann, C. J. (1950). *J. Amer. Chem. Soc.* **72**, 5348.
Ojeda, M., and Wyatt, P. A. H. (1964). *J. Phys. Chem.* **68**, 1857.
O'Reilly, D. E., and Leftin, H. P. (1960). *J. Phys. Chem.* **64**, 1555.
Paul, M. A. (1954). *J. Amer. Chem. Soc.* **76**, 3236.
Paul, M. A., and Long, F. A. (1957). *Chem. Rev.* **57**, 1.
Phillips, J. N. (1961). *Aust. J. Chem.* **14**, 183.
Plattner, P. A., Heilbronner, E., and Weber, S. (1949). *Helv. Chim. Acta*, **32**, 574.
Randles, J. E. B., and Tedder, J. M. (1955). *J. Chem. Soc.* 1218.
Reeves, R. L. (1966). *J. Amer. Chem. Soc.* **88**, 2240.
Robinson, E. A., and Quadri, S. A. A. (1967a), *Canad. J. Chem.* **45**, 2385.
Robinson, E. A., and Quadri, S. A. A. (1967b). *Canad. J. Chem.* **45**, 2391.
Rosenthal, D., and Dwyer, J. S. (1962). *J. Phys. Chem.* **66**, 2687.
Rosenthal, D., and Dwyer, J. S. (1963a). *Canad. J. Chem.* **41**, 80.
Rosenthal, D., and Dwyer, J. S. (1963b). *Anal. Chem.* **35**, 161.
Ryabova, R. S., Medvetskaya, I. M., and Vinnik, M. I. (1966). *Russ. J. Phys. Chem.* **40**, 182.
Schubert, W. M., and Quacchia, R. H. (1962). *J. Amer. Chem. Soc.* **84**, 3778.
Schubert, W. M., and Quacchia, R. H. (1963). *J. Amer. Chem. Soc.* **85**, 1278.
Schulze, J., and Long, F. A. (1964). *J. Amer. Chem. Soc.* **86**, 322.
Schwabe, K. (1967). *Electrochim. Acta*, **12**, 67.
Sherrill, M. S., and Lyons, E. H. (1932). *J. Amer. Chem. Soc.* **54**, 979.
Staples, P. J. (1966). *J. Inorg. Nuclear Chem.* **28**, 2209.
Stewart, R., and Mathews, T. (1960). *Canad. J. Chem.* **38**, 602.
Stewart, R., Granger, M. R., Moodie, R. B., and Muenster, L. J. (1963). *Canad. J. Chem.* **41**, 1065.
Strehlow, H., and Wendt, H. (1961). *Z. phys. Chem. N.F.* **30**, 141.
Sweeting, L. M., and Yates, K. (1966). *Canad. J. Chem.* **44**, 2395.
Taft, R. W. (1960). *J. Amer. Chem. Soc.* **82**, 2965.
Taft, R. W., and Levins, P. L. (1962). *Anal. Chem.* **34**, 436.
Turney, T. A., and Wright, G. A. (1958). *J. Chem. Soc.* 2415.
Vetešník, P., Bielavský, J., and Večeřa, M. (1968). *Coll. Czech. Chem. Comm.* **33**, 1687.
Vinnik, M. I., and Librovich, N. B. (1966). *Tetrahedron*, **22**, 2945.
Vinnik, M. I., Ryabova, R. S., and Chirkov, N. M. (1959a). *Russ. J. Phys. Chem.* **33**, 253.
Vinnik, M. I., Ryabova, R. S., and Chirkov, N. M. (1959b). *Russ. J. Phys. Chem.* **33**, 575.
Wasif, S. (1967). *J. Chem. Soc. A*, 142.
Weller, A. (1952). *Z. Elektrochem.* **56**, 662.

Westheimer, F. H., and Kharasch, M. S. (1946). *J. Amer. Chem. Soc.* **68**, 7871.
Whipple, E. B., Chiang, Y., and Hinman, R. L. (1963). *J. Amer. Chem. Soc.* **85**, 26.
Wyatt, P. A. H. (1957). *Discuss. Faraday Soc.* **24**, 162.
Yates, K., and Riordan, J. G. (1965). *Canad. J. Chem.* **43**, 2329.
Yates, K., and Stevens, J. B. (1965). *Canad. J. Chem.* **43**, 529.
Yates, K., Stevens, J. B., and Katritzky, A. R. (1964). *Canad. J. Chem.* **42**, 1957.
Yates, K., and Wai, H. (1965). *Canad. J. Chem.* **43**, 2131.
Yates, K., and Wai, H. (1968). Personal communication.
Zalewski, R. I., and Dunn, G. E. (1968). *Canad. J. Chem.* **46**, 2469.

CHAPTER 4

The Rates of Reactions in Concentrated Aqueous Acid Solutions. Theoretical Approaches

Hammett and Deyrup (1932) defined the H_0 acidity function and noted that the rates of certain reactions correlated with the acidity function. Thus the observed rate constants k, given by equation (4.a), where C_S is the concentration of the reacting organic substrate, were found to be consistent with equation (4.b).

$$k = -(1/C_S)(dC_S/dt) \tag{4.a}$$

$$\log_{10} k + H_0 = \text{constant} \tag{4.b}$$

This is exemplified by the results for the hydrolysis of sucrose shown in Fig. 4.1 (Hammett and Paul, 1934; Hammett, 1935). It was soon realized that not all acid catalysed reactions conformed with this relation. Thus Zucker and Hammett (1939) found that the rates of iodination of acetophenone in aqueous perchloric or sulphuric acid were directly proportional to the acid concentration (Fig. 4.2). A clear distinction between the acidity dependence of different reactions is therefore apparent. Zucker and Hammett proposed that this was associated with precise differences in mechanism between reactions for which $\log_{10} k$ follows $-H_0$ and those for which $\log_{10} k$ follows $\log_{10} C_{H^+}$. The test of these alternatives for a given reaction should therefore provide an indication of the mechanism of the reaction. This idea forms the basis of what has become known as the "Zucker–Hammett hypothesis" (Long and Paul, 1957).

As more experimental results became available for reactions catalysed by concentrated acids it was apparent that the Zucker–Hammett criteria of mechanism were not generally applicable. Although the possible breakdown of the treatment was soon recognized it was some years before an alternative approach was suggested. The use of the Zucker–Hammett treatment has been widespread. The basic concepts of this method are therefore developed below. This also serves as an introduction to the basic kinetic equations for acid catalysed reactions and therefore provides a basis on which to build the discussion of more recent approaches to the determination of reaction mechanism.

A discussion of the relative merits of the different methods will be postponed until the end of Chapter 5 in which experimental results for many acid catalysed reactions are presented.

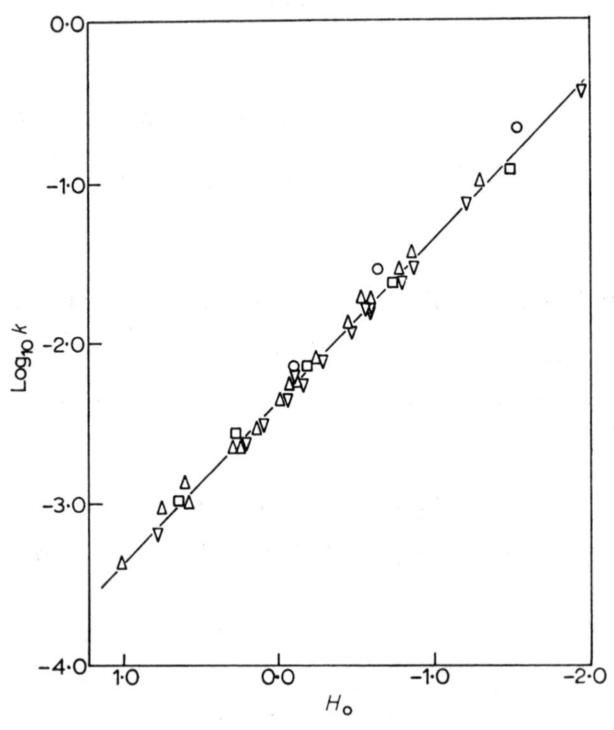

FIG. 4.1. The correlation between H_0 and rate constant for the hydrolysis of sucrose. Catalysing acids were: ○ $HClO_4$; □ H_2SO_4; ▽ HNO_3; △ HCl. The line is drawn with unit slope. (Hammett and Paul, 1934; Hammett, 1935.)

4.1. The Zucker–Hammett Criterion of Mechanism

The A-1 mechanism for the hydrolysis of an organic solute S in acid solution may be written as follows (Ingold, 1963a). The rate determining step is the unimolecular decomposition of the protonated substrate SH^+ to an intermediate A^+ which rapidly reacts with water to give the reaction products. The rate of loss of the sum of the equilibrium concentrations of S and SH^+ is given

$$S + H^+ \rightleftharpoons SH^+ \quad \text{(fast, pre-equilibrium)} \tag{4.1}$$

$$SH^+ \rightarrow A^+ \quad \text{(slow)} \tag{4.2}$$

$$A^+ + H_2O \rightarrow \text{products} \quad \text{(fast)} \tag{4.3}$$

by equation (4.4) where k_0 is the rate constant for the slow step (4.2) and $y^{\ddagger}$ is the activity coefficient of the transition state for that step. The experimental

$$\frac{-d(C_S + C_{SH^+})}{dt} = k_0\, C_{SH^+} \frac{y_{SH^+}}{y^{\ddagger}} \tag{4.4}$$

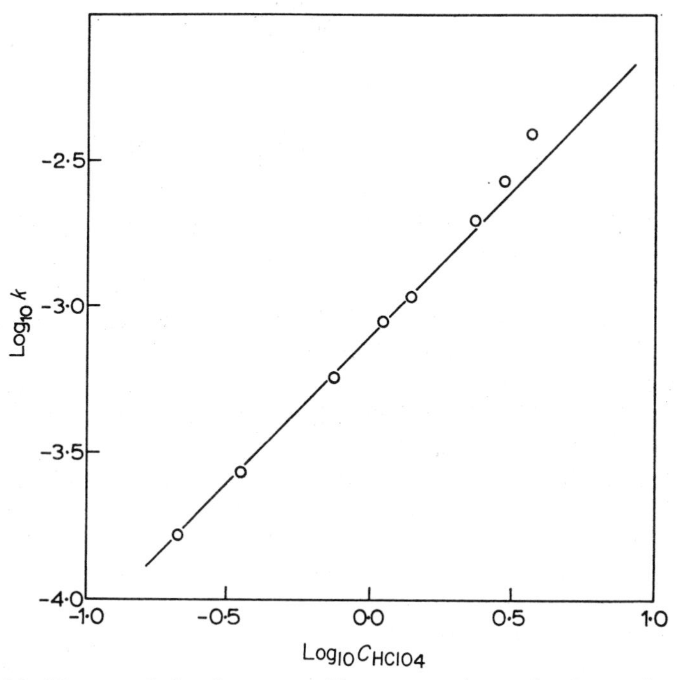

FIG. 4.2. The correlation between acid concentration and rate constant for the iodination of acetophenone in aqueous $HClO_4$ at 25°. The line is drawn with unit slope. (Zucker and Hammett, 1939.)

first order rate constant k_1 is defined by equation (4.5). Combination of equations (4.4) and (4.5) with the expression for the acid ionization constant K_{SH^+} of SH^+ leads to equation (4.6). But for concentrated acid solutions the

$$k_1 = -\left(\frac{1}{C_S + C_{SH^+}}\right)\frac{d(C_S + C_{SH^+})}{dt} \tag{4.5}$$

$$k_1 = \left(\frac{C_S}{C_S + C_{SH^+}}\right)\frac{k_0}{K_{SH^+}}\frac{a_{H^+}\,y_S}{y^{\ddagger}} \tag{4.6}$$

hydrogen ion activity may be equated with the Hammett acidity function via equation (1.29) and therefore

$$k_1 = \left(\frac{C_S}{C_S + C_{SH^+}}\right)\frac{k_0}{K_{SH^+}}\,h_0\,\frac{y_S\,y_{BH^+}}{y_B\,y^{\ddagger}} \tag{4.7}$$

Three situations now arise and depend upon the relative values of h_0 and K_{SH^+}.

Firstly if $h_0 \ll K_{SH^+}$ ($H_0 \gg pK_{SH^+}$) the equilibrium (4.1) lies practically entirely on the left-hand side. Thus $c_{SH^+} \ll c_S$ and equation (4.7) becomes

$$k_1 = \frac{k_0}{K_{SH^+}} h_0 \frac{y_S \, y_{BH^+}}{y_B \, y^{\ddagger}} \tag{4.8}$$

which on taking logarithms is equivalent to

$$\log_{10} k_1 = -H_0 + \log_{10}\left(\frac{k_0}{K_{SH^+}}\right) + \log_{10}\left(\frac{y_S \, y_{BH^+}}{y_B \, y^{\ddagger}}\right) \tag{4.9}$$

Equations (4.b) and (4.9) are identical providing the activity coefficient term in the latter is constant over the range of acid concentrations studied. It is a requirement of the Zucker–Hammett hypothesis that this should be so. The equality of the activity coefficient ratios ($y_S/y^{\ddagger}$) and (y_B/y_{BH^+}) would lead to this result. For any reaction for which this requirement is obeyed $\log_{10} k_1$ will be a linear function $-H_0$ with unit slope. The acid catalysed hydrolysis of sucrose conforms with this prediction and is therefore in accord with the Zucker–Hammett hypothesis. However the requirement that ($y_S/y^{\ddagger}$) and (y_B/y_{BH^+}) should be the same for a particular acid concentration is analogous to the original acidity function concept that (y_B/y_{BH^+}) and hence H_0 should be independent of the chemical structure of the base B. The evidence disproving this idea has been discussed in Chapter 3. It is therefore unlikely that all reactions conforming to mechanism (4.1)–(4.3) in concentrated acid solutions will show similar acidity function behaviour. The Zucker–Hammett postulate that the fitting of equation (4.b) is a useful criterion of mechanism must be suspect. Thus, for example, although for many reactions $\log_{10} k$ is a linear function of $-H_0$ the slopes of the graphs often differ significantly from 1.

Secondly if $h_0 \approx K_{SH^+}$ then appreciable quantities of both S and SH^+ exist in the pre-equilibrium mixture. The concentrations C_S and C_{SH^+} are related via K_{SH^+}. Hence substitution for C_{SH^+} in equation (4.6) leads to equation (4.10) which combined with equation (1.29) gives equation (4.11).

$$k_1 = \frac{k_0}{K_{SH^+}}\left(1 + \frac{a_{H^+} \, y_S}{K_{SH^+} \, y_{SH^+}}\right)^{-1} \frac{a_{H^+} \, y_S}{y^{\ddagger}} \tag{4.10}$$

$$k_1 = \frac{k_0}{K_{SH^+}}\left(1 + \frac{h_0}{K_{SH^+}} \frac{y_S \, y_{BH^+}}{y_{SH^+} \, y_B}\right)^{-1} h_0 \frac{y_S \, y_{BH^+}}{y_B \, y^{\ddagger}} \tag{4.11}$$

The test of this equation relies on the assumptions that

$$\frac{y_S \, y_{BH^+}}{y_{SH^+} \, y_B} \approx \frac{y_S \, y_{BH^+}}{y_B \, y^{\ddagger}} \approx 1 \tag{4.12}$$

whence (Schubert and Latourette, 1952; Dayagi, 1961)

$$\log_{10} k_1 = -H_0 - \log_{10}\left(1 + \frac{h_0}{K_{SH^+}}\right) + \log_{10}\frac{k_0}{K_{SH^+}} \tag{4.13}$$

Finally if $h_0 \gg K_{SH^+}$ ($H_0 \ll pK_{SH^+}$) then all the added substrate S is protonated in the pre-equilibrium step. Equation (4.11) therefore reduces to

$$k_1 = k_0 \frac{y_{SH^+}}{y^{\ddagger}} \tag{4.14}$$

Thus if $(y_{SH^+}/y^{\ddagger})$ = ca. 1 for all acid concentrations then the observed reaction rate will be independent of acid concentration and will stay approximately constant.

The A-2 mechanism for acid catalysed hydrolysis has a bimolecular rate determining step and may be represented by reactions (4.15) and (4.16).

$$S + H^+ \rightleftharpoons SH^+ \quad \text{(fast, pre-equilibrium)} \tag{4.15}$$

$$SH^+ + H_2O \rightarrow \text{products} \quad \text{(slow)} \tag{4.16}$$

(Ingold, 1963). The rate of loss of the sum of the equilibrium concentrations of S and SH^+ is given by equation (4.17) in which k_2 is the rate constant for step

$$\frac{-d(C_S + C_{SH^+})}{dt} = k_2\, C_{SH^+} \frac{a_w\, y_{SH^+}}{y^{\ddagger}} \tag{4.17}$$

(4.16). Combined with the definition (4.5) of the experimental rate constant and on substitution for C_{SH^+} via the equilibrium constant expression for reaction (4.15) this leads to equation (4.18).

$$k_1 = \left(\frac{C_S}{C_S + C_{SH^+}}\right)\frac{k_2}{K_{SH^+}}\, C_{H^+} \frac{y_S\, y_{H^+}\, a_w}{y^{\ddagger}} \tag{4.18}$$

As for equation (4.7) three possibilities now arise depending upon the relative values of h_0 and K_{SH^+}.

For $h_0 \ll K_{SH^+}$, $C_{SH^+} \ll C_S$ and equation (4.18) becomes, taking logarithms

$$\log_{10} k_1 = \log_{10} C_{H^+} + \log_{10}\frac{k_2}{K_{SH^+}} + \log_{10}\left(\frac{y_S\, y_{H^+}\, a_w}{y^{\ddagger}}\right) \tag{4.19}$$

The use of this equation as a criterion of mechanism is exemplified by the results in Fig. 4.2 for the iodination of acetophenone. These may be taken as fitting equation (4.19) providing the identity (4.20) is accepted.

$$\log_{10}\left(\frac{y_S\, y_{H^+}\, a_w}{y^{\ddagger}}\right) \approx 0 \tag{4.20}$$

It is a basic postulate of the Zucker–Hammett hypothesis that equation (4.20) is obeyed by all A-2 reactions whose mechanism is in accord with equations

(4.15) and (4.16). If the Zucker–Hammett ideas are generally applicable, then correlation of $\log_{10}k_1$ with either $-H_0$ (equation 4.9) or $\log_{10}C_{H^+}$ (equation 4.19) would provide a clear criterion of mechanism for acid catalysed hydrolyses.

For $h_0 \approx K_{SH^+}$ appreciable concentrations of both S and SH^+ exist at equilibrium and the appropriate expression for k_1 is (Rosenthal and Taylor, 1957)

$$\log_{10}k_1 = \log_{10}C_{H^+} - \log_{10}\left(1 + \frac{h_0}{K_{SH^+}}\right) + \log_{10}\frac{k_2}{K_{SH^+}} + \log_{10}\left(\frac{y_S\, y_{H^+}\, a_w}{y^{\ddagger}}\right) \quad (4.21)$$

Alternatively when $h_0 \gg K_{SH^+}$, $C_{SH^+} \gg C_S$ and equation (4.22) is applicable. However the Zucker–Hammett approach is mainly concerned with reactions

$$\log_{10}k_1 = \log_{10}C_{H^+} + H_0 + \log_{10}k_2 + \log_{10}\left(\frac{y_S\, y_{H^+}\, a_w}{y^{\ddagger}}\right) \quad (4.22)$$

for which in the pre-equilibrium $C_{SH^+} \ll C_S$. Thus providing the appropriate assumptions concerning activity coefficient terms are valid the reaction rates should either fit equation (4.9) or equation (4.19) and this should indicate an A-1 or A-2 mechanism respectively. In order to distinguish between these alternatives a reaction of unknown mechanism must be studied in concentrations of catalysing acid for which C_{H^+} and h_0 diverge appreciably.

A further mechanistic possibility which is relevant to the present discussion is that in which protonation of the substrate molecule is the rate-determining step in the reaction (Long and Paul, 1957). The observed rate constant k_1 is given by equations (4.23)–(4.25) in which k is the rate constant for the protonation step.

$$-\frac{dC_S}{dt} = kC_S\, C_{H^+}\frac{y_S\, y_{H^+}}{y^{\ddagger}} \quad (4.23)$$

$$k_1 = -\frac{1}{C_S}\frac{dC_S}{dt} = kC_{H^+}\frac{y_S\, y_{H^+}}{y^{\ddagger}} = kh_0\frac{y_S\, y_{BH^+}}{y_B\, y^{\ddagger}} \quad (4.24)$$

$$\log_{10}k_1 = -H_0 + \log_{10}k + \log_{10}\left(\frac{y_S\, y_{BH^+}}{y_B\, y^{\ddagger}}\right) \quad (4.25)$$

Equation (4.25) is of the same form as equation (4.9). It is logical to expect that for such a reaction $\log_{10}k_1$ should be linear in $-H_0$ with unit slope. The Zucker–Hammett approach does not provide a means of distinguishing between reactions of this type and those conforming to the A-1 mechanism (4.1)–(4.3).

Zucker and Hammett (1939) proposed their ideas at a time when it was thought that there might be a unique acidity function scale for aqueous solutions of a particular acid. The above treatment has therefore been considered entirely in terms of the Hammett H_0 acidity function defined using primary amines as indicators. Tests of the treatment have been almost exclusively based on use of H_0 as the appropriate acidity function. The now established fact (Chapter 3) that changes in indicator structure can lead to the need for significantly different acidity functions leaves no doubt that the Zucker–Hammett approach will not be generally applicable. Thus, for example, why should B in equation (4.25) be an amine rather than an amide or an olefin? Should $\log_{10}k_1$ be a linear function with unit slope of $-H_0$ or $-H_A$ or $-H'_R$? The most likely acidity function to give the required correlation would presumably be the one defined by indicators whose protonated form most resembles in structure the transition state in the slow step of the reaction under investigation. These doubts are borne out by the majority of results obtained for acid catalysed reactions. Thus the acidity function dependence of the reaction rates, as with the protonation equilibria of neutral bases, is dependent upon the chemical structure of the reacting substrate molecules. An *exact* parallel of $\log_{10}k_1$ with either $-H_0$ or $\log_{10}C_{H^+}$ in accord with the extreme interpretation of Zucker and Hammett's ideas is therefore probably quite fortuitous

It is interesting to note here that Arnett and Mach (1966) have shown that for some reactions $(\log_{10}k_1 + H_0)$ is a linear function of the concentration of the catalysing acid. Examples include the results for the iodination of acetophenone given in Fig. 4.2. This implies that the activity coefficient term in equation (4.9) may be represented by the combination of four Setschenow equations (Long and McDevit, 1952). The possibility that the slopes of these graphs may be a useful criterion of mechanism has not been investigated in any detail.

4.2. Correlation of Reaction Rates with the J_0 Acidity Function

The use of the J_0 acidity function for the determination of reaction mechanisms was first proposed to explain the rates of certain nitration reactions in concentrated sulphuric acid solutions. Possible nitrating agents which may be formed in strongly acidic solutions are the nitracidinium ion $H_2NO_3^+$ and the nitronium ion NO_2^+. These would be formed by the ionization of nitric acid according to equilibria (4.26) and (4.27) respectively.

$$HONO_2 + H_2SO_4 \rightleftharpoons H_2\overset{+}{O}NO_2 + HSO_4^- \tag{4.26}$$

$$HONO_2 + 2H_2SO_4 \rightleftharpoons NO_2^+ + H_3O^+ + 2HSO_4^- \tag{4.27}$$

Cryoscopic studies of nitric acid in H_2SO_4 suggested that the latter was predominant (Hantzsch, 1907, 1908). Westheimer and Kharasch (1946) found that $\log_{10}k_1$ for the nitration of nitrobenzene and $\log_{10}(C_{R^+}/C_{ROH})$ for the ionization of 4,4′,4″-trinitrotriphenylcarbinol were parallel functions of the sulphuric acid concentration. This strengthens the conclusion that NO_2^+ is the nitrating agent because equation (4.27) is analogous to equation (3.1) for the ionization of a triarylcarbinol.

This idea was developed by Lowen *et al.* (1950; Williams and Lowen, 1950) and Gold and Hawes (1951) who defined the J_0 acidity function. The rates of aromatic nitration reactions may then be correlated directly with the J_0 acidity function as follows. Equation (4.29) follows from equation (4.28) in which S is the species being nitrated, k is the rate constant for the rate determining step in the reaction, and k_1 is the experimental rate constant. These equations are written on the assumption that for equilibrium (4.27) $C_{NO_2^+} \ll C_{HNO_3}$, and that the nitration proceeds via a slow electrophilic attack of substrate S by

$$\frac{-dC_S}{dt} = kC_{NO_2^+} C_S \frac{y_{NO_2^+} y_S}{y^{\ddagger}} = k_1 C_S C_{HNO_3} \quad (4.28)$$

$$k_1 = k \frac{C_{NO_2^+}}{C_{HNO_3}} \frac{y_{NO_2^+} y_S}{y^{\ddagger}} \quad (4.29)$$

NO_2^+ (Ingold, 1963b). Combination of equation (4.29) with equation (4.30) and the definition of J_0 ($= -\log_{10} j_0$) equation (3.4) leads to equation (4.31) for k_1.

$$K_{HNO_3} = \frac{C_{NO_2^+} y_{NO_2^+} a_w}{C_{HNO_3} y_{HNO_3} a_{H^+}} \quad (4.30$$

$$k_1 = kK_{HNO_3} j_0 \left(\frac{y_{R^+} y_{HNO_3}}{y_{NO_2^+} y_{ROH}}\right)\left(\frac{y_S y_{NO_2^+}}{y^{\ddagger}}\right) \quad (4.31)$$

If the ionization of nitric acid in aqueous sulphuric acid solutions parallels the J_0 acidity function for the latter, then equations (4.32) and (4.33) are

$$J_0 = -pK_{HNO_3} - \log_{10}(C_{NO_2^+}/C_{HNO_3}) \quad (4.32)$$

$$\left(\frac{y_{R^+} y_{HNO_3}}{y_{NO_2^+} y_{ROH}}\right) = 1 \quad (4.33)$$

applicable. Hence taking logarithms, equation (4.31) becomes

$$\log_{10} k_1 = -J_0 + \log_{10}(kK_{HNO_3}) + \log_{10}\left(\frac{y_S y_{NO_2^+}}{y^{\ddagger}}\right) \quad (4.34)$$

Thus, providing the activity coefficient term in equation (4.34) is approximately zero, a plot of $\log_{10} k_1$ against $-J_0$ should be linear with unit slope. If the nitracidium ion were the nitrating species the Zucker–Hammett treatment would indicate that $\log_{10} k_1$ should be a linear function of $-H_0$ with unit slope. Because of the wide divergence (Fig. 3.2) between J_0 and H_0 for aqueous sulphuric acid solutions these are clearly distinguishable alternatives. The parallel between rates of nitration of benzene derivatives and the ionization ratios of triarylcarbinols confirms that NO_2^+ is the active species in the nitration process. Correlations with the J_0 acidity function can therefore provide a decisive test of the mechanism of certain reactions. Although the equations have been developed for nitration reactions the applicability of the use of J_0 as a criterion of mechanism is general. Other examples will be presented in the following chapter.

4.3. The Bunnett Criteria of Mechanism

4.3.1. *w and w* Parameters*

The *exact* requirement of the Zucker–Hammett hypothesis that $\log_{10} k_1$ should be a linear function with *unit* slope of either $-H_0$ or $\log_{10} C_{H^+}$ is rarely obeyed. This is exemplified by the results given in Table 4.1. Thus the use of the

TABLE 4.1

The slopes of the plots of $\log_{10} k_1$ against $-H_0$ and the Bunnett w, w^* and ϕ parameters for some acid catalysed reactions at 25° (Bunnett, 1961a; Bunnett and Olsen, 1966b)

Reaction	Acid	Slope	w	w^*	ϕ
Methylal, hydrolysis	HCl	1·27	−5·26		−0·70
Methoxymethyl acetate, hydrolysis	HCl	1·19	−2·39		−0·37
Trioxane, depolymerization	HCl	1·20	−2·10		−0·35
Trioxane depolymerization	H_2SO_4	1·08	−1·02		−0·13
D-sec-Butyl hydrogen sulphate, racemization	H_2SO_4	1·00	−0·02		−0·006
Ethyl acetate, hydrolysis	HCl	0·37	+4·15	−0·60	+0·87
Isopropyl acetate, hydrolysis	HCl	0·42	+4·62	−0·71	+0·86
4-Hydroxy-4-phenyl-2-butanone, dehydration	H_2SO_4	0·54	+3·17	+0·45†	+0·66
Cyanamide, hydrolysis	HNO_3	0·95	+0·80		+0·12
Acetone, iodination	HCl		+6·66	−1·42	+0·83
Acetophenone, iodination	H_2SO_4	0·59	+3·63	0·00†	+0·62
Acetophenone, iodination	$HClO_4$		+6·35	−1·62	+0·80
2,4-Dinitrobenzyl alcohol, sulphation	H_2SO_4	0·82	+0·31		+0·19
4-Nitrophenol-2-*d*, exchange	H_2SO_4	0·65	+0·40		+0·37
Sucrose, hydrolysis	HCl	1·02	−0·43		−0·04

† $w_A \sim w^* + 1·6$, see text.

Zucker–Hammett treatment as a criterion of mechanism is limited particularly when the dependence of rate with acidity falls somewhere between the two possibilities allowed by the theory. An alternative approach was therefore desirable.

Bunnett (1961a) noted that for many reactions $(\log_{10} k_1 + H_0)$ was a linear or approximately linear function of $\log_{10} a_w$. An example is given in Fig. 4.3. Equation (4.35) is therefore applicable in which w is the slope of the graph and C_1 (and C_2, C_3, ... in the following equations) is a constant.

$$(\log_{10} k_1 + H_0) = w \log_{10} a_w + C_1 \tag{4.35}$$

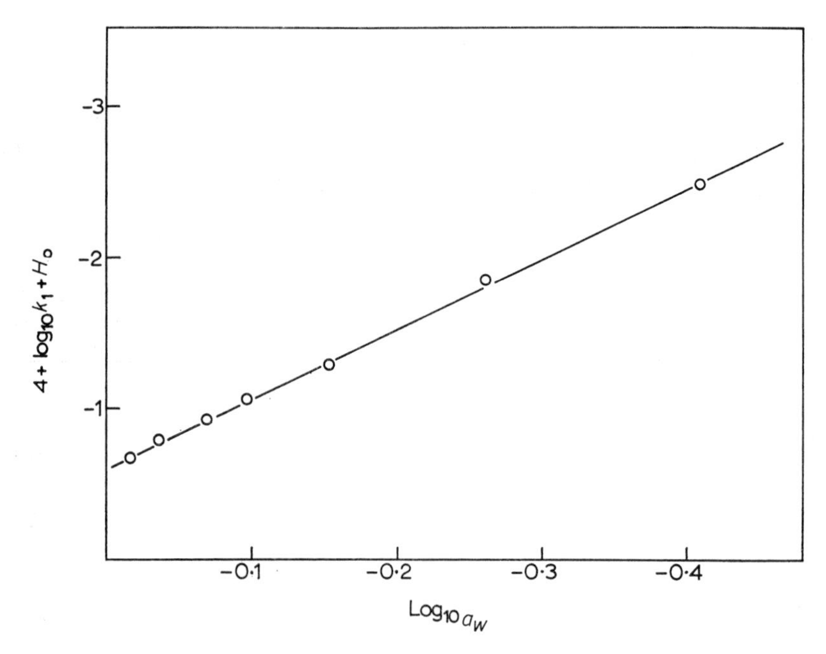

FIG. 4.3. Test of equation (4.35) for the hydrolysis of nicotinamide (at 76°) in aqueous HCl. Hence $w = 4{\cdot}8$.

In accord with the difference between equations (4.9) and (4.13), equation (4.36) is a more general form of equation (4.35). Equation (4.35) is only valid

$$\log_{10} k_1 - \log_{10}\left(\frac{h_0}{h_0 + K_{SH+}}\right) = w \log_{10} a_w + C_2 \tag{4.36}$$

when $C_{SH+} \ll C_S$ in the pre-equilibrium step. Similarly equation (4.37) is applicable when $C_{SH+} \gg C_S$ at equilibrium. Equations (4.35), (4.36) and (4.37)

$$\log_{10} k_1 = w \log_{10} a_w + C_3 \tag{4.37}$$

are therefore analogous to the three cases considered in Section 4.1 and which led to equations (4.9), (4.13) and (4.14) respectively. Consideration of these latter three equations on the Zucker–Hammett hypothesis would require that $w = 0$ for all reactions conforming to mechanism (4.1)–(4.3).

Table 4.1 list w factors for a few acid catalysed reactions. The best agreement with the requirement $w = 0$ is for the racemization of D-sec-butyl hydrogen sulphate in aqueous H_2SO_4. Even for the hydrolysis of sucrose, which apparently closely fits the Zucker–Hammett ideas (Fig. 4.1), w is small but significantly different from zero. The w factor is much more sensitive to changes of reacting substrate than the slopes of the conventional Zucker–Hammett plots. Bunnett (1961b, 1961d) has therefore suggested that the value of w for an acid catalysed reacion may provide a useful criterion for the mechanism of that reaction.

For certain reactions the plots appropriate to the determination of w are not linear. However in these cases one of the three equations (4.38)–(4.40) is obeyed and therefore enables the deduction of a parameter w^*. These equations

$$\log_{10} k_1 - \log_{10} C_{\text{acid}} = w^* \log_{10} a_w + C_4 \qquad (4.38)$$

$$\log_{10} k_1 - \log_{10}\left(\frac{C_{\text{acid}}}{h_0 + K_{\text{SH}^+}}\right) = w^* \log_{10} a_w + C_5 \qquad (4.39)$$

$$\log_{10} k_1 - \log_{10} C_{\text{acid}} - H_0 = w^* \log_{10} a_w + C_6 \qquad (4.40)$$

are related to equations (4.19), (4.21) and (4.22) respectively. The w^* factor is also a useful criterion of reaction mechanism. For many reactions both w and w^* can be measured (Table 4.1). This arises because over not too wide ranges of acid concentration $(H_0 + \log_{10} C_{\text{acid}})$ varies linearly with $\log_{10} a_w$ (Bunnett, 1961a). Thus equation (4.41) obeyed. Over wide ranges of acid concentration

$$H_0 + \log_{10} C_{\text{acid}} = (w - w^*) \log_{10} a_w + C_7 \qquad (4.41)$$

this equation is invalid and either the plots for the determination of w or those for the determination of w^* must be curved. It follows that w and/or w^* will be some function of the particular acid concentration range which has been studied. This must be borne in mind when considering the mechanistic implications of w and w^* values. Reactions with mechanism (4.15), (4.16) should give $w^* = 0$ if the Zucker–Hammett postulates are strictly obeyed. The corresponding value of w (usually $> +5$) will depend on the nature and concentration range of the catalysing acid.

It is difficult to determine w^* values for reactions catalysed in aqueous H_2SO_4 for which 0·5 mole litre$^{-1} < C_{H_2SO_4} < 3$ mole litre^{-1} because of the uncertainty with which the extent of ionization of HSO_4^- in these solutions is known. Bunnett (1961a) proposed the definition of a parameter w_A by equation (4.42)

$$\log_{10} k_1 - \log_{10} k_A = w_A \log_{10} a_w + C_8 \qquad (4.42)$$

in which k_A (interpolated from the results of Zucker and Hammett, 1939) is the observed rate constant for the enolization of acetophenone at the same sulphuric acid concentration. Plots of $(\log_{10} k_1 - \log_{10} k_A)$ against $\log_{10} a_w$ are frequently linear. It follows that by definition $w_A = 0$ for the enolization of acetophenone. Since $w^* = -1{\cdot}62$ for this reaction in aqueous $HClO_4$ it would seem that $w_A \sim (w^* + 1{\cdot}6)$. However this identity is not necessarily true because w values for the same reaction catalysed by different acids are sometimes different. Thus if w is positive it is fairly general that $w(HClO_4) > w(H_2SO_4)$ by about 1 or 2 w units. That w^* is appreciably different from zero for the enolization of acetophenone contrasts with the supposition (see Fig. 4.2) that it conforms with one extreme of the Zucker–Hammett treatment. This demonstrates the greater sensitivity of w^* over the slope of the $\log_{10} k_1$ against $\log_{10} C_{H^+}$ plot to small deviations from the Zucker–Hammett criteria.

The use of w and w^* as criteria of mechanism is based on a detailed analysis by Bunnett (1961a; Bunnett and Olsen, 1966b) of results for many reactions. A consideration of reactions for which independent evidence for particular mechanisms was available showed that certain classes of mechanisms usually gave w values in a particular range. These classes were characterized by the mode of involvement of water in the rate determining step. The reactions were divided into three groups: those in which water does not participate in the rate-determining step, those in which water acts as a nucleophile in the rate-determining step, and those in which water acts as a proton transfer agent in the rate-determining step. The slow steps of the three classes may therefore be summarized by equations (4.43), (4.44) and (4.45) respectively. The last class is

$$SH^+ \rightarrow \text{products} \tag{4.43}$$

$$SH^+ + H_2O \rightarrow H_2O.SH^+ \tag{4.44}$$

$$SH^+ + H_2O \rightarrow S + H_3O^+ \tag{4.45}$$

also exemplified by reactions in which in the rate-determining step the H_3O^+ ion transfers a proton to a reacting substrate molecule. This is analogous to the reverse of reaction (4.45). The difference is whether the water is in the process of gaining or losing a proton in the transition state. The summary of the w values characteristic of each class are given in Table 4.2. Bunnett (1961b) has proposed that this table should be taken as the basis for the interpretation of w and w^* values as criteria of mechanism.

The effect of temperature on w and w^* is small. Thus although it would be preferable to correlate rate constants for a particular temperature with H_0 and water activities for the same temperature the lack of available data makes this impossible except for 25°. It is therefore necessary to correlate k_1 for any temperature with H_0 and water activities at 25°. The results in Table 4.3 give an indication of how much w may vary with temperature. The variation for trioxane depolymerization is greater than that for sucrose hydrolysis. However for both reactions the w values remain inside the limits of one of the criteria of mechanism given in Table 4.2.

The deviations between the acidity functions defined by different weakly basic indicators are at least in part associated with the different solvation

TABLE 4.2

Values of w and w^* characteristic of the way in which water is involved in the rate determining step of acid catalysed reactions (Bunnett, 1961b).

Involvement of water	w values	w^* values
For substrates protonated on oxygen or nitrogen		
Not involved	−2·5 to 0·0	
Nucleophile	1·2 to 3·3	<-2†
Proton transfer agent	$>3{\cdot}3$	>-2†
For substrates which are hydrocarbon-like bases		
Proton transfer agent	ca. 0	

† This distinction is not exactly defined.

TABLE 4.3

Values of w deduced from correlation of rate constants at different temperatures with H_0 and water activities at 25°. Catalysing acid was HCl. (Bunnett, 1961a)

Trioxane depolymerization		Sucrose hydrolysis	
Temperature	w	Temperature	w
25	−2·10	0	−0·43
30	−0·76	10	−0·39
40	−0·82	15	−0·80
50	−1·11	20	−0·68
60	−2·46	25	−0·43

requirements of the neutral bases and conjugate acids of each class of base (Chapter 3). Bunnett (1961c) adopted a similar approach to explain the significance of w and w^* values. The ideas are considered in terms of the generalized mechanism (4.46), (4.47) for measurably slow acid catalysed reactions. Here

$$S(H_2O)_s + H(H_2O)_n{}^+ \rightleftharpoons SH(H_2O)_p{}^+ + (s+n-p)H_2O \quad (4.46)$$

$$SH(H_2O)_p{}^+ + (t-p)H_2O \rightleftharpoons \ddagger(H_2O)_t{}^+ \rightarrow \text{products} \quad (4.47)$$

s, n, and p, represent the hydration numbers of S, H^+ and SH and t is the number of water molecules associated (hydrating and reacting) with the transition state. Overall the solvated transition state contains $(t-n-s)$ more water molecules than the solvated reactants $(S + H^+)$. The expressions (4.6) and

(4.18) for the observed rate constant may now be written in the more general form (4.48) in which it has been assumed that $H_0 \gg K_{SH^+}$ ($C_{SH^+} \ll C_S$) and k is

$$k_1 = \frac{k}{K_{SH^+}} C_{H(H_2O)_n^+} a_w^{(t-s-n)} \frac{y_{S(H_2O)_s} y_{H(H_2O)_n^+}}{y^{\ddagger}_{(H_2O)t^+}} \tag{4.48}$$

the rate constant for the rate determining step. Equation (4.49) follows and may be compared with equation (4.38). Thus $w^* = (t - s - n)$ providing the

$$\log_{10} k_1 - \log_{10} C_{H(H_2O)_n^+} = (t - s - n)\log_{10} a_w + \log_{10} \frac{k}{K_{SH^+}} + \log_{10}\left(\frac{y_{S(H_2O)_s} y_{H(H_2O)_n^+}}{y^{\ddagger}_{(H_2O)_t^+}}\right) \tag{4.49}$$

activity coefficient term in equation (4.49) is a constant independent of acid concentration.

The interpretation of w depends not only upon the solvation requirements of the species involved in the particular reaction under investigation but also on the solvation requirements of the primary amines and their conjugate acid used to define the H_0 scale. Thus equation (2.34) is more correctly written as equation (4.50) which leads to equation (4.51) for h_0. This equation is similar

$$B(H_2O)_b + H(H_2O)_n^+ \rightleftharpoons BH(H_2O)_a^+ + (b + n - a)H_2O \tag{4.50}$$

$$h_0 = C_{H(H_2O)_n^+} a_w^{(a-b-n)} \frac{y_{B(H_2O)_b} y_{H(H_2O)n^+}}{y_{BH(H_2O)_a^+}} \tag{4.51}$$

to equation (2.35) but as discussed in Chapter 3 the solvation requirements of B and BH^+ as well as of H^+ are considered to contribute to the observed acidity function dependence for a particular base. Substitution for $C_{H(H_2O_n)^+}$ from equation (4.51) in equation (4.48) and taking logarithms leads to equation (4.52). Comparison of this equation with equation (4.35) shows that

$$\log_{10} k_1 + H_0 = (t - s + b - a)\log_{10} a_w + \log_{10} \frac{k}{K_{SH^+}} + \log_{10}\left(\frac{y_{S(H_2O)_s} y_{BH(H_2O)_a^+}}{y_{B(H_2O)_b} y^{\ddagger}_{(H_2O)_t^+}}\right) \tag{4.52}$$

$w = (t - s + b - a)$ providing the activity coefficient term in equation (4.52) is a constant independent of the acid concentration.

The above ideas represent an extreme interpretation of w and w^* parameters. From a consideration of equation (4.41) Bunnett (1961c) concluded that the assumption that the activity coefficient terms in equations (4.49) and (4.52) are constant is not completely true. However solvation changes do play a significant part in determining the variation of reaction rate with acid concentration in acid solutions. Added evidence for this given by the approximately linear correlation between w and entropies of activation $\Delta S^{\ddagger}$ for several acid catalysed reactions. From the slope (–4·1 e.u. per w unit) the entropy loss per water molecule incorporated in the transition state would appear to be ca. 4·1 cal deg^{-1} mole^{-1}. This is of the right order of magnitude (Bunnett, 1961c). The evaluation of w and w^* for reactions following Bunnett's treatment may therefore give valuable information about the solvation requirements of the transition states in acid catalysed reactions.

It is important to remember that the solvation numbers of species in concentrated solutions will probably decrease with increasing acid concentration. Thus over wide ranges of acid concentration and particularly in most concentrated solutions both w and w^* should not be constant. Several treatments related to equilibrium measurements for concentrated acid solutions have tried to take this into account (Wyatt, 1957; Högfeldt, 1960; Perrin, 1964; Robertson and Dunford, 1964). It would be rather ambitious to attempt to make a quantitative assessment of this effect on reaction rates. However it must be accepted as a further complicating factor.

McTigue (1964) has formulated a theory for reaction rates in concentrated electrolyte solutions. The activity coefficients in the rate equation are equated with the Glueckauf (1955, 1959) expression for activity coefficients. The electrostatic contribution to the chemical potentials of ions of the same charge are taken as equal. The equations require a knowledge of the hydration number and apparent molal volume of the swamping electrolyte (the acid or an added salt). Appropriate plots give a parameter h which is interpreted as being the difference between the hydration numbers of the proton and the transition state in the reaction. Effectively the approach is a more rigorous treatment than that which interprets w^* as $(t-s-n)$ in equation (4.49). However the interpretation of h and w^* is similar. McTigue's calculations give an important theoretical approach to reaction rates in concentrated solutions. Bunnett's correlations are more generally applicable and easier to test and are therefore probably adequate for most mechanistic studies.

4.3.2. *Linear Free Energy Relationships*

Bunnett and Olsen (1965, 1966a) have investigated the validity of the linear free energy relationship (3.22) which correlates the protonation behaviour of any weak base B with the Hammett acidity function. They (1966b) have also extended this approach to measurably slow acid catalysed reactions. Thus equation (4.53) is applicable providing $C_{SH^+} \ll C_S$ in the fast pre-equilibrium

$$\log_{10} k_1 + H_0 = \phi(H_0 + \log_{10} C_{H^+}) + \log_{10}(k/K_{SH^+}) \qquad (4.53)$$

protonation step of the reaction for which k_1 is the experimental rate constant. Graphs of $(\log_{10} k_1 + H_0)$ against $(H_0 + \log_{10} C_{H^+})$ are often linear and hence allow deduction of (k/K_{SH^+}) and the proportionality slope parameter ϕ. This is exemplified by the results for the hydrolysis of isonicotinamide shown in Fig. 4.4. Bunnett and Olsen tested these plots for about 100 reactions, some of them at several temperatures. In cases where the protonation pre-equilibrium gave appreciable concentrations of SH^+ in solution appropriate corrections to the left-hand side of equation (4.53) were made (cf. equations (4.35)–(4.37)). If the equilibrium concentration of SH^+ can be measured by visible or ultraviolet spectroscopy then it is better to plot $\log_{10}[k_1(C_S+C_{SH^+})/C_{SH^+}]$ rather

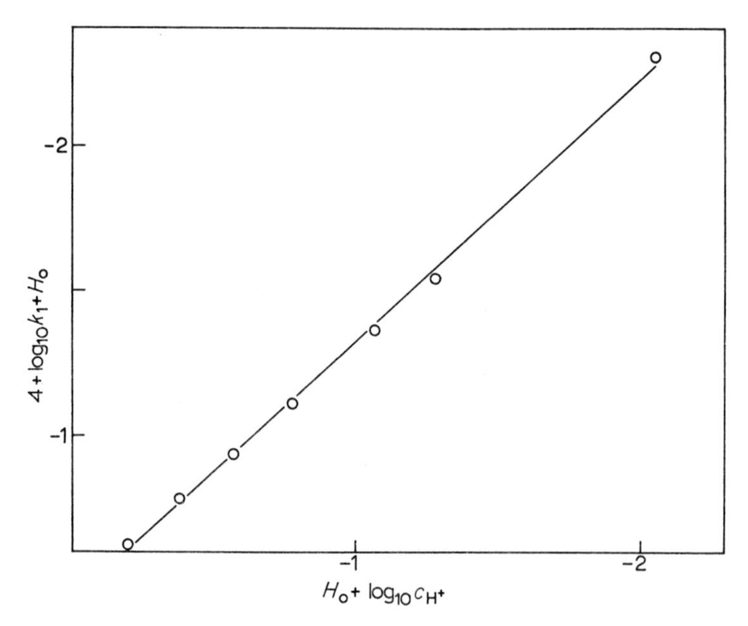

FIG. 4.4. Test of the linear free energy relationship (4.53) for the hydrolysis of isonicotinamide in aqueous HCl at 68° (H_0 at 25°). Reproduced by permission from Bunnett and Olsen (1966b).

than the left-hand side of equation (4.36) against $(H_0 + \log_{10}C_{H^+})$. This eliminates the ambiguity of making the assumption given in equation (4.12) that (y_S/y_{SH^+}) equals (y_B/y_{BH^+}).

Values of ϕ for some reactions are included in Table 4.1. It is a general result that the linear free energy relationship graphs are more clearly linear over wider ranges of acid concentration than either the conventional $\log_{10}k_1$ against H_0 plots or the graphs appropriate to the determination of w and w^*. Thus ϕ can usually be measured more accurately than the w and w^* parameters. It is therefore logical to assume that ϕ will probably provide a more reliable criterion for mechanism. However as would be expected the results in Table 4.1 do show that there is an approximately linear relationship between ϕ and w.

The plot of w against ϕ for all the reactions considered by Bunnett (1961a; Bunnett and Olsen, 1966b) is slightly curved. From this curve values of ϕ indicative of particular mechanisms can be deduced from the corresponding w values given in Table 4.2. Thus for substrates protonated on nitrogen or oxygen, the ranges of ϕ for different modes of involvement of water in the rate determining step are $-0{\cdot}34 < \phi < 0$ for water not involved, $0{\cdot}18 < \phi < 0{\cdot}47$ for water acting as a nucleophile, and $\phi > 0{\cdot}47$ for water acting as a proton transfer agent. For hydrocarbon-like basic substrates $\phi =$ ca. 0 if water is acting as a proton transfer agent.

Despite the correlation between certain ranges of ϕ and certain ranges of w as criteria of mechanism, the overall value of ϕ for a particular reaction is not of prime mechanistic importance. Thus ϕ is composed of two parts, ϕ_e which relates to the equilibrium protonation step (4.46) and ϕ_r which relates to the slow rate determining step (4.47) in the reaction. It is the latter which should be most sensitive to changes in mechanism. Equation (3.22) applied to the equilibrium (4.46) leads to equation (4.54) for ϕ_e.

$$\log_{10}(C_{SH^+}/C_S) + H_0 = \phi_e(H_0 + \log_{10} C_{H^+}) + pK_{SH^+} \qquad (4.54)$$

The corresponding linear free energy relationship for the slow reaction (4.47) may be written as equation (4.55).

$$\log_{10} k' = \phi_r(H_0 + \log_{10} C_{H^+}) + \log_{10} k \qquad (4.55)$$

The rate constant k' for the slow step is related to the overall experimental rate constant by equation (4.56) which when combined with equations (4.54) and (4.55) leads to equation (4.57). Equation (4.57) is written specifically for the

$$\log_{10} k_1 = \log_{10} k' + \log_{10} [C_{SH^+}/(C_S + C_{SH^+})] \qquad (4.56)$$

$$\log_{10} k_1 + H_0 = (\phi_e + \phi_r)(H_0 + \log_{10} C_{H^+}) + \log_{10}(k/K_{SH^+}) \qquad (4.57)$$

situation where $C_{SH^+} \ll C_S$ in the pre-equilibrium of S and SH^+. Comparison of equations (4.53) and (4.57) shows that $\phi = (\phi_e + \phi_r)$. The deduction of ϕ_r from ϕ therefore requires a knowledge of ϕ_e. If direct measurement of ϕ_e is not possible it may be reasonably assumed that the appropriate value is similar to ϕ_e for bases of related structure for which ϕ_e may be determined.

Bunnett and Olsen (1966b) have given examples which illustrate the additive nature of ϕ values for reactions composed of a sequence of steps. A complete analysis of ϕ_r values for reactions of known mechanism has not been made. Thus the mechanistic significance of particular values of ϕ_r is not clear. It should be noted that when $C_{SH^+} \gg C_S$ in the pre-equilibrium (that is for the more basic substrates) then $k_1 = k'$ and $\phi = \phi_r$.

4.4. Kresge's α-Coefficient

Kresge *et al.* (1965) have suggested an approach to mechanistic studies which does not rely on the use of the H_0 acidity function. The idea was outlined by consideration of the rates of hydrogen exchange for trihydroxybenzene and for trimethoxybenzene. The mechanisms of the two reactions should be similar. However the slopes of the appropriate $\log_{10}k_1$ against $-H_0$ plots were 0·80 ($w=+3$) for the former and 1·14 ($w=-2$) for the latter. Kresge *et al.* drew attention to the similar difference (see Section 3.2.5) between the slopes of 1·10 and 1·95 respectively (see however Schubert and Quacchia, 1962, 1963) for the plots of $\log_{10}(C_{SH^+}/C_S)$ against $-H_0$ relevant to the corresponding protonation equilibria.

The mechanism of the exchange reaction is represented by the equilibrium (4.58) from which it follows that the experimental rate constant will be given by equation (4.59).

$$H'Ar + H^+ \rightleftharpoons H'ArH^+ \rightleftharpoons H'^+ + HAr \tag{4.58}$$

$$k_1 = \frac{kC_{H^+}}{K_{SH^+}} \frac{y_{H^+}y_{HAr}}{y^{\ddagger}} \tag{4.59}$$

The basic concept of this approach is that the activity coefficient $y^{\ddagger}$ of the transition state can be represented by equation (4.60) in which α represents the extent to which the transition state resembles $H'ArH^+$. Combination of

$$y^{\ddagger} = (y_{HAr}y_{H^+})^{1-\alpha}y^{\alpha}_{HArH^+} \tag{4.60}$$

equations (4.59) and (4.60) with the ionization ratio $I = (C_{HArH^+}/C_{HAr})$ and the acid dissociation constant K_{SH^+} of $HArH^+$ leads to equation (4.61). Thus the

$$\log_{10}(k_1/C_{H^+}) = \alpha\log_{10}(I/C_{H^+}) + \log_{10}(kK^{\alpha-1}_{SH^+}) \tag{4.61}$$

parameter α can be determined from the slope of the linear graph of $\log_{10}(k_1/C_{H^+})$ against $\log_{10}(I/C_{H^+})$. Hence $\alpha = 0{\cdot}48$ for trihydroxybenzene and $\alpha = 0{\cdot}44$ for trimethoxybenzene. Thus although the acidity function dependences of the two reactions are appreciably different the values of α are similar and this confirms that the mechanisms are identical.

Comparison of equations (4.53), (4.54) and (4.61) shows that α is related to the Bunnett ϕ and ϕ_e parameters by equation (4.62).

$$\alpha = (\phi - 1)/(\phi_e - 1) \tag{4.62}$$

Whether α or $(\phi - \phi_e)$ is the best criterion for the mechanism of acid catalysed reactions has still to be tested (Bunnett and Olsen, 1966b).

REFERENCES

Arnett, E. M., and Mach, G. W. (1966). *J. Amer. Chem. Soc.* **88**, 1177.
Bunnett, J. F. (1961a). *J. Amer. Chem. Soc.* **83**, 4956.

Bunnett, J. F. (1961b). *J. Amer. Chem. Soc.* **83**, 4968.
Bunnett, J. F. (1961c). *J. Amer. Chem. Soc.* **83**, 4973.
Bunnett, J. F. (1961d). *J. Amer. Chem. Soc.* **83**, 4978.
Bunnett, J. F., and Olsen, F. P. (1965). *Chem. Comm.* 601.
Bunnett, J. F., and Olsen, F. P. (1966a). *Canad. J. Chem.* **44**, 1899.
Bunnett, J. F., and Olsen, F. P. (1966b). *Canad. J. Chem.* **44**, 1917.
Dayagi, S. (1961). *Bull. Res. Council Israel Sect. A*, **10**, 152.
Glueckauf, E. (1955). *Trans. Faraday Soc.* **51**, 1235.
Glueckauf, E. (1959). "The Structure of Electrolytic Solutions" (W. J. Hamer, ed.), p. 97. Wiley, New York.
Gold, V., and Hawes, B. W. V. (1951). *J. Chem. Soc.* 2102.
Hammett, L. P. (1935). *Chem. Rev.* **16**, 67.
Hammett, L. P., and Deyrup, A. J. (1932). *J. Amer. Chem. Soc.* **54**, 2721.
Hammett, L. P., and Paul, M. A. (1934). *J. Amer. Chem. Soc.* **56**, 830.
Hantzsch, A. (1907). *Z. phys. Chem.* **61**, 257.
Hantzsch, A. (1908). *Z. phys. Chem.* **65**, 41.
Högfeldt, E. (1960). *Acta Chem. Scand.* **14**, 1627.
Ingold, C. K. (1963). "Structure and Mechanism in Organic Chemistry", (a) Ch. XIV, (b) Ch. VI. Bell, London.
Kresge, A. J., More O'Ferrall, R. A., Hakka, L. E., and Vitullo, V. P. (1965). *Chem. Comm.* 46.
Long, F. A., and Paul, M. A. (1957). *Chem. Rev.* **57**, 935.
Long, F. A., and McDevit, W. F. (1952). *Chem. Rev.* **51**, 119.
Lowen, A. M., Murray, M. A., and Williams, G. (1950). *J. Chem. Soc.* 3318.
McTigue, P. T. (1964). *Trans. Faraday Soc.* **60**, 127.
Perrin, C. (1964). *J. Amer. Chem. Soc.* **86**, 256.
Robertson, E. B., and Dunford, H. B. (1964). *J. Amer. Chem. Soc.* **86**, 5080.
Rosenthal, D., and Taylor, T. I. (1957). *J. Amer. Chem. Soc.* **79**, 2684.
Schubert, W. M., and Latourette, H. K. (1952). *J. Amer. Chem. Soc.* **74**, 1829.
Schubert, W. M., and Quacchia, R. H. (1962). *J. Amer. Chem. Soc.* **84**, 3778.
Schubert, W. M., and Quacchia, R. H. (1963). *J. Amer. Chem. Soc.* **85**, 1278.
Westheimer, F. H., and Kharasch, M. S. (1946). *J. Amer. Chem. Soc.* **68**, 7871.
Williams, G., and Lowen, A. M. (1950). *J. Chem. Soc.* 3312.
Wyatt, P. A. H. (1957). *Discuss. Faraday Soc.* **24**, 162.
Zucker, L., and Hammett, L. P. (1939). *J. Amer. Chem. Soc.* **61**, 2791.

CHAPTER 5

The Rates of Reactions in Concentrated Aqueous Acid Solutions. Experimental Results

Measurements of the rates of reactions in concentrated acid solutions can give valuable information about the mechanisms of the reactions. The current status of the means whereby acidity function correlations may provide a criterion for mechanism have been presented in the previous chapter. Many acid catalysed reactions have been studied. For some of these there is independent evidence for a particular mechanism whereas for others only the acidity function criteria are available. It is the object of the present chapter to present the experimental results for these reactions. The relative merits of the different approaches described in Chapter 4 may then be assessed. For clarity each type of reaction is considered in turn and the general conclusions are summarized in Section 5.21.

It must be emphasized that throughout this chapter the Zucker–Hammett hypothesis is being considered in its approximate form in which the linearity of the $\log_{10} k_1$ against $\log_{10} C_{\text{acid}}$ or $-H_0$ plots is being taken as the criterion for a bimolecular or unimolecular rate determining reaction respectively.

5.1. The Hydrolysis of Carboxylic Esters

In Day and Ingold's (1941) classification there are four mechanisms possible for the acid catalysed hydrolysis of esters. These are differentiated by the molecularity of the slow step in the reactions and the position of bond fission in the ester. Either "acyl–oxygen fission" or "alkyl–oxygen fission" can occur and may be represented by

$$\mathrm{R.CO} \vdots \mathrm{OR'} \quad \text{and} \quad \mathrm{R.CO.O} \vdots \mathrm{R'}$$

respectively. The unimolecular and bimolecular acyl–oxygen fission hydrolyses are designated $A_{AC}1$ and $A_{AC}2$ and the corresponding alkyl–oxygen fission reactions are $A_{AL}1$ and $A_{AL}2$ (Ingold, 1953).

The use of ^{18}O in the solvent water was first used by Polanyi and Szabo (1934) to distinguish between acyl- and alkyl–oxygen fission in the base catalysed hydrolysis of amyl acetate. Whether the ^{18}O appears in the carboxylic acid or the alcohol product determines the bond which is broken. This method has been extended to acid catalysed hydrolyses (Datta *et al.* 1939; Long and Friedman, 1950) and to the reverse esterification reaction (Roberts and Urey, 1938). In general for simple esters acyl–oxygen fission occurs.

Several criteria enable a distinction to be drawn between the $A_{AC}1$ and $A_{AC}2$ mechanisms. Experiments using small concentrations of water in an inert solvent test whether a reaction is first ($A_{AC}2$) or zero ($A_{AC}1$) order in the water concentration. Friedman and Elmore (1941) showed that methyl acetate is hydrolysed via the $A_{AC}2$ mechanism. The effects of steric hindrance produced by bulky substituents in the ester should have a serious retarding effect for the $A_{AC}2$ mechanism. Steric effects will be negligible for the unimolecular decomposition of a protonated ester. The bimolecular reaction is characterized by the small electronic effect of different substituents in the acyl group. The rate of heterolysis of the acyl–oxygen bond in the $A_{AC}1$ mechanism is very sensitive to changes in the electronic structure of the protonated ester. The $A_{AC}2$ mechanism should show specific hydrogen-ion cataylsis and not general acid catalysis (Bell, 1941). The experimental evidence suggests that the bimolecular mechanism is the commonest for the hydrolysis of simple esters in aqueous acid solutions.

Equations (5.1)–(5.3) represent the generalized mechanism for an acid catalysed $A_{AC}2$ reaction (Ingold, 1953).

$$R.CO.OR' + H^+ \rightleftharpoons R.CO.\overset{+}{O}HR' \qquad \text{(fast)} \qquad (5.1)$$

$$RCO.\overset{+}{O}HR' + H_2O \rightleftharpoons R.CO.\overset{+}{O}H_2 + HOR' \qquad \text{(slow, } A_{AC}2\text{)} \qquad (5.2)$$

$$R.CO.\overset{+}{O}H_2 \rightleftharpoons R.CO.OH + H^+ \qquad \text{(fast)} \qquad (5.3)$$

For such a reaction the Zucker–Hammett hypothesis would predict that the experimental rate constant should be a linear function of the catalysing acid concentration with unit slope. This exact requirement is never obeyed. However there is a much closer correlation between $\log_{10} k_1$ and $\log_{10} C_{\text{acid}}$ than between $\log_{10} k_1$ and $-H_0$ for the hydrolysis of many esters. These include methyl formate and ethyl acetate in HCl, ethyl acetate in H_2SO_4 (Bell *et al.* 1955), methyl acetate in HCl (Duboux and de Sousa, 1940), methyl benzoate, α-glyceryl monobenzoate and α-glyceryl monoanisoate in $HClO_4$ (Chmiel and Long, 1956), potassium ethoxycarbonylmethanesulphonate in HCl and H_2SO_4 (Bell and Rawlinson, 1958), isopropyl acetate in HCl (Salomaa, 1959), n-butyl acetate in H_2SO_4 (Librovich and Vinnik, 1968), and methylene diacetate and ethylidene diacetate in HCl (Salomaa, 1957e; Bell and Lukianenko, 1957). The Zucker–Hammett hypothesis appears to be approximately applicable for these reactions (Long and Paul, 1957). Vinnik and Librovich (1967) concur with the proposed A-2 mechanism for ethyl acetate hydrolysis. The A-2 mechanism is also consistent with the similarity between the volumes of activation for the hydrolyses of methyl acetate and ethyl acetate in dilute and concentrated aqueous HCl solutions (Whalley, 1964).

The hydrolysis of simple esters in dilute aqueous H_2SO_4 is an A-2 reaction but two molecules of water (one being the nucleophile) are considered to be involved in the transition states (Lane, 1964; Jaques, 1965; Yates and McClelland, 1967). This follows since $r =$ ca. 2 in equation (5.3a) for the hydrolyses of several esters. In equation (5.3a) $m =$ ca. 0·62 is the slope of plots of

$$\log_{10} k_1 + mH_0 = r \log_{10} a_w + \text{constant} \qquad (5.3a)$$

$\log_{10}(C_{SH^+}/C_S)$ against $-H_0$ for esters S which are hydrolysed sufficiently slowly for equilibrium measurements to be able to be made. At high ($> 70\%$) concentrations of H_2SO_4 r becomes slightly negative (Yates and McClelland, 1967)

TABLE 5.1

The w,w^* and ϕ parameters and the slopes of the $\log_{10} k_1$ against $-H_0$ plots for the acid catalysed hydrolyses of some esters (Bunnett, 1961a; Bunnett and Olsen, 1966b)

Ester	Acid	Temp.	w	w^*	ϕ	Slope
Methyl benzoate	$HClO_4$	90	7·02	−0·13	0·98	
α-Glyceryl monobenzoate	$HClO_4$	90	6·09	−0·06	0·99	
Methyl acetate	HCl	25	5·83	−1·47	0·80	
α-Glyceryl monoanisoate	$HClO_4$	90	5·72		0·96	
$C_2H_5OOCCH_2SO_3K$	H_2SO_4	25	4·88	+1·41†	0·88	
Isopropyl acetate	HCl	25	4·62	−0·71	0·86	0·42
$C_2H_5OOCCH_2SO_3K$	HCl	25	4·52	−0·63	0·86	0·36
Ethyl acetate	H_2SO_4	25	4·50	+1·12†	0·84	
Methyl formate	HCl	25	4·21	−1·50	0·74	
Methylene diacetate	HCl	25	4·17	−1·41	0·75	0·50
Ethyl acetate	HCl	25	4·15	−0·60	0·87	0·37
Ethylidene diacetate	HCl	25	4·11	−1·46	0·74	0·51
Ethylidene diacetate	HCl	25	3·95	−1·22	0·77	0·46
α-Glyceryl mono-3,4,5-trimethoxybenzoate	H_2SO_4	50	2·91	−0·83	0·76	
Methoxymethyl formate	HCl	25	1·24		0·15	
Methoxymethyl acetate	H_2SO_4	25	−0·67		−0·06	
Methyl mesitoate	H_2SO_4	90	−1·10		−0·25	
tert-Butyl acetate	HCl	25	−1·17		−0·21	1·13
Methoxymethyl acetate	HCl	25	−2·39		−0·37	1·19
Methyl mesitoate	$HClO_4$	90	−2·47		−0·42	1·30
Ethoxymethyl acetate	HCl	25	−3·57		−0·44	

† w_A value; see Table 4.1.

in accord with the proposal (Jaques, 1965) that an A-1 mechanism becomes predominant as the H_2SO_4 concentration is increased. The mechanism may then be $A_{AC}1$ or $A_{AL}1$ depending on the structure of the particular ester being hydrolysed.

The w, w^* and ϕ parameters for the hydrolyses of several esters in concentrated aqueous acid are given in Table 5.1. For the first fourteen esters the Zucker–Hammett approach would indicate a bimolecular mechanism. The values of w, w^* (see Table 4.2) and ϕ are also consistent with this conclusion. However these parameters have quite a large range ($2{\cdot}91 \leqslant w \leqslant 7{\cdot}02$) of values for the series of esters. This may in part arise from differences in the solvation requirements of the protonated esters and the transition states in the reactions produced by changes in either R or R′ (Bunnett, 1961c). Bunnett (1961d) has also distinguished between rate determining processes in which either water is attacking a carbon atom or an alkoxy group is leaving a tretrahedral intermediate complex. Typical w values would be 2 to 3 for the former and about 6 for the latter. The observed w factors are between these two possibilities and therefore reflect the extent to which each contributes to the rate determining process.

For the remaining esters in Table 5.1 $\log_{10}k_1$ is a linear function of $-H_0$ with slope approximating to 1. The Zucker–Hammett hypothesis and the observed w factors suggest that these hydrolyses all proceed via a unimolecular mechanism. Ingold (1953) has stressed that the unimolecular mechanism will be favoured when the bimolecular mechanism is retarded by steric hindrance. Thus although the hydrolysis of methyl benzoate in aqueous $HClO_4$ is an $A_{AC}2$ reaction the hydrolysis of methyl mesitoate in either $HClO_4$ or H_2SO_4 follows the $A_{AC}1$ mechanism (Chmiel and Long, 1956). In its general form this is represented by reactions (5.4)–(5.7). Treffers and Hammett (1937) and

$$\mathrm{RCO.OR'} + \mathrm{H^+} \rightleftharpoons \mathrm{R.CO.\overset{+}{O}HR'} \qquad \text{(fast)} \tag{5.4}$$

$$\mathrm{R.CO.\overset{+}{O}HR'} \rightleftharpoons \mathrm{R.\overset{+}{C}O} + \mathrm{R'OH} \qquad (\text{slow, } A_{AC}1) \tag{5.5}$$

$$\mathrm{R.\overset{+}{C}O} + \mathrm{H_2O} \rightleftharpoons \mathrm{R.CO.\overset{+}{O}H_2} \qquad \text{(fast)} \tag{5.6}$$

$$\mathrm{R.CO.\overset{+}{O}H_2} \rightleftharpoons \mathrm{R.CO.OH} + \mathrm{H^+} \qquad \text{(fast)} \tag{5.7}$$

Newman (1941) have shown that mesitoic acid in anhydrous H_2SO_4 exists predominantly as the corresponding acylium cation $\mathrm{R.\overset{+}{C}O}$ and this supports the proposed mechanism for the hydrolysis reaction. It is interesting to note that the hydrolysis of methyl benzoate in nearly anhydrous H_2SO_4 is an A-1 (probably $A_{AC}1$) reaction (Leisten, 1956).

In dilute aqueous acid tert-butyl acetate is hydrolysed by an A-1 mechanism predominantly via alkyl–oxygen bond fission (Bunton and Wood, 1955). This is in accord with the observed acidity function dependence of the reaction rates (Salomaa, 1959), The alkyl–oxygen fission mechanism is as follows

$$Me.CO.OCMe_3 + H^+ \rightleftharpoons Me.C\overset{+}{O}.OHCMe_3 \quad \text{(fast)} \tag{5.8}$$

$$Me.CO.\overset{+}{O}HCMe_3 \rightleftharpoons Me.CO.OH + \overset{+}{C}Me_3 \quad \text{(slow, } A_{AL}1\text{)} \tag{5.9}$$

$$\overset{+}{C}Me_3 + H_2O \rightleftharpoons CMe_3.OH + H^+ \quad \text{(fast)} \tag{5.10}$$

The hydrolyses of methoxymethyl acetate and ethoxymethyl acetate are also $A_{AL}1$ reactions. This is consistent with the acidity function data in Table 5.1 (Salomaa, 1957a) and with the dependence of the reaction rates on solvent composition (Salomaa, 1957b, 1957c) and temperature (Salomaa, 1957d). For methoxymethyl formate the acidity function result (Salomaa, 1957a, 1957b) is intermediate between that expected for an A-1 or an A-2 mechanism. This ester probably undergoes concurrent $A_{AL}1$ and $A_{AC}2$ hydrolyses in aqueous acid solutions. The curved Arrhenius plot for the hydrolysis reaction (Salomaa, 1957d) supports this conclusion.

In general it may be concluded that for acid catalysed ester hydrolyses the Zucker–Hammett and the Bunnett approaches give a good indication of the molecularity of the rate determining step in the reactions. Where independent evidence is available it supports the acidity function conclusions. Values of w, w^* and ϕ for a particular general mechanism will be influenced by small differences in the solvation requirements of the reactants and in the solvation requirement and extent of bond making or breaking at the transition state in the slow step. Anomalous acidity function behaviour may indicate competing mechanisms (Salomaa and Linnantie, 1958).

5.2. The Hydrolysis of Amides

The experimental rate constants are approximately proportional to the HCl concentration for the hydrolyses in concentrated aqueous HCl solutions of isonicotinamide, picolinamide (Jellinek and Urwin, 1953), nicotinamide (Jellinek and Gordon, 1949), piperazine-2,5-dione (Edward and Meacock, 1957b), and acethydrazide (Edward *et al.* 1955). The Zucker–Hammett approach would therefore indicate a bimolecular rate determining step. This is confirmed by the w^* values which are in the range $0{\cdot}35 \geqslant w^* \geqslant -0{\cdot}63$ and the w factors which are about 5 (Bunnett, 1961a). Most amides hydrolyse by an $A_{AC}2$ mechanism in which the slow step is the bimolecular addition of water to the protonated amide. The acidity function behaviour is consistent with this mechanism. Independent evidence is provided by the observed effects on reaction rate of substituents in the benzene nucleus of benzamide (Reid, 1899, 1900; Leisten, 1959). Electron withdrawing *meta* or *para* substituents increase the reaction rate whereas *ortho* substituents, independent of their polar effects, retard the reaction through steric hindrance. Although neutral amides are predominantly protonated on oxygen (Fraenkel and Franconi, 1960) the slow hydrolysis reaction probably occurs mainly through the nitrogen protonated conjugate acid (Huisgen and Brade, 1957).

For the hydrolysis of many amides in concentrated acid solutions the experimental rate constant has a maximum value at a particular acid concentration (for example: Taylor, 1930; Krieble and Holst, 1938). At higher concentrations the reaction rate decreases with increasing acid concentration. This considerably complicates tests of acidity function behaviour as a criterion of mechanism. The results are explicable if the amides are protonated to an appreciable extent in the rapid pre-equilibrium which precedes the rate determining step of the reaction. For the expected $A_{AC}2$ mechanism the Zucker–Hammett hypothesis may be tested using equation (4.21). The evaluation of w and w^* requires use of equations (4.36) and (4.39) respectively. Rosenthal and Taylor (1957) found that equation (4.21) fitted the experimental results for the hydrolysis of thioacetamide in aqueous HCl and $HClO_4$ providing the activity coefficient term was written as a linear function of electrolyte concentration. Bunnett (1961a) deduced values of w via equation (4.36) for the hydrolyses of benzamide in HCl, benzamide and 4-nitrobenzamide in H_2SO_4 (Edward and Meacock, 1957a), acetamide in HCl (Rosenthal and Taylor, 1957; Edward *et al.* 1955), propionamide in HCl (Rabinovitch and Winkler, 1942), and acetylgylcine in HCl (Edward and Meacock, 1957b). However because of the uncertainty of the K_{SH^+} values used in the correlations, the w factors which were deduced are all incorrect. It is a difficulty inherent in the use of equations (4.21), (4.36) and (4.39) that the equilibrium constant for the pre-equilibrium protonation must be known. This can be avoided if the equilibrium protonation is studied separately by spectrophotometry (Moodie *et al.* 1963; Armstrong *et al.* 1968). Recent measurements for amides are summarized in Section 3.2.2. Using these results Bunnett and Olsen (1966b) deduced more reliable w parameters for the hydrolysis reactions.

Direct knowledge of ionization ratios (C_{SH^+}/C_S) as a function of acid concentration also eliminates the need to make the assumption $(y_{SH^+}/y_S) = (y_{BH^+}/y_B)$ which is given in equation (4.12). Comparison of the ionization behaviour of amines (B) and amides (S) clearly shows that this identity is not acceptable when considering the hydrolysis of amides. An alternative way to avoid this problem is to use the amide acidity function H_A rather than H_0 in the appropriate equations (Yates and Stevens, 1965; Yates and Riordan, 1965). Thus for example equation (4.36) becomes

$$\log_{10} k_1 - \log_{10}\left(\frac{h_A}{K_{SH^+} + h_A}\right) = r \log_{10} a_w + \text{constant} \tag{5.11}$$

where r has a similar interpretation to that for w. This equation gave linear plots of the left-hand side against $\log_{10}a_w$ for the hydrolyses of benzamide and 4-nitrobenzamide in aqueous H_2SO_4. Values of r were 2·6 and 2·7 respectively. These figures compare favourably with that of $b = 3·3$ deduced by Moodie *et al.* (1963) who fitted Edward and Meacock's (1957a) data for the hydrolysis of benzamide in aqueous H_2SO_4 to equation (5.12). This equation is consistent

$$\log_{10}k_1 + \log_{10}\left(\frac{C_S + C_{SH^+}}{C_{SH^+}}\right) = \log_{10}\left(\frac{k\,y_{SH^+}}{y^{\ddagger}}\right) + b\log_{10}a_w \qquad (5.12)$$

with reaction (5.13) as the slow step in the hydrolysis (compare equation 4.47). It has the advantage that it does not involve any acidity function but does however need precise measurements of the equilibrium ionization ratios of the reacting substrate in the catalysing acid solutions. Despite the difference in approach between the use of equations (5.11) and (5.12). Moodie *et al.* (1963)

$$SH^+(H_2O)_p + bH_2O \rightarrow \ddagger(H_2O)_{p+b} \quad \text{(slow)} \qquad (5.13)$$

and Yates and Stevens (1965) came to the same conclusion that the transition state in the acid catalysed hydrolysis of benzamide could be represented as follows:

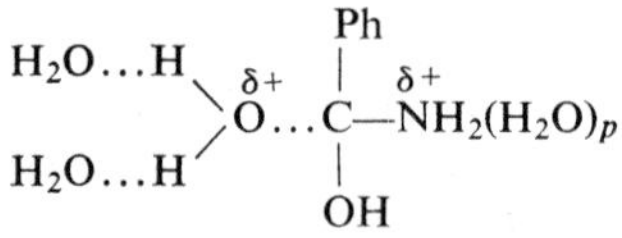

```
               Ph
H2O...H   δ+   |   δ+
       \       |
        O...C—NH2(H2O)p
       /       |
H2O...H        |
               OH
```

A similar transition state has been proposed for the hydrolysis of other simple amides (Yates and Riordan, 1965). However this is not consistent with the idea that the slow reaction predominantly goes through nucleophilic attack of water on the nitrogen-protonated conjugate acid of amides (Ingold, 1963; Huisgen and Brade, 1957; Bunnett, 1961b).

Bunton *et al.* (1967) have shown that the rates of hydrolysis of benzamide, *N*-methylbenzamide and *N,N*-dimethylbenzamide in aqueous HCl are not consistent with any of the above equations which correlate rates with either H_0 or H_A. They conclude that the hydrolyses proceed along two different competing mechanistic paths, one which involves the slow addition of H_3O^+ to the neutral amide and a second in which the protonated amide undergoes nucleophilic attack by water. The former has an *N*-protonated amide and the latter an *O*-protonated amide participating in the transition state.

The rates of cyanamide hydrolysis in aqueous nitric acid correlate better with H_0 than with the nitric acid concentration (Hammett, 1935; Sullivan and Kilpatrick, 1945). The w value (ca. 1·5) is correspondingly low for the reaction (Bunnett, 1961a, 1961d). However this figure does fall within the limits set by Bunnett's criteria (Table 4.2) for an A-2 reaction. The Zucker–Hammett approach is therefore misleading and gives an incorrect prediction of the mechanism. The reaction is complicated in certain acid media by the competition between water and other nucleophiles in their attack on the protonated amide (Kilpatrick, 1947). In this context Vinnik and Medvetskaya (1967) have suggested that the hydrolysis of 2,4-dinitroacetanilide in aqueous H_2SO_4 involves nucleophilic attack by H_2O in predominantly aqueous solutions and by H_2SO_4 in more concentrated solutions. In non-aqueous or in very concentrated aqueous H_2SO_4 solutions several *N*-alkyl substituted amides apparently hydrolyse by a unimolecular ($A_{AL}1$) mechanism (Lacey, 1960; Dayagi, 1961).

5.3. The Hydrolysis of Lactones

The rates of hydrolysis of β-propiolactone (Long and Purchase, 1950), β-butyrolactone (Olson and Miller, 1938), and β-isovalerolactone (Liang and Bartlett, 1958) in concentrated aqueous acid solutions correlate with the H_0 acidity function (Fig. 5.1). The w and ϕ values (Bunnett, 1961a; Bunnett and Olsen, 1966b) are $w = 0{\cdot}39$, $\phi = 0{\cdot}10$ for β-propiolactone in H_2SO_4, $w = -1{\cdot}18$, $\phi = -0{\cdot}22$ for β-propiolactone in $HClO_4$, and $w = 0{\cdot}06$, $\phi = 0{\cdot}01$ for β-isovalerolactone in $HClO_4$ (all at 25°). There are insufficient results to be able to deduce w and ϕ for β-butyrolactone hydrolysis. The Zucker–Hammett hypothesis and the Bunnett criteria of mechanism (Table 4.2) indicate that these hydrolyses are all A-1 reactions. Olson and Hyde (1941) studied the hydrolysis of β-butyrolactone in $H_2{}^{18}O$ isotopically enriched water and showed that acyl–oxygen fission occurs. The reactions therefore conform to the $A_{AC}1$ mechanism (Ingold, 1953). In dilute acid solution there are appreciable contributions to the overall rates of hydrolysis from the water catalysed $B_{AL}2$ reaction (Long and Purchase, 1950; Ingold, 1963).

γ-Butyrolactone also undergoes acyl–oxygen fission in acid solutions (Long and Friedman, 1950). However the rates of reaction correlate much more closely with the acid concentration (HCl or $HClO_4$) than with the Hammett acidity function (Long *et al.* 1951a). The figures $w = 6{\cdot}11$, $w^* = -0{\cdot}91$, $\phi = 0{\cdot}87$ for HCl at 0° and $w = 8{\cdot}50$, $w^* =$ ca. 0, $\phi = 1{\cdot}04$ for $HClO_4$ at 5° confirm this and suggest that the $A_{AC}2$ mechanism is operative (Bunnett, 1961a; Bunnett and Olsen, 1966b).

The lactonization of γ-hydroxybutyric acid is the reverse reaction to the hydrolysis of γ-butyrolactone and has also been studied (HCl at 0° and $HClO_4$ at 5°) by Long *et al.* (1951a). The rates contrast with those for the hydrolysis in that they correlate better with H_0 than with the acid concentration. This is in accord with the following reaction scheme

$$\underset{\displaystyle\text{OH}}{\text{H}_2\text{C}}\text{CH}_2\text{CH}_2\text{COOH} \rightleftharpoons \text{H}_2\text{O} + \text{H}_2\text{C}\ \text{CH}_2\text{CH}_2\text{CO}\ (\text{ring closed by }-\text{O}-)$$

$$\text{H}_2\text{C}(\text{OH})\text{CH}_2\text{CH}_2\text{COOH} \xrightarrow{2} \text{transition state} \xleftarrow{3} \text{lactone}$$

On the Zucker–Hammett hypothesis the forward reaction would be expected to follow H_0 and the reverse reaction should follow the acid concentration. This is what is observed. However both reactions must have a common mechanism and a common transition state. It follows that the deviation

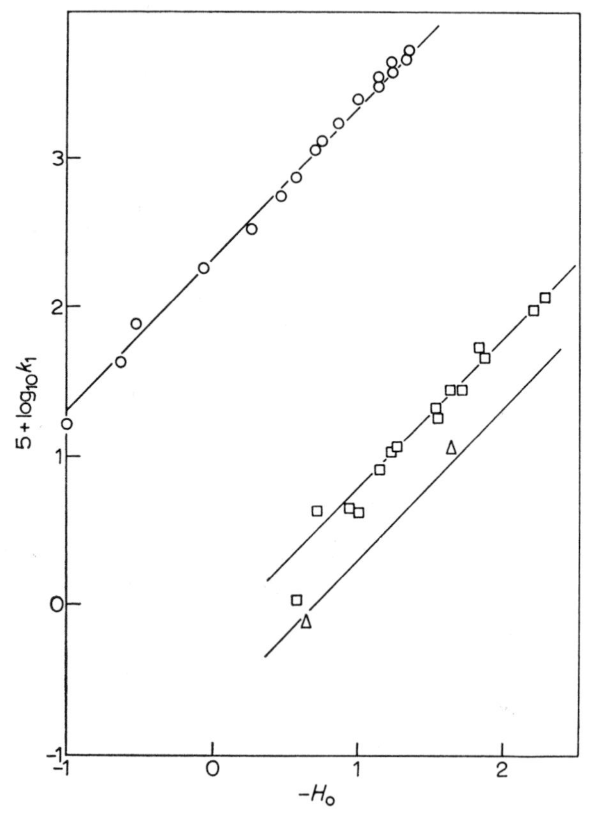

FIG. 5.1. Linear relationship between $\log_{10} k_1$ and $-H_0$ for the acid catalysed hydrolyses of β-butyrolactone (△), β-propiolactone (□) and β-isovalerolactone (○). Reproduced by permission from Liang and Bartlett (1958).

between the acidity function behaviour of the lactonization and the hydrolysis reactions must result from differences in the activity coefficient behaviour of the γ-lactone and γ-hydroxybutyric acid (Taft *et al.* 1958; Bunnett, 1961d). The additivity of ϕ parameters is demonstrated by considering the above reaction scheme in conjunction with the experimental results obtained for the reactions in aqueous HCl at 0°. For the overall equilibrium $\phi_1 = -0{\cdot}57$ (Long *et al.* 1951b; Bunnett and Olsen, 1966a), for the lactonization

$\phi_2 = 0{\cdot}32$, and for the hydrolysis $\phi_3 = 0{\cdot}87$ (Bunnett and Olsen, 1966b). Thus $\phi_1 \approx (\phi_2 - \phi_3)$ in accord with Bunnett and Olsen's proposed additivity principle.

Long *et al.* (1951b; Long *et al.* 1951a) have made a detailed study of the salt effects of NaCl and $NaClO_4$ on the γ-butyrolactone/γ-hydroxybutyric acid system. The salt effects for the hydrolysis and lactonization reactions are

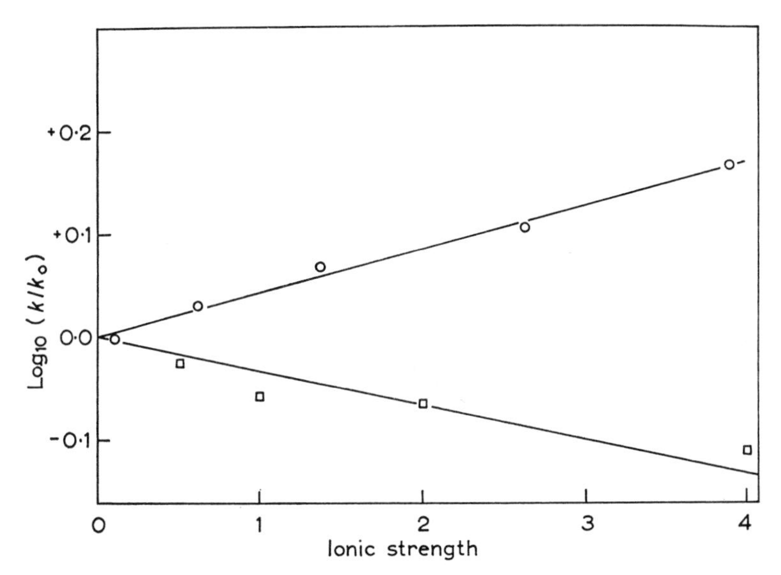

FIG. 5.2. Effect of added NaCl (○) and $NaClO_4$ (□) on the rate of hydrolysis of γ-butyrolactone in aqueous solution at 25°. Catalysing acid HCl (0·100 mole litre^{-1}) or $HClO_4$ (0·104 mole litre^{-1}). Reproduced by permission from Long *et al.* (1951a).

shown in Figs 5.2. and 5.3. These are consistent with the variation of the γ-butyrolactone/γ-hydroxybutyric acid equilibrium concentration ratio with increasing NaCl or $NaClO_4$ concentration. The equilibrium concentration ratio may in turn be correlated with the variation of activity coefficients with salt concentration (NaCl or $NaClO_4$) for γ-butyrolactone and γ-hydroxybutyric acid. The activity coefficients were evaluated from the results of distribution measurements. The conclusion mentioned above that the reaction rates are influenced largely by the activity coefficient behaviour of the reacting substrate (γ-butyrolactone or γ-hydroxybutyric acid) is therefore also applicable to the present data. The linearity of the plots in Figs. 5.2 and 5.3 shows that the logarithm of the activity coefficient ratio ($y_S y_{H_3O^+}/y^{*}$) in the rate equation is directly proportional to the ionic strength I at least up to $I =$ ca. 4 mole litre^{-1}.

5.4. The Hydrolysis of Epoxides

Brønsted *et al.* (1929) found that the acid catalysed hydrolysis of epoxides was susceptible only to specific hydrogen ion catalysis. Isotope studies of the hydrolysis of isobutylene oxide and propylene oxide in water enriched with $H_2{}^{18}O$ showed that the C—O bond fission occurs entirely at the most substituted carbon for the former and predominantly so for the latter (Long and Pritchard, 1956). Pritchard and Long (1956a) have measured the rates of

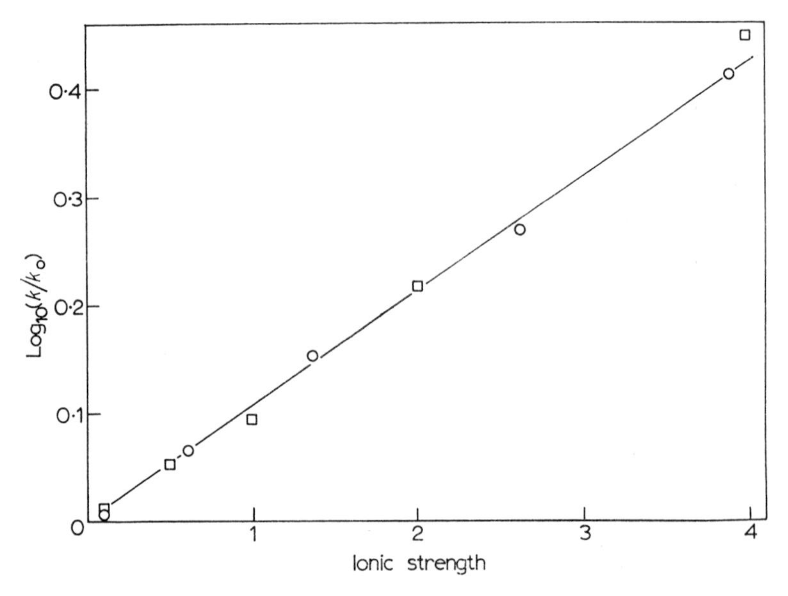

Fig. 5.3. Effect of added NaCl (○) and $NaClO_4$ (□) on the lactonization of γ-hydroxybutyric acid in aqueous solution at 25°. Catalysing acid HCl (0·100 mole litre^{-1}) or $HClO_4$ (0·104 mole litre^{-1}). Reproduced by permission from Long *et al.* (1951a).

hydrolysis of ten epoxides in aqueous $HClO_4$ at 0°. The experimental rate constants correlated with the H_0 acidity function better than with the $HClO_4$ concentration. The slopes of the $\log_{10} k_1$ against $-H_0$ plots are given in Table 5.2 together with the measured entropies (Long *et al.* 1957; Koskikallio and Whalley, 1959a) and volumes (Koskikallio and Whalley, 1959a) of activation where known. Data for the related compound trimethylene oxide are also included (Pritchard and Long, 1958; Long *et al.* 1957). Where the slopes of $\log_{10} k_1$ against $\log_{10} C_{HClO_4}$ graphs are quoted the reactions are too fast to be studied in acid concentrations for which $\log_{10} C_{HClO_4}$ and $-H_0$ diverge appreciably. The rates of hydrolysis in H_2O and D_2O have been compared for

TABLE 5.2

Experimental data for the acid catalysed hydrolyses of trimethylene oxide and ten epoxides. References are given in the text

Oxide	Slope (0°)	(k_{D_2O}/k_{H_2O}) (25°)	$\Delta S^{\ddagger}(25°)$ (cal mole^{-1} deg^{-1})	$\Delta V^{\ddagger}(0°)$ (cm^3 mole^{-1})
Epibromohydrin	0·86†			
Epichlorohydrin	0·87†	1·89		
β-Methylepichlorohydrin	0·89†			
Glycidol	0·89†			
β-Methylglycidol	0·89†	2·05		
Ethylene oxide	1·06†	2·20	−6·1(−7·5)	−5·9§
Propylene oxide	1·06†		−4·3	−8·4
trans-2,3-Epoxybutane	1·01‡			
cis-2,3-Epoxybutane	0·95‡			
Isobutylene oxide	0·98‡		−4	−9·2
Trimethylene oxide	1·00†	2·21	−3·9	

† Plot $\log_{10} k_1$ against $-H_0$.
‡ Plot $\log_{10} k_1$ against $\log_{10} C_{HClO_4}$.
§ $\Delta V^{\ddagger} = -7{\cdot}9$ cm^3 mole^{-1} at 25°.

three epoxides (Pritchard and Long, 1956b) and trimethylene oxide (Pritchard and Long, 1958).

The (k_{D_2O}/k_{H_2O}) values for epoxide hydrolysis indicate that there is rapid proton transfer pre-equilibrium step in the reaction. On the basis of the Zucker–Hammett hypothesis the linear correlations between $\log_{10} k_1$ and $-H_0$ were interpreted as evidence for an A-1 mechanism not involving a water molecule in the rate-determining step (Pritchard and Long, 1956a; Long and Paul, 1957). The (k_{D_2O}/k_{H_2O}) ratios and entropies of activation were taken as supporting this conclusion although for both the observed values are somewhere between those typical of either A-1 or A-2 reactions. Koskikallio and Whalley (1959a) have argued that the volumes of activation support the A-2 mechanism. Thus although for ethylene oxide and propylene oxide plots of $\log_{10} k_1$ against $-H_0$ are linear with slope 1·06 (k_1 at 0°, H_0 at 25°) independent evidence points to an A-2 mechanism. The Zucker–Hammett criterion for mechanism therefore breaks down when applied to the hydrolysis of epoxides. The conflict between the conclusions drawn from the volumes of activation and from the Zucker–Hammett hypothesis have been discussed by Whalley (1964). Schaleger and Long (1963) agree that the present evidence favours the bimolecular mechanism. However Long and Paul's (1957) counsel that "further study is desirable" would seem to remain appropriate.

For the first five epoxides listed in Table 5.2 volumes of activation have not been measured. Although $\log_{10} k_1$ is a linear function of $-H_0$ Bunnett (1961d) has pointed out that the appropriate w values are about 2–3. On his (1961b) classification this would indicate a bimolecular reaction with water acting as a nucleophile in the rate-determining step. Thus where the Zucker–Hammett approach is unsatisfactory the Bunnett approach is apparently successful. However $w = $ ca. 0 (slightly negative) for the hydrolyses of ethylene oxide and propylene oxide and on Bunnett's criteria this would require that the reactions are A-1. The Bunnett approach also is therefore misleading in these cases.

Long *et al.* (1957; Pritchard and Long, 1958) have proposed that trimethylene oxide undergoes A-1 acid catalysed hydrolysis. The acidity function behaviour, (k_{D_2O}/k_{H_2O}) ratio, and entropy of activation for this reaction are all similar to the corresponding values for epoxide hydrolysis. It would be logical to suppose that the mechanisms are the same.

5.5. The Hydrolysis of Acetals and Ketals

The established A-1 mechanism for the hydrolysis of acetals and ketals has been generalized by Kreevoy and Taft (1955a) as follows

$$R_1R_2C(OR_3)(OR_4) + H_3O^+ \rightleftharpoons R_1R_2C(\overset{+}{O}HR_3)(OR_4) + H_2O \quad \text{(fast)}$$

$$R_1R_2C(\overset{+}{O}HR_3)(OR_4) \longrightarrow R_1R_2C{=}\overset{+}{O}R_4 + R_3OH \quad \text{(slow, A-1)}$$

$$R_1R_2C{=}\overset{+}{O}R_4 + 2H_2O \rightleftharpoons R_1R_2C{=}O + R_4OH + H_3O^+ \quad \text{(fast)}$$

Much evidence for this mechanism has been obtained. The hydrolysis of acetals in ^{18}O enriched water gave alcohols of normal ^{18}O content. For the reverse condensation of ^{18}O enriched aldehydes with alcohols the ^{18}O label appeared in the water of condensation (Stasiuk *et al.* 1956). D-(+)-2-Butyl acetal was prepared from D-(+)-2-butanol and D-(+)-2-butyl orthoformate and was then hydrolysed in aqueous phosphoric acid. The alcohol recovered gave the same specific rotation as the starting material and this confirms that the reaction mechanism does not involve an alkyl carbonium ion intermediate (Alexander *et al.* 1952). A similar result was obtained from a study of the hydrolysis of the acetal of D-(+)-2-octanol (O'Gorman and Lucas, 1950). The hydrolysis of acetal in DCl/D_2O is 2·64 times faster than in HCl/H_2O (Orr and Butler, 1937). A similar result was obtained for the hydrolysis of 1,1-dimethoxyethane (Kilpatrick, 1963). Only specific hydrogen ion catalysis was detected for the hydrolysis of acetals (Brønsted and Wynne-Jones, 1929; Brønsted and Grove, 1930). The observed electronic effects of the groups R_1 and R_3 on the rates of hydrolysis and the absence of appreciable steric effects are consistent with the A-1 mechanism (Kreevoy and Taft, 1955b). The entropies (Schaleger and Long, 1963) and volumes of activation (Whalley, 1964) for acetal hydrolysis also support this mechanism. The accumulated evidence has been presented in detail by Cordes (1967).

The acidity function dependence of the rates of hydrolysis of simple acetals is in accord with the above mechanism which involves the unimolecular decomposition of protonated acetal in the rate determining step. Thus $\log_{10}k_1$ is a linear function of $-H_0$ for the hydrolysis of methylal in aqueous HCl, $HClO_4$, H_2SO_4 (McIntyre and Long, 1954), CCl_3COOH, $CHCl_2COOH$ and CF_3COOH (Hamer and Leslie, 1960). Similar results have been obtained for the hydrolyses of ethylal in aqueous HCl (Leininger and Kilpatrick, 1939) and chloroacetal in aqueous HCl and $HClO_4$ (Kreevoy and Taft, 1955a). The slopes of the plots are generally greater than 1 and for methylal show small differences for the six catalysing acids. These have been explained by McIntyre and Long (1954) in terms of the different salt effects of the catalysing acids on the activity coefficients of methylal and the transition state in the reaction. The w values are all negative (Bunnett, 1961a) although there is no simple correlation between w and ΔS^{+} for the hydrolysis of acetals (Bunnett, 1961c).

The hydrolysis of the 4-dimethoxymethylpyridinium ion in 36–50% aqueous $HClO_4$ is a two proton process. The substrate is practically entirely in the form of the pyridinium ion and therefore the unimolecular decomposition of the doubly protonated 4-dimethoxypyridine molecule is the rate determining step in the reaction. This is confirmed (Fuller and Schubert, 1963) by the linearity and slope (1·07) of the plot of $\log_{10}k_1$ against $-H_0$ (Fig. 5.4). Furthermore $w = -0{\cdot}34$ and the entropy of activation is 4·5 e.u. All these values are in accord with an A-1 mechanism. Strictly as the pre-equilibrium in the reaction is the protonation of a pyridinium cation the reaction rates should be correlated with H_+ rather than H_0. The results shown in Fig. 5.4 are added evidence for the supposition that H_+ and H_0 are parallel functions of acid concentration. The agreement between the acidity function behaviour for this reaction and that for the hydrolysis of neutral acetals is further confirmation of this point.

Long and McIntyre (1954) have measured the increase in rate of hydrolysis of methylal in aqueous HCl caused by the addition of eight salts to the reaction solutions. The addition of three of the salts to aqueous HCl causes a decrease in the acidity of the solutions as indicated by the ionization of 4-nitroaniline (Paul, 1954). There is therefore no direct parallel between the salt effects on reaction rate and those on the H_0 acidity function. The appropriate relationship between the experimental rate constants and H_0 is equation (4.9) where S is methylal and B is 4-nitroaniline. The lack of correlation between $\log_{10}k_1$ and

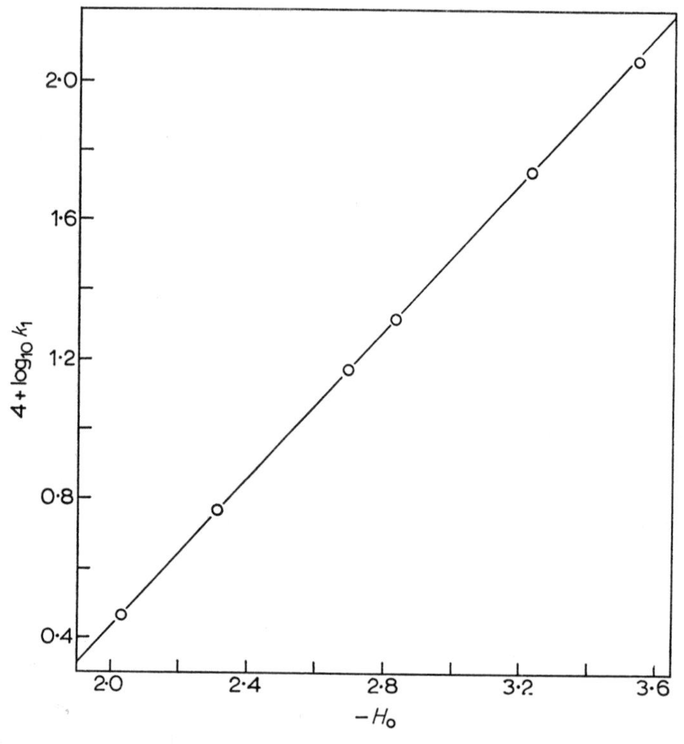

FIG. 5.4. The hydrolysis of the 4-dimethoxymethylpyridinium ion in aqueous $HClO_4$ at 25°. Reproduced by permission from Fuller and Schubert (1963).

$-H_0$ indicates that the identity (5.14) is not valid for the solutions. Measurements of y_S and y_B as a function of salt concentration shows that this arises

$$(y_S/y^{\ddagger}) = (y_B/y_{BH^+}) \qquad (5.14)$$

primarily because of differences between the salt effects on the activity coefficients y_S of methylal and y_B of 4-nitroaniline.

5.6. The Hydrolysis of Glycosides

The hydrolyses of four glucopyranosides in aqueous $HClO_4$ were shown by Bunton *et al.* (1955b) to proceed via an A-1 mechanism which involves hexose–oxygen bond fission. This confirmed the mechanism suggested by Edward

(1955). Since then a vast amount of evidence has been accumulated which supports the general conclusion that most glycosides undergo acid catalysed A-1 hydrolysis. The (k_{D_2O}/k_{H_2O}) ratios of 2·05 for sucrose hydrolysis (Wiberg, 1955) and 2.5 for the hydrolysis of α-2-deoxy-D-glucopyranoxide (Armour *et al.* 1961) are typical of the corresponding values for the A-1 hydrolyses of acetals and orthoesters (Cordes, 1967). Entropies of activation are in the range 9 e.u. $< \Delta S^{+} <$ 17 e.u. (Armour *et al.* 1961; Timell, 1964; Semke *et al.* 1964) which is again typical for A-1 reactions (Schaleger and Long, 1963; Whalley, 1964). Reactions studied include the hydrolyses of sixteen glucopyranosides, two galactopyranosides, two methyl α-D-mannopyranoside, three glycuronides and eight disaccharides. The volume of activation $\Delta V^{+} = 5{\cdot}1$ cm^3 $mole^{-1}$ (Whalley, 1964) for the hydrolysis of methyl α-D-glucopyranoside at 100° conforms with expectation for an A-1 reaction. Many other studies have been made supporting these conclusions although only those which have made use of acidity function correlations are mentioned here. In all cases $\log_{10} k_1$ for the hydrolyses were linear functions of the Hammett acidity function. In fact the hydrolysis of sucrose was one of the first reactions for which the rates were correlated with H_0 (Hammett and Paul, 1934). Although the slopes of the plots are often different from 1 the Zucker–Hammett hypothesis indicates that the A-1 mechanism is probable. This agrees with the other evidence and the Zucker–Hammett approach therefore gives an acceptable indication of mechanism.

The rates of hydrolysis of cellulose, xylan and laminarin in concentrated aqueous acid also follow $-H_0$ (Novikova and Konkin, 1959). However the slopes of the $\log_{10} k_1$ against $-H_0$ graphs for the hydrolysis of cellubiose are quite different for HCl and H_2SO_4 as the catalysing acids. Similar results were obtained for the hydrolysis of methyl α-D-glucopyranoside (Fig. 5.5). Although the slope of the graph was ca. 1 for HCl it falls to ca. 0·6 for H_3PO_4 (Timell, 1964). This contrasts with the data shown in Fig. 4.1 for the hydrolysis of sucrose where the results for HCl, $HClO_4$, H_2SO_4 and HNO_3 all fall on the same line. The exact reasons for the differences in behaviour is not certain although it is possible that salt effects (Long and McIntyre, 1954) make a significant contribution to the deviations between the four plots for methyl α-D-glucopyranoside. For this reaction the Bunnett w factor varies enormously with changes in the catalysing acid. On the Bunnett criterion for mechanism (Table 4.2) the w values for the hydrolysis in $HClO_4$, H_2SO_4, H_3PO_4 would suggest an A-2 bimolecular mechanism.

Bunnett (1961a; 1961d) has listed w values for the hydrolysis of several glycosides. Although in some cases $w < 1$ in accord with the A-1 mechanism there are also several instances where $w > 1{\cdot}2$. On the Bunnett (1961b) criteria for mechanism this indicates an A-2 reaction. The w values are given in Table 5.3 together with the entropies of activation where known for the reactions.

Schaleger and Long (1963) have argued that the similarity in the ΔS^{+} values for all these reactions excludes the possibility that any of them are A-2 hydrolyses. The positive volume of activation for methyl α-D-glucopyranoside is strong evidence against the A-2 mechanism for that compound (Whalley,

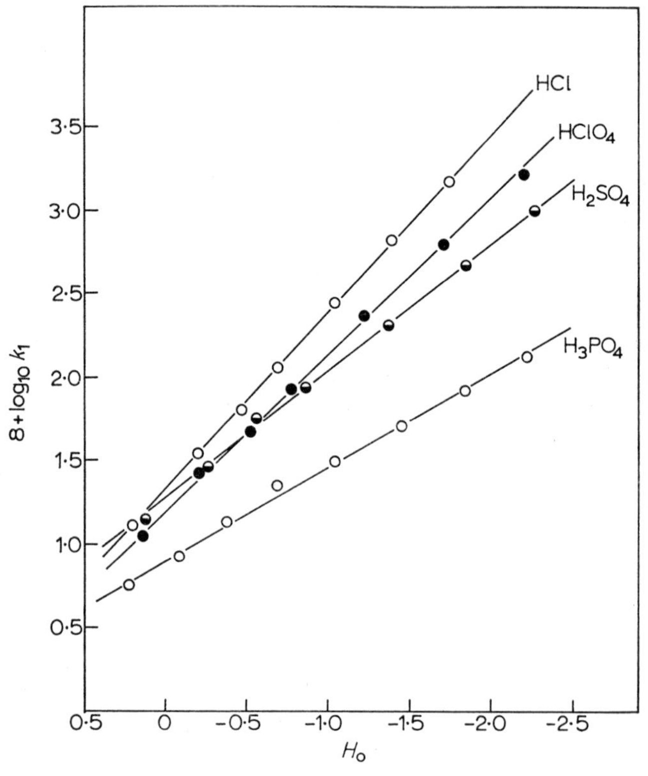

FIG. 5.5. Variation of reaction rate with H_0 for the hydrolysis of methyl α-D-glucopyranoside in aqueous solutions of four acids. Reproduced by permission from Timell (1964).

1964). It must be concluded that the Bunnett (1961b) criteria for mechanism are not valid for the hydrolysis of certain glycosides. Overend *et al.* (1962) consider that some furanosides may be hydrolysed via an A-2 mechanism. These reactions are characterized by low or negative entropies of activation.

Most glycopyranosides are hydrolysed by the following mechanism which is similar to that given in the previous section for the hydrolysis of acetals and ketals. The slow heterolysis of the hexose–oxygen bond in the protonated glycoside leads to a pyranosyl carbonium ion which rapidly combines with

TABLE 5.3

The w values (Bunnett, 1961d) and entropies of activation for the hydrolysis of some glycosides in aqueous acid

Glycoside	w	$\Delta S^{\ddagger}$ (cal deg^{-1} $mole^{-}$)	Reference†
t-Butyl β-D-glucopyranoside	−5·0		
Methyl β-2-deoxy-D-glucopyranoside	−1·6	13·7	a
Methyl α-2-deoxy-D-glucopyranoside	−1·8	16·7	a
Sucrose	−0·6	7·9	c
Methyl α-D-mannopyranoside	+0·4	10·4	c
Methyl α-D-glucopyranoside	+1·7	13·8, 14·8	a, c
Methyl β-D-glucopyranoside	+1·6	16·5	c
Phenyl α-D-glucopyranoside	+3·0	13·2, 13·3	a, c
Phenyl β-D-glucopyranoside	+1·4	10·0, 10·8	b, c
Lactose	+3·0	14·9	d
Maltose	+2·0	14·0	d

† Reference: a. Armour *et al.* 1961. b. Semke *et al.* 1964. c. Overend *et al.* 1962. d. Timell, 1964.

water to give a sugar (Bunton *et al.* 1955b; Semke *et al.* 1964; Timell, 1964). Details of alternative mechanisms are given in the references cited in this section. The acid catalysed hydrolysis of glycosides has been reviewed by BeMiller (1967).

O OR, H + H_3O^+ ⇌ H, O OR (+) + H_2O (fast)

H, O OR (+), H ⟶ O, + H ⟷ O+, H + ROH (slow, A-1)

O, + H + $2H_2O$ ⟶ O, H,OH + H_3O^+ (fast)

5.7. The Hydrolysis of Ethers

For the hydrolysis of diethyl ether in aqueous acid the volume of activation is –8·5 cm^3 $mole^{-1}$ and the entropy of activation is –9·0 e.u. (Koskikallio and Whalley, 1959b). Both these figures are consistent with the A-2 hydrolysis reaction given by equations (5.15)–(5.17).

$$(C_2H_5)_2O + H_3O^+ \rightleftharpoons (C_2H_5)_2\overset{+}{O}H + H_2O \qquad \text{(fast)} \tag{5.15}$$

$$(C_2H_5)_2\overset{+}{O}H + H_2O \rightleftharpoons C_2H_5OH + C_2H_5\overset{+}{O}H_2 \qquad \text{(slow, A-2)} \tag{5.16}$$

$$C_2H_5\overset{+}{O}H_2 + H_2O \rightleftharpoons C_2H_5OH + H_3O^+ \qquad \text{(fast)} \tag{5.17}$$

The variation of reaction rate (at 120° and 50 atm) shows a closer parallel with h_0 than with acid concentration. The Zucker–Hammett approach would indicate an A-1 mechanism and therefore the Zucker–Hammett hypothesis is not applicable in this case. However the Bunnett w parameter is +2·7 for the reaction (Bunnett, 1961a) and this is in accord with the above mechanism (Bunnett, 1961b). Jaques and Leisten (1964) have measured the rates of hydrolysis of diethyl ether in aqueous $HClO_4$ and H_2SO_4. They have suggested that in concentrated acid solution the A-1 mechanism becomes predominant. However other interpretations would also fit their data and further study is clearly desirable.

The acid catalysed hydrolyses of several aryl alkyl ethers also conform to the A-2 mechanism. For azoaryl ethers the w values are greater than +2·8 (Bunnett and Buncel, 1961; Bunnett *et al.* 1962). Schubert and Quacchia (1963) have studied the hydrolysis of 1-methoxy-3,5-dihydroxybenzene and 1,3,5-trimethoxybenzene in aqueous $HClO_4$. For the former $\Delta S^{\ddagger} = -15{\cdot}2$ e.u. in accord with a bimolecular rate determining step, although as this is a two proton hydrolysis the conclusion is less certain than it would be for a one proton process. Oxygen-18 tracer measurements with isotopically enriched water gave the normal abundance ^{18}O in the product methanol formed from the hydrolysis of 1,3,5-trimethoxybenzene. This proves that aryl–oxygen rather than alkyl–oxygen fission occurs.

The two proton mechanism for the hydrolysis of 1-methoxy-3,5-didihydroxybenzene is as follows

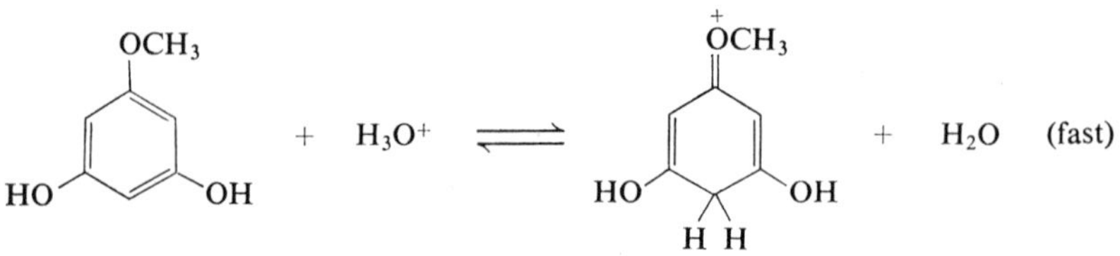

OCH$_3$ / HO OH (1-methoxy-3,5-dihydroxybenzene) + H_3O^+ $\rightleftharpoons$ $\overset{+}{O}CH_3$ / HO OH / H H + H_2O (fast)

$$\text{HO–C}_6\text{H}_2(\text{H}_2)(=\overset{+}{\text{O}}\text{CH}_3)\text{–OH} + H_3O^+ \rightleftharpoons \text{HO–C}_6\text{H}_2(\text{H}_2)(=\overset{+}{\text{O}}\text{CH}_3)\text{–}\overset{+}{\text{O}}\text{H}_2 + H_2O \quad \text{(fast)}$$

$$\text{HO–C}_6\text{H}_2(\text{H}_2)(=\overset{+}{\text{O}}\text{CH}_3)\text{–}\overset{+}{\text{O}}\text{H}_2 + H_2O \longrightarrow \text{HO–C}_6\text{H}_2(\text{H}_2)(H_2O^+)(\text{OCH}_3)\text{–}\overset{+}{\text{O}}\text{H}_2 \quad \text{(slow, A-2)}$$

$$\text{HO–C}_6\text{H}_2(\text{H}_2)(H_2O^+)(\text{OCH}_3)\text{–}\overset{+}{\text{O}}\text{H}_2 + 2H_2O \longrightarrow \text{HO–C}_6\text{H}_3(\text{OH})\text{–OH} + CH_3OH + 2H_3O^+ \quad \text{(fast)}$$

Strictly the reaction rates for such a mechanism should be correlated with H_+ and not H_0. However H_+ has not been investigated in detail (see Section 3.3) and therefore H_0 has been used. This adds a further ambiguity to the interpretation of the acidity function correlations.

5.8. The Hydrolysis of Acetic Anhydride

On the basis of a linear correlation between $\log_{10} k_1$ and H_0 for the hydrolysis of acetic anhydride in aqueous HCl, $HClO_4$, H_2SO_4 and H_3PO_4 Gold and Hilton (1955a, 1955b) deduced that an $A_{AC}1$ mechanism was operating. The appropriate w factors are $-0{\cdot}4$ for $HClO_4$ and ca. 0 for high concentrations of HCl and H_2SO_4. These figures seemingly support the A-1 hypothesis. However the plots for the determination of w are curved (Bunnett, 1961a). Thus for dilute H_2SO_4 $w \sim +12$ and $w = +3{\cdot}9$ for $4{\cdot}6$ mole litre^{-1} $< C_{HCl} <$ $5{\cdot}6$ mole litre^{-1}. These two figures would be consistent with an A-2 mechanism on Bunnett's interpretation of w. Strong evidence in favour of the A-2 mechanism comes from the values of $\Delta S^{\ddagger} =$ ca. -20 e.u. and $\Delta V^{\ddagger} = -17{\cdot}1$ cm^3 mole^{-1} for the entropy and volume of activation respectively (Koskikallio *et al.* 1959). The Zucker–Hammett hypothesis is therefore probably invalid when applied

$$\begin{matrix} CH_3\text{–C}\overset{\nearrow \overset{+}{O}H}{\searrow} \\ \quad O \\ CH_3\text{–C}\overset{\nearrow}{\searrow} O \end{matrix} \longrightarrow \begin{matrix} CH_3 \\ | \\ C^+ \\ \| \\ {}^+OH \end{matrix} + CH_3CO_2^- \quad \text{(A-1)} \qquad (5.18)$$

to the hydrolysis of acetic anhydride. The possibility does arise that if the unimolecular reaction (5.18) is the slow step in the hydrolysis then this would be consistent with the observed $\Delta V^{\neq}$ and $\Delta S^{\neq}$. However Koskikallio *et al.* (1959) consider this unlikely. The Bunnett approach is ambiguous in this case and gives no indication of whether the unimolecular or the bimolecular mechanism is more likely.

5.9. The Depolymerization of Paraldehyde and Trioxane

The entropies of activation for the acid catalysed hydrolyses of paraldehyde and trioxane are ca. 20 e.u. (Skrabal *et al.* 1934) and 2·4 e.u. (Brice and Lindsay, 1960) respectively. The volumes of activation are 3·0 cm³ mole⁻¹ for paraldehyde and −1·8 cm³ mole⁻¹ for trioxane (Withey and Whalley, 1963a). If the reactions are catalysed by the acids HCl, $HClO_4$, H_2SO_4 and HNO_3 the reaction rates follow the Hammett acidity function (Paul, 1950; Bell *et al.* 1956; Bell and Brown, 1954; Brice and Lindsay, 1960; Hamer and Leslie, 1960). The slopes of the $\log_{10}k_1$ against $-H_0$ plots are ca. 1·2 for both polymers. The w parameters are therefore negative (Bunnett, 1961a). All these facts are consistent with the following A-1 mechanism for the depolymerization reactions (R = H, trioxane; R = CH_3, paraldehyde)

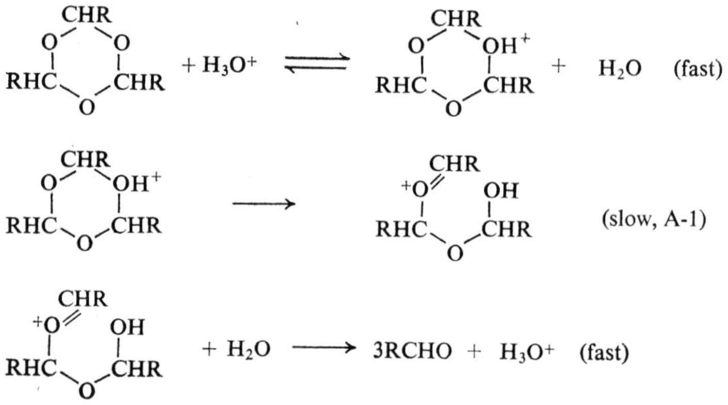

$$\text{(cyclic } (RCHO)_3\text{: CHR, O, O, RHC, CHR, O)} + H_3O^+ \rightleftharpoons \text{(ring with } OH^+\text{)} + H_2O \quad \text{(fast)}$$

$$\text{(ring with } OH^+\text{)} \longrightarrow \text{(opened: } {}^+O{=}CHR \ldots OH\text{, RHC–O–CHR)} \quad \text{(slow, A-1)}$$

$$\text{(opened cation)} + H_2O \longrightarrow 3RCHO + H_3O^+ \quad \text{(fast)}$$

Bell and Brown (1954) have measured the rates of depolymerization of paraldehyde catalysed by $KHSO_4$, CCl_3COOH and $CHCl_2COOH$. The rates increase more steeply with increasing acid concentration than would be expected from the linear correlation of $\log_{10}k_1$ with H_0 found for catalysis by strong acids. Similar results were obtained by Hamer and Leslie (1960) for the depolymerization of trioxane by CF_3COOH, CCl_3COOH and $CHCl_2COOH$. These are shown in Fig. 5.6. Bell and Brown considered that their results were caused by general acid catalysis by the acid molecules. However Long and Paul (1957) preferred an explanation which invoked a salting-in effect by the large acid molecules of the neutral indicators used to measure H_0. The discrepancy is then not that the reaction rates are too fast but

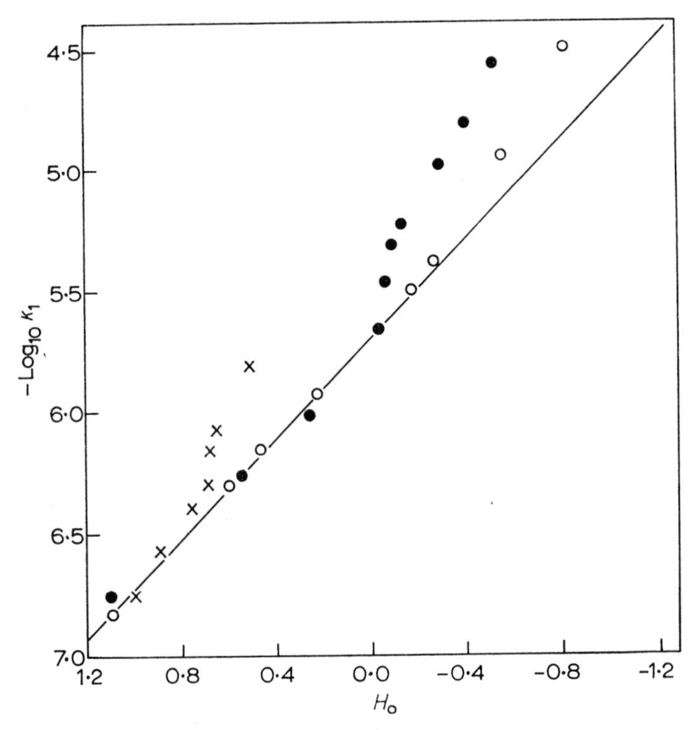

FIG. 5.6. Rates of depolymerization of trioxane at 25°. Catalysing acids were ○ = CF_3COOH, ● = CCl_3COOH, × = $CHCl_2COOH$, continuous line (slope 1·0) H_2SO_4. Reproduced by permission from Hamer and Leslie (1960).

rather that the acidity functions do not decrease sufficiently rapidly with increasing acid concentration. Hamer and Leslie (1960) also studied the hydrolysis of methylal catalysed by CF_3COOH, CCl_3COOH and $CHCl_2COOH$. The deviations of $\log_{10}k_1$ from a linear plot against H_0 were considerably smaller than those for the trioxane depolymerization reaction. Thus although Long and Paul's explanation is probably correct for methylal hydrolysis some other effect must be predominant for the depolymerization reactions. Hamer and Leslie favour Bell and Brown's conclusion that catalysis by undissociated acids occurs.

Paul (1950) has shown that the rate of decomposition of trioxane in aqueous $HClO_4$ is accelerated by the addition of sodium perchlorate. However the changes in rate parallel the corresponding changes in the Hammett acidity function for the solutions. The effect of adding salts (concentration 1 mole litre^{-1}) on the rate of depolymerization of trioxane in 4 mole litre^{-1} aqueous HCl was investigated by Brice and Lindsay (1960). The salts all increased the rate of reaction the relative magnitudes of the increase being LiCl > NaCl > KCl = NH_4Cl > C_5H_5NHCl > $N(CH_3)_4Cl$. In general the smaller the cations the greater the salt effect as would be expected. Equation (5.19) describes the

$$\log_{10} k_1 = -1{\cdot}16 H_0 + 0{\cdot}320 C_{\mathrm{LiCl}} - 6{\cdot}318 \quad (5.19)$$

effect of LiCl (concentration C_{LiCl}) on the rate of depolymerization of trioxane in HCl/LiCl mixtures at 40°. The value of H_0 is that for the solution of acid alone.

5.10. The Hydrolysis of Non-carboxylic Esters

5.10.1. *Sulphites*

The acid catalysed hydrolysis of several cyclic sulphites proceeds via sulphur–oxygen bond fission (Bunton *et al.* 1958a). For ethylene sulphite and trimethylene sulphite in 1 mole litre^{-1} $HClO_4$ the entropies of activation are −15·8 e.u. and −13·0 e.u. respectively (Bunton *et al.* 1958c). These values are indicative of a bimolecular A-2 mechanism (Schaleger and Long, 1963; Whalley, 1964). The reaction rates in aqueous acid do not correlate with either C_{acid} or H_0. Thus a plot (Fig. 5.7) of $\log_{10} k_1$ against $-H_0$ for the hydrolysis of ethylene sulphite in aqueous $HClO_4$ at 44·6° was linear (slope 0·68) over a limited range only. Despite the linearity at low acid concentrations the slope is too low to be acceptable as an indication of a unimolecular rate-determining step. The effect of adding $NaClO_4$ to reaction solutions is also shown on the graph. Although the acidity function decreases appreciably on addition of $NaClO_4$ (McIntyre and Long, 1954) the reaction rates are increased by a comparatively small amount. The conclusion is that the reaction rates approximately correlate with the acid concentration and not h_0. A bimolecular A-2 mechanism is therefore indicated. This indirect application of the Zucker–Hammett hypothesis is however rather unsatisfactory. In contrast the Bunnett approach is much more successful and gives a direct indication of the bimolecular mechanism. Thus for ethylene, trimethylene, tetramethylethylene, (−)-2,3-butylene and *meso*-2,3-butylene sulphites the interpolated w factors are 4·21, 3·93, 4·35, 3·47 and 4·19 respectively (Bunnett, 1961a). For tetramethylethylene sulphite the entropy of activation is −17·8 e.u. (Bunton *et al.* 1958b).

Bunton (1958c) also showed that the hydrolyses of ethylene sulphite and trimethylene sulphite are catalysed by added salts, particularly by NaCl and NaBr. The observed catalytic effects are in the order of the nucleophilic power of the anions of the salts, viz. $Br^- > Cl^- \gg HSO_4^-$ ($> ClO_4^-$). The mechanism therefore probably involves nucleophilic attack by the negative ion (X^-) on the sulphur atom of the protonated sulphite. The overall reaction mechanism is

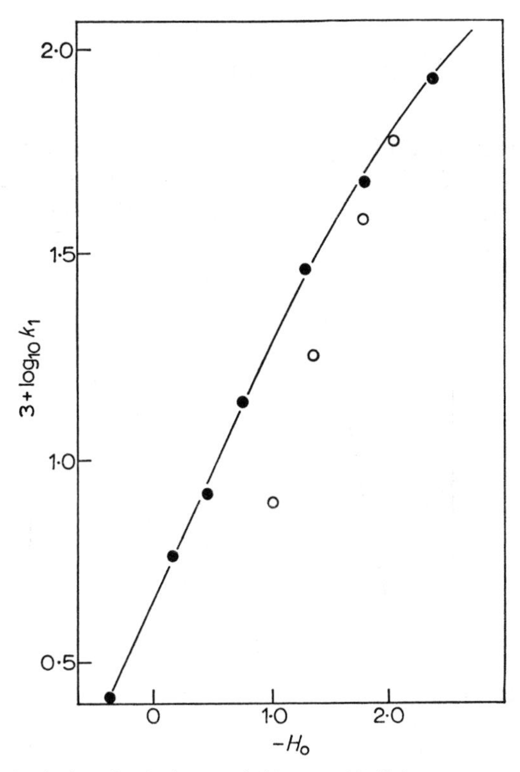

FIG. 5.7. Hydrolysis of ethylene sulphite at 44·6° in aqueous $HClO_4$ (●) and $HClO_4 + NaClO_4$(○). Reproduced by permission from Bunton *et al.* (1958c).

given by equations (5.20)–(5.24) where equations (5.21) and (5.22) involve H_2O as the nucleophile in the slow step and equations (5.23) and (5.24) involve X^- as the nucleophile.

$$(CH_2)_n\langle{}^{O}_{O}\rangle SO + H_3O^+ \rightleftharpoons (CH_2)_n\langle{}^{\overset{H}{O^+}}_{O}\rangle SO + H_2O \quad (5.20)$$

$$(CH_2)_n\langle{}^{\overset{H}{O^+}}_{O}\rangle SO + H_2O \longrightarrow (CH_2)_n\langle{}^{OSO_2H}_{OH} + H^+ \quad \text{(slow, A-2)} \quad (5.21)$$

$$(CH_2)_n\langle{}^{OSO_2H}_{OH} \longrightarrow (CH_2)_n\langle{}^{OH}_{OH} + SO_2 \quad (5.22)$$

$$(CH_2)_n\begin{matrix}\diagup \overset{H}{O^+} \diagdown \\ \diagdown O \diagup\end{matrix} SO + X^- \longrightarrow (CH_2)_n\begin{matrix}\diagup OSOX \\ \diagdown OH\end{matrix} \quad \text{(slow)} \tag{5.23}$$

$$(CH_2)_n\begin{matrix}\diagup OSOX \\ \diagdown OH\end{matrix} + H_2O \longrightarrow (CH_2)_n\begin{matrix}\diagup OH \\ \diagdown OH\end{matrix} + SO_2 + HX \tag{5.24}$$

The lack of direct correlation between $\log_{10}k_1$ and $\log_{10}C_{\text{acid}}$ in aqueous solutions of strong acids is predominantly caused by the concurrent catalysis via reaction (5.23). This confuses direct mechanistic diagnosis using acidity function methods.

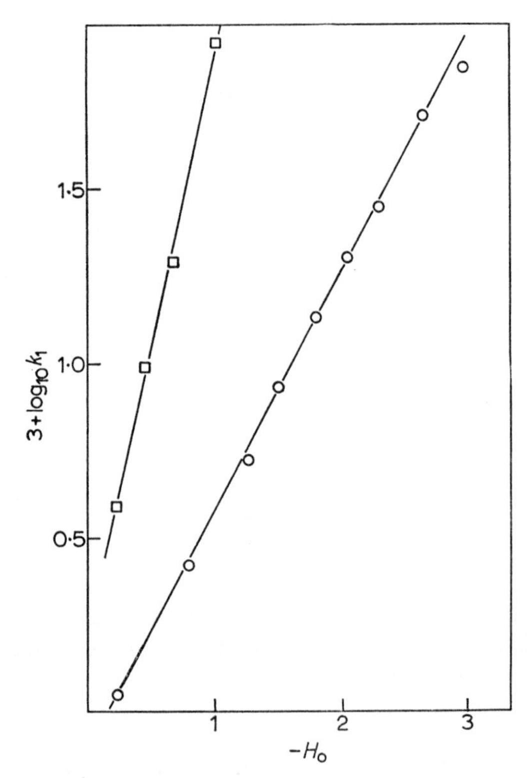

FIG. 5.8. The hydrolysis of diethyl sulphite at 0°. Catalysing acids are □ = HCl, ○ = $HClO_4$. Reproduced by permission from Bunton *et al.* (1959).

The hydrolyses of diethyl sulphite and dimethyl sulphite gave results similar to those for cyclic sulphites (Bunton *et al.* 1959). For diethyl sulphite in aqueous HCl and $HClO_4$ the plots of $\log_{10}k_1$ against $-H_0$ show two quite different lines (Fig. 5.8). For the $HClO_4$ catalysed reaction the graph has 0·69 slope. The deviation between the lines is consistent with the different salt effects of NaCl and $NaClO_4$ on the reaction rate (Fig. 5.9). The entropy of activation in 1 mole litre^{-1} $HClO_4$ is $-8{\cdot}2$ e.u. These results indicate a bimolecular A-2 hydrolysis reaction consisting of two concurrent slow steps in which water or

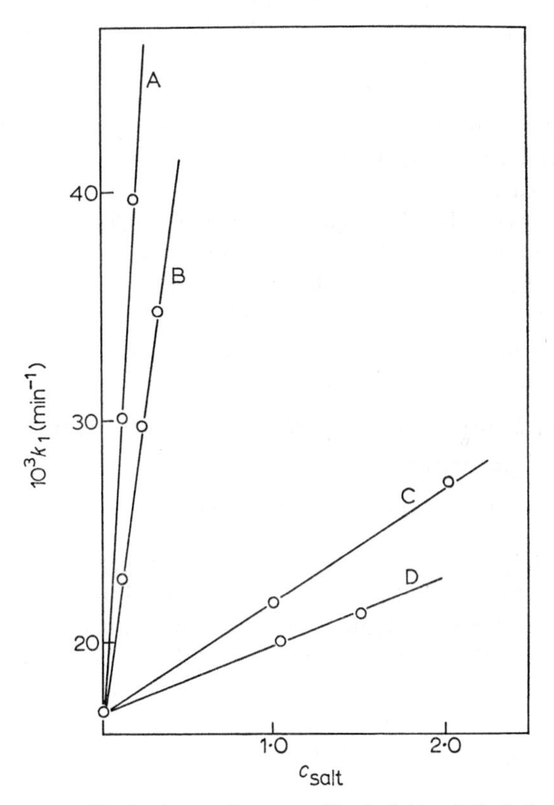

FIG. 5.9. Effect of added salts on the rate of hydrolysis of diethyl sulphite in 0·27 mole litre^{-1} $HClO_4$ at 35°. Salts were A, NaBr; B, NaCl; C, $NaClO_4$; D, $C_7H_7SO_3Na$. Reproduced by permission from Bunton *et al.* (1959).

the anion nucleophile derived from the strong acid or added salt compete for attack on the sulphur atom of the protonated sulphite. The two competing reactions are analogous to equations (5.21) and (5.23). The calculated w values are 2·15 for diethyl sulphite and 2·36 for dimethyl sulphite in $HClO_4$ at 0° (Bunnett, 1961a). These figures may be directly interpreted as indicating a straightforward A-2 hydrolysis in accord with equations (5.20)–(5.22) (Bunnett 1961d). However this seems to be an oversimplification of the true state of affairs.

5.10.2. *Phosphates*

The acid catalysed hydrolysis of α-D-glucose-1-phosphate proceeds via carbon–oxygen bond fission. The reaction rates parallel h_0 closer than the acid concentration (Barnard *et al.* 1955; Bunton *et al.* 1958f). The Zucker–Hammett approach would therefore indicate an A-1 mechanism. Strong support for this conclusion is provided by the entropy of activation $\Delta S^{\ddagger} = +14{\cdot}8$ e.u. and the volume of activation $\Delta V^{\ddagger} = 4{\cdot}3$ cm^3 mole^{-1} for the reaction (Osborn and Whalley, 1961). However the w factors are 3·34 for the hydrolysis in H_2SO_4 and 1·75 for $HClO_4$ (Bunnett 1961a). These values clearly conflict with the other data and therefore the Bunnett approach is apparently unreliable in this case.

The hydrolyses of t-butyl phosphate in aqueous $HClO_4$ also involves C—O bond fission, although P—O bond fission occurs in neutral solutions (Lapidot *et al.* 1964). The reaction rate follows the Hammett acidity function; the $\log_{10} k_1$ against $-H_0$ plot was linear with unit slope. This implies that w = ca. 0 and therefore both the Zucker–Hammett and the Bunnett criteria are in accord with the following unimolecular mechanism.

$$\mathrm{Bu^tOPO_3H_2 + H_3O^+ \rightleftharpoons [Bu^tOP(OH)_3]^+ + H_2O} \quad \text{(fast)}$$

$$\mathrm{[Bu^tOP(OH)_3]^+ \rightleftharpoons Bu^{t+} + H_3PO_4} \quad \text{(slow, A-1)}$$

$$\mathrm{Bu^{t+} + 2H_2O \rightarrow Bu^tOH + H_3O^+} \quad \text{(fast)}$$

The acid catalysed hydrolysis of methyl dihydrogen phosphate proceeds mainly by C—O fission but partly (27% in 5 mole litre^{-1} $HClO_4$) by P—O bond fission (Bunton *et al.* 1958e). The reaction rate is proportional to acid concentration and the reaction is therefore probably bimolecular. Formulations for the transition states are (I) for the attack of water on the carbon atom of the protonated phosphate (C—O bond fission) and (II) for the attack of water at the phosphorus atom (P—O bond fission).

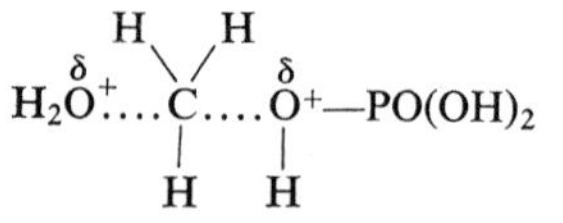

```
  H   H                                  O
δ+ \ /   δ                          δ+   ||    δ+
H2O....C....O+—PO(OH)2             H2O....P....OMe
       |    |                            / \    \
       H    H                          HO   OH   H

      (I)                                 (II)
```

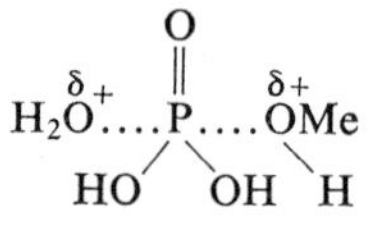

Although this reaction involves nucleophilic attack by water in the slow step $w^* = -0{\cdot}1$ (Bunnett, 1961a). This value is much higher than that required ($w^* < -2$) by the Bunnett criteria for water acting as a nucleophile. Bunnett (1961d) has invoked an unusually high solvation requirement for the transition state to explain this discrepancy. For dimethyl hydrogen phosphate (Bunton *et al.* 1960b) $w^* = -1{\cdot}9$, a figure which is almost in the range of the Bunnett criterion for water acting as a nucleophile. The better agreement may be ascribed to the replacement of one OH group by OMe in the transition state, the solvation requirement of which is thereby reduced (Bunnett, 1961d).

The rates of hydrolysis of 4-acetylphenyl and 4-nitrophenyl dihydrogen phosphates exhibit maxima in concentrated acid solutions (Barnard *et al.* 1966). Diphenyl and triphenyl phosphates behave similarly. Phosphorus–oxygen bond fission occurs. It is believed that the falling off in rate at high concentrations is the result of the sensitivity of water activity and the activity coefficients of the reactants and transition states to changes at high ionic strengths. Neopentyl dihydrogen phosphate also hydrolyses by acid catalysed P—O bond fission although no maximum in rate was observed for catalysis by up to $C_{HCl} = 8$ mole litre^{-1} (Bunton 1966). For this reaction $\log_{10} k_1$ was a linear function of $-H_0$. However the slope of 0·65 is too low for the linearity to be interpreted as indicating an A-1 mechanism on the Zucker–Hammett hypothesis. The w parameter is consistent with expectation for an A-2 reaction with water acting as a nucleophile. It is interesting that the slope of the $\log_{10} k_1$ against H_0 graph is nearly identical to that for the similar plots for the hydrolysis of organic sulphites in aqueous $HClO_4$. These undergo A-2 hydrolysis with concurrent hydrolysis involving nucleophilic attack by the anions of the catalysing acids (see previous Section 5.10.1). Nucleophilic attack by halide ions on the conjugate acid of methyl dihydrogen phosphate also contributes to the hydrolysis rates of that ester (Bunton *et al.* 1966).

5.10.3. *Fluorides*

For the acid catalysed ($HClO_4$, 20·5°) hydrolysis of 2-bromo-2-deoxy-β-arabinosyl fluoride and β-arabinosyl fluoride $\log_{10} k_1$ is a linear function (slope ca. 1) of $-H_0$ (Kent and Barnett, 1964). This suggests an A-1 mechanism. The entropies of activation are $-1{\cdot}1$ and $+0{\cdot}45$ e.u. respectively. These are rather low, but are nevertheless still consistent with an A-1 hydrolysis. The low values for ΔS^{+} may be associated with high hydration of the transition states caused by the presence of a strongly hydrogen-bonding fluoride moiety. An A-1 mechanism is also probable for the H_2SO_4 catalysed hydrolysis of benzylidyne trifluoride for which k_1 is proportional to h_0 (Coombes *et al.* 1969).

5.10.4. *Isopentyl Nitrate*

Isopentyl alcohol is almost quantitatively esterified to its nitrate ester by a mixture of nitric acid and sulphuric acid at 0°. The reaction rate correlates with the J_0 acidity function (Section 4.2) (Bonner and Frizel, 1959a). The reverse hydrolysis reaction of isopentyl nitrate correlates with H_0 (Bonner and Frizel, 1959b). Thus in aqueous $HClO_4$ and H_2SO_4 as catalysing acids the slopes of the $\log_{10} k_1$ against $-H_0$ linear graphs were 0·92 and 0·85 respectively. Bonner and Frizel concluded by consideration of the Zucker–Hammett hypothesis that the rate determining step in the hydrolysis is a unimolecular decomposition of the protonated isopentyl nitrate.

5.11. The Hydrolysis of *N*-Cyano Cations

The equilibrium concentrations of the *N*-cyanoisoquinolinium ion (III) and its "pseudo base" (IV) in aqueous perchloric acid at 22° are given by

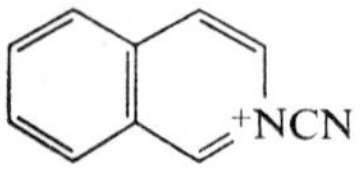

$\overset{+}{\mathrm{N}}\mathrm{CN}$ NCN H OH $\overset{+}{\mathrm{N}}.\mathrm{CONH_2}$

(III) 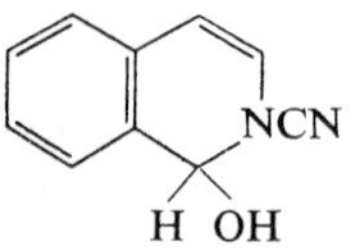 (IV) 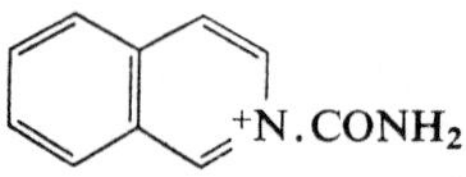 (V)

equation (5.25) (Huckings and Johnson, 1964). Hence pK_{ROH} = ca. −2·0 for

$$\log_{10}(C_{III}/C_{IV}) = -0{\cdot}81J_0 - 1{\cdot}62 \qquad (5.25)$$

the ion (III). A similar equation holds for the corresponding equilibrium for the *N*-cyanoquinolinium ion where ca. $-0{\cdot}9 = pK_{ROH}$. The rates of hydrolysis of both ions were found to correlate with H_0. The experimental rate constants showed maximum values at an acid concentration (both for $HClO_4$ and H_2SO_4) for which the J_0 value approximately equalled pK_{ROH} for the ions. This is because although the reaction rate follows H_0 the pre-equilibrium between (III) and (IV) is governed by J_0. J_0 changes faster than H_0 with increasing acid concentration and therefore the decrease in equilibrium concentration of (IV) (the reactive substrate in the hydrolysis) causes an overall decrease in the observed rate at high acid concentrations. The *N*-carbamoylisoquinolinium ion (V) is an isolatable intermediate in the hydrolysis of the *N*-cyanoisoquinolinium ion. The rates of hydrolysis in aqueous H_2SO_4 are faster than those in aqueous $HClO_4$ because catalysis by the HSO_4^- ion also occurs in the former.

5.12. The Hydrolysis of Hydroxylamine Sulphonates

The rates of hydrolysis of hydroxylamine trisulphonate $(SO_3)_2NOSO_3^{3-}$, hydroxylamine *N*,*O*-disulphonate $SO_3NHOSO_3^{2-}$, and hydroxylamine *N*,*N*-disulphonate $(SO_3)_2NOH^{2-}$ all approximately correlate with the H_0 acidity function. An exact parallel is even less likely than for neutral substrates since the reactants are anions and correlation with either H_{2-} or H_{3-} would be more appropriate. However a unimolecular mechanism is indicated for all three reactions. This is confirmed by the entropies of activation which are 11, 15 and 4 e.u. respectively (Candlin and Wilkins, 1961, 1965).

For hydroxylamine *N*-monosulphonate the observed rate constants were proportional to the acid concentration. This evidence for a bimolecular mechanism is supported by the entropy of activation $\Delta S^{+} = -12$ e.u. However for the Zwitterion hydroxylamine *O*-monosulphonate $NH_3{}^{+}OSO_3{}^{-}$ the graphs of $\log_{10} k_1$ against $-H_0$ were linear with approximately unit slope. Candlin and Wilkins (1965) deduced that the transition state did not contain a molecule of water. The rate of hydrolysis was slower in D_2O than in H_2O and this could imply a rate determining slow proton transfer. For the other hydroxylamine sulphonates $(k_{D_2O}/k_{H_2O}) > 2$ in accord with a rapid proton transfer pre-equilibrium. The entropy of activation is $\Delta S^{+} - 9$ e.u., a figure which is rather unusual for an A-1 reaction. However other examples of large negative entropies of activation for A-1 reactions have been cited (Salomaa and Kankaanperå, 1961; Noyce and Jorgenson, 1961).

5.13. The Hydration of Olefins

In dilute solutions the rate of hydration of olefins is proportional to acid concentration (Lucas and Eberz, 1934; Lucas and Liu, 1934). In more concentrated solutions the rate of hydration of isobutene was shown by Taft (1952) to be proportional to h_0. On the Zucker–Hammett hypothesis the transition state must contain only an isobutene molecule and a proton and no (apart from solvating) water molecule. The earlier measurements of Lucas and Eberz (1934) and Ciapetta and Kilpatrick (1948) were shown to be consistent with the H_0 correlation. Levy *et al.* (1953) studied the hydration of two isomeric olefins trimethylethylene and asym-methylethylethylene. At 50% reaction of each there was no appreciable conversion to the other isomer. Because protonation of either of the isomers leads to the same carbonium ion it follows that the general mechanism represented by equations (5.26) and (5.27) is not applicable to the hydration of olefins. Both reactions give t-amyl alcohol as product.

$$\text{olefin} + H_3O^{+} \rightleftharpoons \text{carbonium ion} + H_2O \quad \text{(fast)} \tag{5.26}$$

$$\text{carbonium ion} \rightarrow \text{products} \quad \text{(slow)} \tag{5.27}$$

Linear correlations of $\log_{10} k_1$ with $-H_0$ were also obtained for the hydration of trimethylethylene, methylenecyclobutane, and 2,2-dimethylbutene in aqueous HNO_3 (Taft *et al.* 1955). The proportionality factors (rate constants defined in terms of the dissolved concentration of gaseous olefin) were 0·98, 1·11 and 0·99 respectively (1·07 for isobutene). Entropies of activation are 0, +1 and −8 e.u. respectively. From the relative rates of hydration of 2-methyl-2-butene and 1-methyl-1-cyclopentene in H_2O and D_2O Purlee and Taft (1956) concluded that a rapid equilibrium proton transfer preceded the slow step of the hydration reactions. However this conclusion has been criticized by Long and Paul (1957). The mechanism proposed by Taft (1952) is as follows:

$$\text{>C=C<} + H_3O^+ \rightleftharpoons \left[\text{>C}\underset{\underset{H}{\vdots}}{=}\text{C<}\right]^+ + H_2O \quad \text{(fast)}$$

(π complex)

$$\pi \text{ complex} \longrightarrow -\underset{\underset{H}{|}}{C}-\overset{+}{C}- \quad \text{(slow)}$$

(carbonium ion)

$$\text{carbonium ion} + 2H_2O \longrightarrow \text{products} \quad \text{(fast)}$$

The pre-equilibrium is regarded as resulting in a π complex which rearranges to the carbonium ion in a rate determining unimolecular step. This is in accord with the observed small entropies of activation for the hydration reactions. Kwart and Weisfeld (1958) favour this mechanism whereas de la Mare *et al.* (1954) have suggested an alternative.

Long and Paul (1957) suggested that the experimental evidence was also consistent with a mechanism in which the slow step is a rate determining proton transfer to the olefin. Evidence that this is correct for the hydration of styrenes has been presented by Schubert *et al.* (1964). They consider that their results definitely exclude the Taft mechanism. The acidity dependence of the reaction rates for the hydration of styrenes lies roughly half way between the H_0 and the H_R' (Section 3.2.5) acidity functions. This has been attributed to the relatively small solvation requirements of the transition states in the proton transfer reaction. Evidence for the slow proton transfer include the appreciable solvent isotope effects ($k_{H_2O}/k_{D_2O} = \text{ca. } 2$) and one case of general acid catalysis. It is relevant to compare these results with those from a detailed study of the hydration of alkynyl ethers and thioethers (Drenth and Hogeveen, 1960; Hogeveen and Drenth, 1963a, 1963b; Stamhuis and Drenth, 1963a, 1963b). The reactions underwent general acid catalysis, gave $k_{H_2O} > k_{D_2O}$, and the entropies of activation were in the range $0 > \Delta S^{\ddagger} > -6$ e.u. For vinylthioethyne in aqueous $HClO_4$ at 25° a slope of 1·07 was deduced from a linear plot of $\log_{10} k_1$ against $-H_0$. The Bunnett w parameter was –0·9 for this reaction. Drenth and coworkers concluded that the hydration of alkynyl ethers and thioethers proceeds via a slow proton-transfer rate determining reaction. The addition of a water molecule is subsequent to the slow addition of a proton to the neutral ether. Equation (4.25) relates the Hammett acidity function to the observed rate constants for a slow proton transfer to the reacting substrate.

The values of the activity coefficient term determine the deviation from unity of the slope of a $\log_{10}k_1$ against $-H_0$ graph.

The mechanisms discussed above do not involve a molecule of water in the transition state of the reactions. Baliga and Whalley (1964) have suggested that a molecule of water must be included in the transition state during the $HClO_4$ catalysed hydration of propylene and isobutene. No indication of the structure of the transition state is given by the results and it may therefore be conveniently represented as [olefin.H^+.H_2O]‡. The arguments in favour of this are based on the measured entropies and volumes of activation for the reactions. These were $\Delta S^{\ddagger} = -5{\cdot}4$ e.u., $\Delta V^{\ddagger} = -9{\cdot}6$ cm^3 mole^{-1} for propylene, and $\Delta S^{\ddagger} = -7{\cdot}7$ e.u., $\Delta V^{\ddagger} = -11{\cdot}5$ cm^3 mole^{-1} for isobutene. These figures, particularly the volumes, are only consistent with a mechanism which includes water in the transition state (Whalley, 1966). For the hydration of ethylene $\Delta S^{\ddagger} = -5{\cdot}7$ e.u. and $\Delta V^{\ddagger} = -15{\cdot}5$ cm^3 mole^{-1} and therefore the same argument is also applicable (Baliga and Whalley, 1965).

Clearly the acidity function correlations and the volumes of activation give conflicting evidence as to the nature of the transition state in the hydration of olefins. However it is more informative to compare the rates of hydration with H'_R rather than with H_0. Plots of $\log_{10}k_1$ against H'_R for the hydration of styrenes have slopes around 0·6–0·7 (Schubert *et al.* 1964). Slopes of this magnitude for an H_0 plot are typical of bimolecular mechanisms in which water is a component of the transition state. It may be that for olefin hydration the present result is indicative of the involvement of water in the transition state. Comparison of the rates with H'_R is more logical since it gives information about the activity coefficient behaviour of the transition states in the slow hydration relative to that of the carbonium ions formed by equilibrium protonation of olefins. If the correlation between activity coefficient behaviour and solvation requirements is approximately valid (Bunnett, 1961c) then the indication is that the transition states are more heavily hydrated than the corresponding carbonium ions. That one molecule of water is actually reacting, and not merely hydrating, is plausible. The acidity function dependence of the reaction rates is therefore consistent with Whalley's proposals even though the Zucker–Hammett approach leads to a different conclusion. If these arguments are valid both the Taft (1952) and Long and Paul (1957) mechanisms which are discussed above are incorrect. The Bunnett (1961b) criteria for mechanism maintained a clear distinction between reacting substrates which are protonated on oxygen or nitrogen and those which are protonated on carbon. Insufficient data are available for carbon bases to allow satisfactory criteria for mechanism to be proposed. It would seem more appropriate to use the alkene or azulene (Section 3.2.5) acidity function scales as the standard for such criteria.

The rates of hydration of crotonaldehyde and β,β-dimethylacraldehyde are approximately proportional to the catalysing acid concentration (Winstein and Lucas, 1937; Lucas *et al.* 1944). Bell *et al.* (1962) have made more detailed studies of the hydration of crotonaldehyde and mesityl oxide. The latter compound was investigated more extensively. The rates of hydrolysis correlated with neither $-H_0$ or the acid concentration (Fig. 5.10). Furthermore

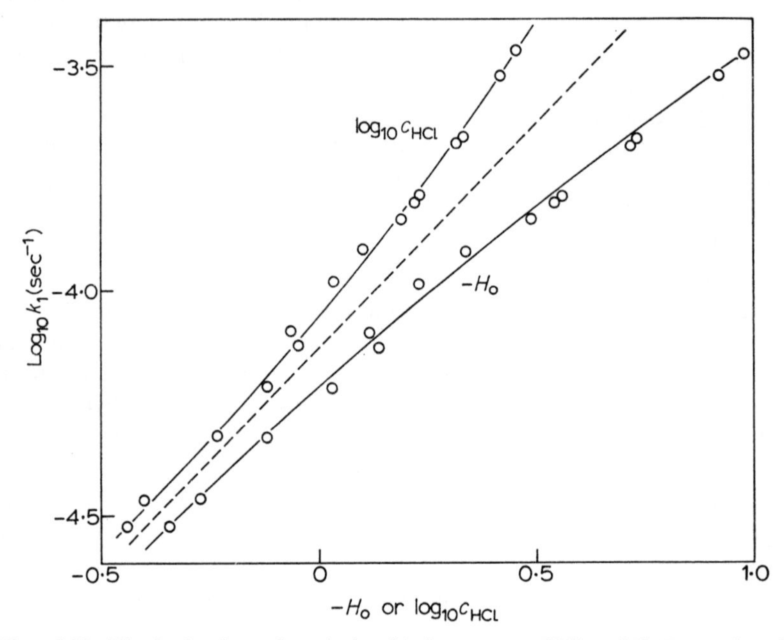

FIG. 5.10. The hydration of mesityl oxide in aqueous HCl at 25°. Reproduced by permission from Bell *et al.* (1962).

catalysis by different acids gave a series of separate $\log_{10} k_1$ against $-H_0$ plots (Fig. 5.11). Thus the reacton undergoes general acid catalysis. The results for catalysis by D_2SO_4/D_2O and DCl/D_2O included in Fig. 5.11 give an indication of the large isotope effect. The ratio $k_{H_2O}/k_{D_2O} = 3{\cdot}5$ for HCl and 4·0 for H_2SO_4. The following mechanism (Noyce and Reed, 1958) is consistent with a large isotope effect, general acid catalysis and the observed acidity function dependence.

$$\mathrm{Me_2C{=}CHCOMe + H_2O \rightleftharpoons HOCMe_2CH{=}CMeOH} \quad \text{(fast)} \quad (5.27a)$$

$$\mathrm{HOCMe_2CH{=}CMeOH + Acid \rightleftharpoons HOCMe_2CH_2CMe{=}\overset{+}{O}H + Base} \quad \text{(slow)} \quad (5.28)$$

$$\mathrm{HOCMe_2CH_2CMe{=}\overset{+}{O}H + Base \rightleftharpoons HOCMe_2CH_2COMe + Acid} \quad \text{(fast)} \quad (5.29)$$

The pre-equilibrium (5.27a) lies well over to the left. It is interesting to compare the present results with those for the hydration of styrenes, alkynyl ethers and alkynyl thioethers for which general acid catalysis and similarly large isotope effects have been found. The marked difference between the acidity function behaviour for the hydration of styrenes and mesityl oxide may be dependent on the same factors that determine the similar deviation between the H_R' and

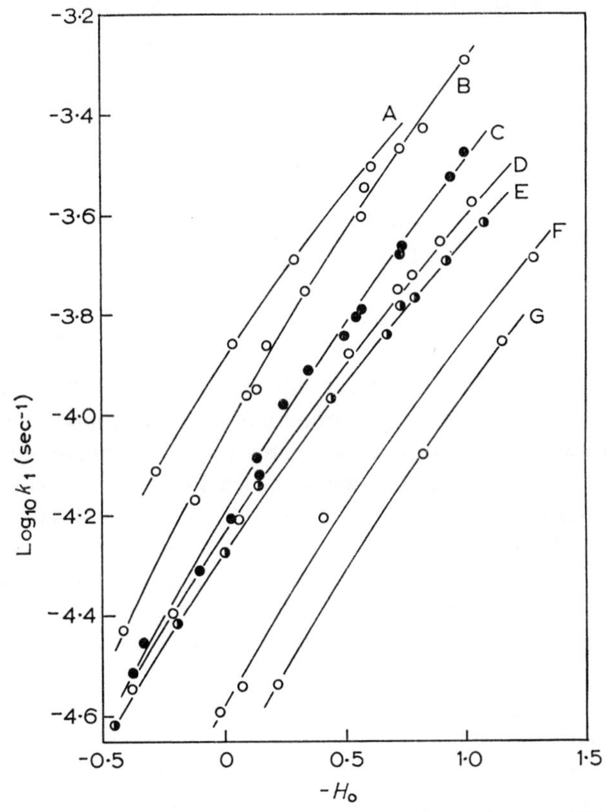

FIG. 5.11. The hydration of mesityl oxide in acid solutions at 25°. A, H_3PO_4; B, H_2SO_4; C, HCl; D, HNO_3; E, $HClO_4$; F, D_2SO_4; G, DCl. Reproduced by permission from Bell *et al.* (1962).

H_B acidity functions (Sections 3.24 and 3.25). Differences in mechanism must obviously be also considered.

It seems probable that the hydration of a simple olefin passes through a transition state which contains a molecule of water. The acidity function dependence of the reaction rates are consistent with this conclusion providing comparison is made with an acidity function scale defined by the protonation equilibria of carbon bases. The use of the Hammett H_0 acidity function is ambiguous and misleading when the rates of hydration of substrates which are carbon bases are being considered.

5.14. Reactions of Alcohols

In aqueous $HClO_4$ the rate of racemization of the optically active form of sec-butyl alcohol is twice the rate of oxygen exchange between the alcohol and the solvent water (Bunton *et al.* 1955a). This implies that inversion occurs during every substitution. Bunton and Llewellyn (1957) obtained a similar result for the reactions in aqueous H_2SO_4. Plots of $\log_{10}k_1$ against $-H_0$ were linear with slopes near unity. Together with the measured entropy of activation $\Delta S^{*} = 12$ e.u. for the racemization the results indicate an A-1 mechanism. The proposed reaction scheme is as follows.

$$\begin{matrix}Me\\Et\end{matrix}\!>\!CH\text{—}\overset{+}{O}H_2 \longrightarrow \begin{matrix}Me\\Et\end{matrix}\!>\!\overset{+}{C}H....OH_2 \qquad \text{(slow, A-1)}$$

$$\begin{matrix}Me\\Et\end{matrix}\!>\!\overset{+}{C}H....OH_2 + H_2{}^{18}O \longrightarrow H_2{}^{18}O^{+}\text{—}CH\!<\!\begin{matrix}Me\\Et\end{matrix} + H_2O \quad \text{(fast)} \qquad (5.30)$$

The occurrence of inversion at every substitution requires that the attacking water molecule is directed to the carbonium ion on the opposite side to that from which a water molecule is leaving. Thus the carbonium ion formed in the slow step cannot be completely freed of the leaving water molecule before the attacking water molecule has committed itself. This shielding of the point of nucleophilic substitution by the leaving water molecule is not always 100% efficient. Thus for 1-phenylethylalcohol in aqueous $HClO_4$ the rate of racemization is only 1·22 times the rate of oxygen exchange (Grunwald *et al.* 1957). Bunnett (1961d) has reasoned that inversion during every substitution is more consistent with an A-2 mechanism and quotes $w = 0{\cdot}8$ for sec-butyl alcohol oxygen exchange and $w = 1{\cdot}3$ for racemization as further evidence for this conclusion. However Withey and Whalley (1963b) have measured the volume of activation to be $-0{\cdot}7$ cm^3 mole^{-1}. This figure is considered to definitely indicate a unimolecular reaction (Whalley, 1964). At high acid concentrations the ratio of rates of racemization and oxygen exchange is less than two because an alternative mechanism becomes significant. This involves an elimination reaction which gives an olefinic intermediate. Addition of water to the olefin gives a racemic mixture of (—)- and (+)-sec-butyl alcohols. For the oxygen exchange reaction of t-butyl alcohol in aqueous H_2SO_4 $w = -2{\cdot}3$ (Bunnett, 1961a). This would suggest that the rate-determining step in the reaction is a slow unimolecular decomposition of the protonated alcohol.

The dehydration of alcohols is the reverse reaction of the hydration of olefins (Section 5.13). The discussion in the previous section is therefore also directly relevant here. The rates of dehydration of 4-(*p*-methoxyphenyl)-4-hydroxy-3-methyl-2-butanone (Noyce and Snyder, 1958) and 4-(*p*-methoxyphenyl)-4-hydroxy-2-butanone (Noyce and Reed, 1958) correlate with the H_0

acidity function. This suggests a carbonium ion mechanism, which for the latter compound may be written as follows (Ar = *p*-methoxyphenyl) (see footnote 21 in Noyce *et al.* 1962c).

$$\mathrm{Ar\underset{|}{\overset{OH}{C}}HCH_2\overset{\overset{O}{\|}}{C}CH_3 + H_3O^+ \rightleftharpoons Ar\underset{|}{\overset{\overset{+}{O}H_2}{C}}HCH_2\overset{\overset{O}{\|}}{C}CH_3 + H_2O} \quad \text{(fast)} \tag{5.31}$$

$$\mathrm{Ar\overset{\overset{+}{O}H_2}{C}HCH_2\overset{\overset{O}{\|}}{C}CH_3 \rightleftharpoons Ar\overset{+}{C}HCH_2\overset{\overset{O}{\|}}{C}CH_3 + H_2O} \quad \text{(fast)} \tag{5.32}$$

$$\mathrm{Ar\overset{+}{C}HCH_2\overset{\overset{O}{\|}}{C}CH_3 \rightarrow ArCH{=}CH\overset{\overset{O}{\|}}{C}CH_3 + H^+} \quad \text{(slow)} \tag{5.33}$$

The suggestion that the rates of a reaction of this mechanism might parallel the J_0 acidity function (Long and Paul, 1957) is reasonable but not apparently applicable in the present cases. In accord with the above mechanism the rates of dehydration of several 1,2-diarylethanols give linear plots (slope ca. 1·3) of $\log_{10}k_1$ against $-H_0$ (Noyce *et al.* 1968b; Noyce *et al.* 1968a).

In contrast to the above the rates of dehydration of β-phenyl-β-hydroxy-propiophenone (Noyce *et al.* 1959), 4-phenyl-4-hydroxy-2-butanone and 4-(*p*-nitrophenyl)-4-hydroxy-2-butanone (Noyce and Reed, 1958) increase less steeply with increasing acid concentration than a correlation with H_0 would require. The w values are 2·8, 3·2 and 3·4 respectively for the reactions in aqueous H_2SO_4. Furthermore Noyce and Reed (1958) have showed that the acidity dependence of the rate of dehydration of 4-phenyl-4-hydroxy-2-butanone parallels exactly the acidity dependence of the rate of enolization of acetophenone (Zucker and Hammett, 1939). The conclusion is that an enolization mechanism is operative for these three dehydration reactions. This may be represented by reactions (5.34)–(5.36) (Noyce and Snyder, 1958). In the

$$\mathrm{Ar\overset{\overset{OH}{|}}{C}HCH_2\overset{\overset{O}{\|}}{C}CH_3 + H_3O^+ \rightleftharpoons Ar\overset{\overset{OH}{|}}{C}H.CH_2\overset{\overset{\overset{+}{O}H}{\|}}{C}CH_3 + H_2O} \quad \text{(fast)} \tag{5.34}$$

$$\mathrm{Ar\overset{\overset{OH}{|}}{C}H.CH_2\overset{\overset{\overset{+}{O}H}{\|}}{C}CH_3 + H_2O \rightarrow Ar\overset{\overset{OH}{|}}{C}HCH{=}\overset{\overset{OH}{|}}{C}CH_3 + H_3O^+} \quad \text{(slow)} \tag{5.35}$$

$$\mathrm{Ar\overset{\overset{OH}{|}}{C}HCH{=}\overset{\overset{OH}{|}}{C}CH_3 \rightarrow ArCH{=}CH\overset{\overset{O}{\|}}{C}CH_3 + H_2O} \quad \text{(fast)} \tag{5.36}$$

slow step water acts as a proton transfer agent. On the Bunnett (1961b) criterion the values of w and w^* ($\approx w_A - 1{\cdot}6$) should be $w > 3{\cdot}3$ and $w^* > -2$. The observed w values given above are on the borderline of Bunnett's distinction between water acting as a nucleophile and as a proton transfer agent.

However in accord with the proposed mechanism $w^* > -2$ for all three dehydration reactions. The w^* values are nearly identical to those for the acid catalysed halogenation (enolization) of ketones (Bunnett, 1961a).

The dehydration of β-phenyl-β-hydroxypropionic acid involves a rate limiting loss of the α proton to give cinnamic acid as product (Noyce and Lane, 1962b). Thus the isotope effect for loss of the appropriate hydrogen (or deuterium in β-phenyl-β-hydroxypropionic-α-d_1 acid) atom is $k_H/k_D = 2{\cdot}9$. The following mechanism has been proposed (Noyce *et al.* 1962c). The rate of

$$\mathrm{Ph\underset{|}{\overset{OH}{C}}HCH_2COOH} + H_3O^+ \rightleftharpoons \mathrm{Ph\underset{|}{\overset{\overset{+}{O}H_2}{C}}HCH_2COOH} + H_2O \quad \text{(fast)} \tag{5.37}$$

$$\mathrm{Ph\underset{|}{\overset{\overset{+}{O}H_2}{C}}HCH_2COOH} \rightleftharpoons \mathrm{Ph\overset{+}{C}HCH_2COOH} + H_2O \quad \text{(fast)} \tag{5.38}$$

$$\mathrm{Ph\overset{+}{C}HCH_2COOH} \rightarrow \mathrm{PhCH{=}CHCOOH} + H^+ \quad \text{(slow)} \tag{5.39}$$

racemization of (+)-β-phenyl-β-hydroxypropionic acid is faster than the rate of dehydration (Noyce and Lane, 1962a). Also the slopes of the $\log_{10}k_1$ against $-H_0$ plots are 1·17 (at 25°) for racemization and 1·39 (at 25°) for dehydration. A slope of 0·98 for a plot of $\log_{10}k_1$ against $H_0 + \log_{10}a_w$ is obtained for the dehydration reaction. However the significance of this result is obscure. The acid catalysed dehydration reactions of β-(p-methoxyphenyl)-β-hydroxypropionic acid and β-(p-chlorophenyl)-β-hydroxypropionic acid have been studied (Noyce *et al.* 1962c). They also conform to the above mechanism. The $\log_{10}k_1$ against $-H_0$ graphs had slopes of 1·34 and 1·24 respectively. The rates do not correlate with J_0 acidity function although the carbonium ion mechanism could be supposed to require such a correlation (Long and Paul, 1957). The effect of substituents in the benzene ring on reaction rate are in accord with the Brown and Okamoto (1958) σ^+ values for the substituents. Slopes of ca. 1·1–1·4 for plots of $\log_{10}k_1$ against $-H_0$ are apparently typical of carbonium ion mechanisms (Noyce and Lane, 1962a). These are exemplified by reactions (5.31)–(5.33) and (5.37)–(5.39).

The rates of isomerization of substituted *cis*-cinnamic acids proceed via an addition elimination sequence which involves the corresponding β-aryl-β-hydroxypropionic acids as intermediates (Noyce *et al.* 1962b; Noyce and Avarbock, 1962; Noyce *et al.* 1962a). The rates of reaction follow the H_0 acidity function but are somewhat greater than the rates of dehydration of the β-aryl-β-hydroxypropionic acids. The isotope ratio $k_{H_2O}/k_{D_2O} = 6$ for the isomerization of *cis*-p-methoxycinnamic acid in 47% aqueous H_2SO_4 at 25°. The reaction rate is governed by a slow proton transfer to the substrate cinnamic acid molecule. Noyce *et al.* (1962a) have pointed out that correlations of reaction rate with the H_0 acidity function cannot be used as a test for or against a rate determining proton transfer step in a reaction.

Deno and Newman (1960) showed that the rates of esterification of several alcohols in aqueous H_2SO_4 correlate with the H_0 acidity function. Williams and Clark (1956) confirmed this result for the sulphation of 2,4-dinitrobenzyl alcohol. For this reaction at 0° $w = 0{\cdot}13$ and at 25° $w = 0{\cdot}31$ (Bunnett and Olsen, 1966b). The esterification of (+)-butan-2-ol by H_2SO_4 proceeds with retention of configuration (Deno and Newman, 1950) and *neo*pentyl alcohol is converted to *neo*pentyl hydrogen sulphate without molecular rearrangement (Whitmore and Rothrock, 1932). It follows that acyl–oxygen fission occurs. The esterification of 2,4-dinitrobenzyl alcohol is faster in D_2O/D_2SO_4 than in H_2O/H_2SO_4 indicating that a rapid pre-equilibrium proton transfer precedes the rate-determining step (Williams and Clark, 1956). A reaction sequence which satisfies the data involves the rapid reversible protonation of an alcohol–bisulphate complex followed by a slow decomposition of the protonated complex. However other mechanisms are possible (Williams and Clark, 1956) and the acceptance of a particular mechanism cannot be made without further study.

The oxidations of cyclohexanol and propan-2-ol by vanadium(V) in aqueous $HClO_4$ are catalysed by hydrogen bromide (Julian and Waters, 1962). The reaction rates are independent of substrate both for these two alcohols and also for crotonic and cumene-*p*-sulphonic acids. The variation of reaction rates with the acidity of the solutions (constant ionic strength = 5 mole litre^{-1}) is shown in Fig. 5.12. The slope of the line is ca. 2 suggesting that a double protonation precedes the slow step and that a water molecule is not involved in the transition state. The reaction is third order in bromide concentration. The rapid pre-equilibria (5.40)–(5.43) are considered to be set up and lead to the rate-determining step (5.44) in the reactions. This step is independent of the organic substrate which is being oxidized.

$$VO_2^+ + H_3O^+ \rightleftharpoons V(OH)_3^{2+} \tag{5.40}$$

$$V(OH)_3^{2+} + Br^- \rightleftharpoons V(OH)_3Br^+ \tag{5.41}$$

$$V(OH)_3Br^+ + H^+ \rightleftharpoons V(OH)_2Br^{2+} + H_2O \tag{5.42}$$

$$V(OH)_2Br^{2+} + Br^- \rightleftharpoons V(OH)_2Br_2^+ \tag{5.43}$$

$$V(OH)_2Br_2^+ + Br^- \rightarrow V^{IV}(OH)_2Br^+ + Br_2^- \tag{5.44}$$

The slow formation of the reactive radical-ion Br_2^- is followed by its rapid interaction with the organic substrate which gives readily oxidizable free radicals. Although the parallel of $\log_{10}k_1$ with $2H_0$ is consistent with this mechanism it would be more rigorous to correlate the rates with the sum $(H_+ + H_{2+})$. In the absence of detailed information about these functions the correlation shown in Fig. 5.12 must suffice.

Polyvinyl alcohol condenses with formaldehyde to form a cyclic acetal with a six-membered ring. The overall reaction is given by equation (5.45) and is acid catalysed. It is first order in the concentrations of formaldehyde and of hydroxyl groups in the alcohol. A plot of $\log_{10}k_1$ against $-H_0$ was linear with slope 1·07 for the reaction catalysed by aqueous H_2SO_4, $HClO_4$ and HCl

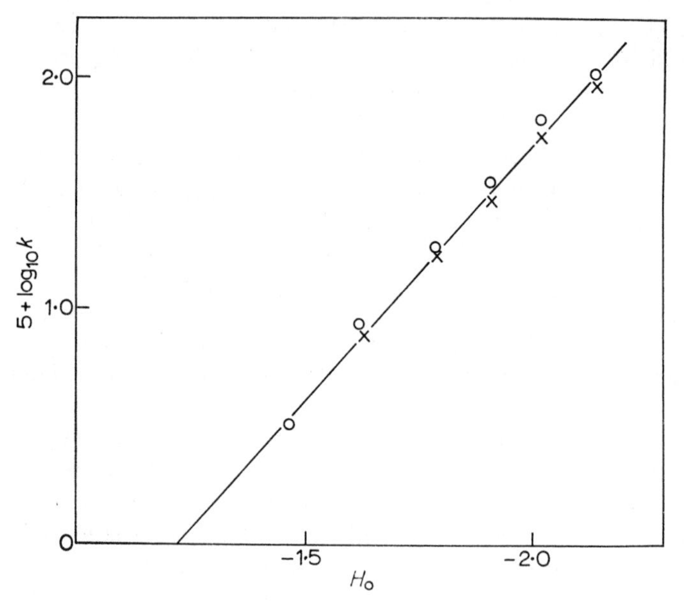

FIG. 5.12. Rates of oxidation of crotonic acid (×) and cyclohexanol (○) in HBr/-vanadium(V)/$HClO_4$ solutions. Initial concentrations of HBr and vanadium were: $c_V = 0{\cdot}045$ mole litre^{-1}, $c_{HBr} = 0{\cdot}16$ mole litre^{-1}.

(Ogata *et al.* 1956). It was concluded that the mechanism involves a rapid pre-equilibrium protonation of formaldehyde, the conjugate acid of which slowly condenses with polyvinyl alcohol in a bimolecular rate-determining step. Rapid cyclization and proton elimination follows. The rate of formation of *N*-t-butylacrylamide from t-butyl alcohol and acrylonitrile is similarly acid

$$\left(\begin{array}{c}\text{—CH—CH}_2\text{—CHCH}_2\text{—} \\ \;\;|\qquad\qquad\quad | \\ \text{OH}\qquad\quad\;\; \text{OH}\end{array}\right)_n + n\text{HCHO} \longrightarrow$$

$$\left(\begin{array}{c}\text{—CH—CH}_2\text{—CH—CH}_2\text{—} \\ \backslash\qquad\qquad | \\ \text{O—CH}_2\text{—O}\end{array}\right)_n + n\text{H}_2\text{O} \quad (5.45)$$

catalysed. The reaction rate is proportional to the t-butyl alcohol and acrylonitrile concentrations and to h_0. Deno *et al.* (1957) have deduced from the correlation with h_0 that the transition state is composed of acrylonitrile and the conjugate acid of t-butyl alcohol. It is unlikely that the transition state contains a t-butyl cation as a correlation of rate with j_0 would then be more appropriate. A correlation of reaction rate with the J_0 acidity function has been found by Gold and Riley (1962) for the t-butylation of anisole by t-butyl alcohol in aqueous $HClO_4$. This implies that the transition state in the reaction consists of a molecule of anisole and a t-butyl cation.

The acid catalysed rearrangement of α-phenylallyl alcohol to cinnamyl alcohol procedes with a volume of activation $\Delta V^{\neq} = -5{\cdot}0$ cm^3 mole^{-1} at 30° in 45·7 w/w % acetone (Harris and Weale, 1956). Whalley (1959) has concluded that the slow step is either a bimolecular attack of water on the protonated alcohol as represented by equation (5.46) or a unimolecular rearrangement according to equation (5.47).

$$\underset{\overset{|}{\underset{+}{OH_2}}}{PhCH}CH{=}CH_2 + H_2O \rightarrow PhCH{=}CHCH_2\overset{+}{O}H_2 + H_2O \qquad (5.46)$$

$$\underset{\overset{|}{\underset{+}{OH_2}}}{PhCH}CH{=}CH_2 \rightarrow PhCH{=}CHCH_2\overset{+}{O}H_2 \qquad (5.47)$$

However Kwart and Herbig (1963) have measured a w factor of –4·7 for this reaction in H_2SO_4 solutions. On Bunnett's criteria for mechanism this clearly eliminates the bimolecular slow step (5.46). Assuming that a w value may be interpreted as giving information about the solvation requirements of the transition state and the reactants in a reaction (Bunnett, 1961c) Kwart and Herbig argued that the most likely slow step consistent with the w value is reaction (5.48). The carbonium ion is regarded as having a much smaller

$$\underset{\overset{|}{\underset{+}{OH_2}}}{PhCH}CH{=}CH_2 \rightarrow PhCHCH\overset{+}{C}H_2 + H_2O \qquad (5.48)$$

extent of hydration than the oxonium ion. It is difficult to see why the two oxonium ions in equation (5.47) should have appreciably different solvation requirements and therefore the very negative w-factor apparently also eliminates this possibility. However the explanation of the negative $\Delta V^{\neq}$ if equation (5.48) is applicable seems uncertain.

The acid catalysed rearrangement of pinacol to pinacone is faster in D_2O/D_2SO_4 than in aqueous H_2SO_4 (Duncan and Lynn, 1957a). The mechanism probably involves a pre-equilibrium proton transfer to the substrate pinacol. The reaction is catalysed specifically by the H_3O^+ ion and not by other acids

(Duncan and Lynn, 1956c). Plots of $\log_{10}k_1$ against $-H_0$ were linear for the rearrangement catalysed by aqueous $HClO_4$, HNO_3, HCl and toluene 4-sulphonic acid solutions up to $C_{\text{acid}} = 1{\cdot}5$ mole litre^{-1} (Duncan and Lynn, 1956a). This has been confirmed for aqueous H_2SO_4 as the catalysing acid up to $C_{H_2SO_4} = 6{\cdot}3$ mole litre^{-1} and in the temperature range $72{\cdot}6° < t < 154°$ (Bunton *et al.* 1956, 1958d; Duncan and Lynn, 1956b). The slopes were less than unity, for example 0·89 at 72·9° and –0·77 at 100° (Bunton *et al.* 1956). However Deno and Perizzolo (1957a) showed that correlation of the reaction rates at 25° with H_0 at 25° gave $d\log_{10}k_1 = -dH_0$.

The simplest mechanism consistent with these observations is the following.

$$\begin{array}{c}\mathrm{Me_2C{-}CMe_2}\\ |\qquad|\\ \mathrm{HO\quad OH}\end{array} + \mathrm{H_3O^+} \rightleftharpoons \begin{array}{c}\mathrm{Me_2C{-}CMe_2}\\ |\qquad|\\ \mathrm{HO\quad \overset{}{OH_2}}\\ \qquad\ +\end{array} + \mathrm{H_2O}\quad \text{(fast)} \tag{5.49}$$

$$\begin{array}{c}\mathrm{Me_2C{-}CMe_2}\\ |\qquad|\\ \mathrm{OH\quad OH_2}\\ \qquad\ +\end{array} \rightleftharpoons \begin{array}{l}\qquad\quad\ \ +\\ \mathrm{Me_2C{-}CMe_2}\\ \quad\ |\\ \quad\mathrm{OH}\end{array} + \mathrm{H_2O} \tag{5.50}$$

$$\begin{array}{l}\qquad\quad\ \ +\\ \mathrm{Me_2C{-}CMe_2}\\ \quad\ |\\ \quad\mathrm{OH}\end{array} \rightarrow \begin{array}{l}\mathrm{MeC{-}CMe_3}\\ \quad\|\\ \quad\mathrm{OH}\\ \quad\ +\end{array} \tag{5.51}$$

$$\begin{array}{l}\mathrm{MeC{-}CMe_3}\\ \quad\|\\ \quad\mathrm{OH}\\ \quad\ +\end{array} + \mathrm{H_2O} \rightleftharpoons \begin{array}{l}\mathrm{MeC{-}CMe_3}\\ \quad\|\\ \quad\mathrm{O}\end{array} + \mathrm{H_3O^+}\quad \text{(fast)} \tag{5.52}$$

The question arises which of the reactions (5.50) or (5.51) is the slow step in the overall mechanism. An ^{18}O tracer study led to the conclusion that the rates of the backward reaction (5.50) and the forward reaction (5.51) are similar (Bunton *et al.* 1958d). As would be expected, the former (with a water molecule in the transition state) increases less rapidly than the latter (no water molecule involved) with increasing acid concentration. It was also concluded that reaction (5.51) was the rate-limiting step. Duncan and Llynn (1957b) agree with this conclusion although they also propose that subtle differences in mechanism may exist for the rearrangement below and above ca. 60°.

The H_0 correlation apparently leads to the contradictory conclusion that reaction (5.50) is rate determining (Deno and Perizzolo, 1957a, 1957b; Long and Paul, 1957). Thus whereas the rate of reaction (5.50) might be expected to parallel h_0 the rate of reaction (5.51), which involves the rearrangement of a carbonium ion, should show a greater tendency to parallel j_0. Bunnett (1961a) has listed w values for the pinacol rearrangement in aqueous H_2SO_4 which on his criterion for mechanism would suggest a bimolecular rate-determining step. The Bunnett criteria are therefore apparently not applicable here. This may arise because of the large difference between the temperature to which the H_0

scale refers and the temperatures appropriate to the measured rate constants. The relative magnitudes of the rate constants in steps (5.50) and (5.51) are such as to make it difficult to make any unambiguous deduction from the correlation between the measured rate constants and acidity functions.

The rates of rearrangement of 2-methylpropane-1,2-diol (Ley and Vernon, 1957a, 1957b) and 2-methylbutane-2,3-diol (Smith *et al.* 1959) also parallel h_0. The acidity function correlations and the results of several other measurements indicate a similar mechanism to that for pinacol itself. The pinacol rearrangement has been reviewed by Collins (1960).

5.15. Reactions of Aldehydes and Ketones

The rate of halogenation of ketones is independent of the halogen concentration (Lapworth, 1904) and for a particular ketone is the same for iodination and bromination. The reaction is catalysed by undissociated acid molecules (Dawson and Powis, 1913). The accepted mechanism involves a slow proton transfer from the α-carbon atom on the conjugate acid of the protonated ketone. The enol which is formed in the slow step rapidly reacts with halogen

$$>CHC{\lt}\!\!{=}O + H_3O^+ \rightleftharpoons >CHC{\lt}\!\!{=}\overset{+}{O}H + H_2O \quad \text{(fast)} \qquad (5.53)$$

$$>CHC{\lt}\!\!{=}\overset{+}{O}H + H_2O \longrightarrow >C{=}C{\lt}OH + H_3O^+ \quad \text{(slow)} \qquad (5.54)$$

to give products. The measured rate of halogenation is identical to the rate of enolization of the ketone. Since the slow step involves a water molecule the Zucker–Hammett hypothesis would require that the experimental rate constants should be proportional to the concentration of catalysing acid. Figure 4.2 shows the results of Zucker and Hammett (1939) for the iodination of acetophenone in aqueous $HClO_4$. Indeed Zucker and Hammett proposed their basic criteria for mechanism in part by consideration of the results for this reaction of well-established mechanism. The slope of the plot in Fig. 4.2 is greater than unity. Thus $w = 6{\cdot}35$, $w^* = -1{\cdot}62$ (Bunnett, 1961a) and $\phi = 0{\cdot}8$ (Bunnett and Olsen, 1966b). Again since this reaction was known to involve water as a proton transfer agent in the slow step (5.54) Bunnett (1961b) used these values of w and w^* as standards which together with data for many other reactions enabled the setting up of the proposals given in Table 4.2.

Satchell (1957) showed that the rate of iodination of acetone in aqueous HCl was approximately proportional to the HCl concentration. This is in accord with expectation for mechanism (5.53)–(5.54). Archer and Bell (1959) studied the bromination of acetone in aqueous HCl. They argued that if the extent of protonation of acetone via equilibrium (5.53) was taken into account then the reaction rate correlated much more satisfactorily with h_0 than with C_{HCl}. Clearly this indicates a breakdown of the Zucker–Hammett concepts. However Campbell and Edward (1960) have measured $pK_{BH^+} = -7{\cdot}2$ for acetone in water at 25° which is quite different from the figure of $-1{\cdot}58$ (Nagakura *et al.* 1957) used by Archer and Bell. The conclusion reached by

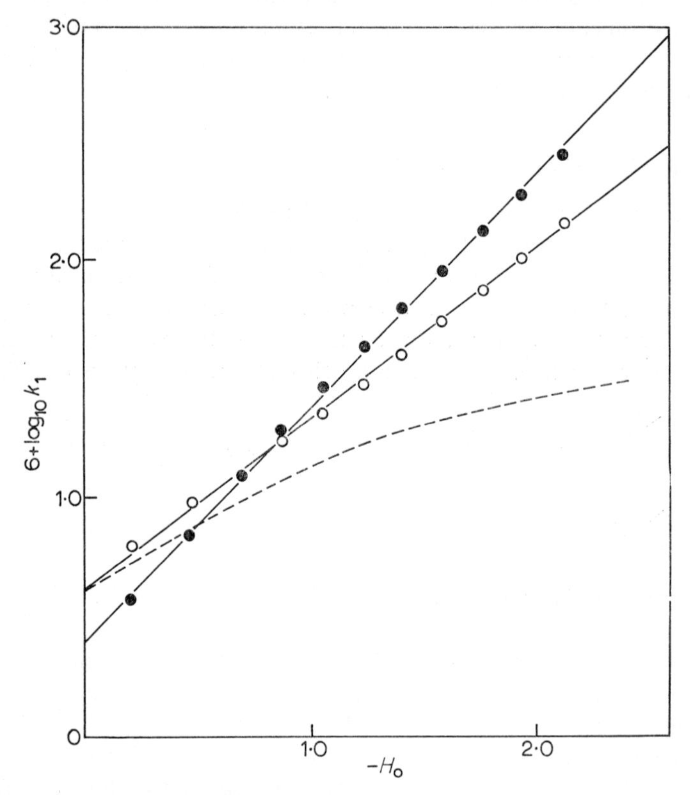

FIG. 5.13. Variation of rate constant with H_0 for the enolization of 1,2-cyclohexanedione (●) and for the reverse ketonization (○) in aqueous HCl at 25°. Dotted line is the theoretical curve for a proportionality between the rate of ketonization and $C_{H_3O^+}$. Reproduced by permission from Long and Bakule (1963).

Campbell and Edward was that Archer and Bell's interpretation was wrong and that the correlation of rate with acid concentration is acceptable. The values of w and w^* for this reaction are consistent with Bunnett's criteria for a reaction involving water as a proton transfer agent in the rate-determining step.

Swain and Rosenberg (1961) have studied the isomerization of D-α-phenylisocaprophenone in 85–94% H_2SO_4. They conclude that not only H_2O but also H_2SO_4 is capable of acting as a base in the slow proton transfer step of the reaction. This has been criticized by Long and Bakule (1963) on the grounds that the rates were correlated directly with the concentrations and not the activities of the reacting species. Medium effects were ignored.

The rates of conversion of the keto and enol forms of 1,2-cyclo-hexanedione to an equilibrium mixture correlate better with H_0 than with the catalyst acid concentration (Long and Bakule, 1963). Typical plots of $\log_{10}k_1$ against $-H_0$ are shown in Fig. 5.13; the slopes of the plots are given in Table 5.4 together

TABLE 5.4

Parameters for the ketonization (subscript f) and enolization (subscript r) reactions of 1,2-cyclohexanedione in concentrated aqueous acids at 25° (Long and Bakule, 1963)

Catalyst acid	$HClO_4$	H_2SO_4	HCl
$-d(\log_{10}k_f/dH_0)$	0·5	0·7	0·7
w_f	~8	~3	~5
w_f^*	−1·1		−3·7
$\Delta S_f^{\ddagger}$	−26		
$-d(\log_{10}k_r/dH_0)$	0·8	0·9	1·0
w_r	1·8	~0	0·0
w_r^*	~−3		~−5
$\Delta S_r^{\ddagger}$	−4		

with values for w, w^* and $\Delta S^{\ddagger}$. Long and Bakule interpreted the results in terms of a keto–enol transformation which is consistent with the accepted general mechanism given in equations (5.53) and (5.54). The difference between the acidity function behaviour for this reaction and that for the enolization of other ketones (for which $k_f \propto C_{H_3O^+}$) is ascribed to the presence of an extra molecule of water in the transition state of the former reaction. It is proposed that the transition state contains the conventional (base) water molecule (equation 5.54) and a monohydrate of the protonated ketone. Further evidence is that 1,2-cyclohexanedione exists predominantly as a monohydrate in water (Bakule and Long, 1963). The abnormal acidity function behaviour for the reaction does not fit the Zucker–Hammett approach which would give an incorrect indication of mechanism. The analysis of the w parameters in Table 5.4 would suggest that water acts as a nucleophile in the enolization reaction and as a proton transfer agent in the ketonization. This is apparently anomalous (Long and Bakule, 1963). The anomaly is probably obviated by a comparison of the ϕ values for the ketonization and enolization reactions with ϕ deduced for the keto–enol equilibrium reaction. The anomaly arises not because of different transition states in the reactions (clearly impossible) but because of the different medium effects for the enol and the keto forms of the substrate in concentrated acid solutions. This is analogous to the analysis presented by Bunnett and Olsen (1966b) for the reversible lactonization of γ-hydroxybutyric acid.

In concentrated acid solutions aromatic aldehydes decompose to carbon monoxide and the corresponding hydrocarbon. The decarbonylation reactions of 2,4,6-trimethyl, 2,4,6-triethyl and 2,4,6-triisopropylbenzaldehydes exhibit rate maxima at particular acidities of aqueous H_2SO_4 (Schubert and Zahler, 1954). At higher H_2SO_4 concentration the rate of decarbonylation decreases. Schubert and Burkett (1956) have proposed the following mechanism to account for the results.

$$\mathrm{ArCHO} + \mathrm{HA} \rightleftharpoons \mathrm{ArCH\overset{+}{O}H} + \mathrm{A} \quad \text{(fast)} \tag{5.55}$$

$$\mathrm{ArCHO} + \mathrm{HA} \rightleftharpoons \overset{+}{\mathrm{Ar}}\begin{smallmatrix}\diagup \mathrm{H} \\ \diagdown \mathrm{CHO}\end{smallmatrix} + \mathrm{A} \tag{5.56}$$

$$\overset{+}{\mathrm{Ar}}\begin{smallmatrix}\diagup \mathrm{H} \\ \diagdown \mathrm{CHO}\end{smallmatrix} + \mathrm{A} \longrightarrow \mathrm{ArH} + \mathrm{CO} + \mathrm{HA} \quad \text{(slow)} \tag{5.57}$$

The reactions are not specifically H_3O^+ catalysed and HA (charge unspecified) represents any general catalysing acid. The decrease in rate at high acidities is attributed to catalysis by other acids (H_2SO_4?) taking over from the H_3O^+ catalysis predominant at lower acid concentrations. This has been questioned by Long and Paul (1957). The formation of high equilibrium concentrations of the non-reactive conjugate acid $ArCHOH^+$ (equation 5.55) is regarded as also contributing to the levelling off and reduction in rate at high acid concentrations. The reaction is bimolecular with reaction (5.57) rate determining. For 2,4,6-trimethylbenzaldehyde in $75\% < C_{H_2SO_4} < 95\%$ the rates fitted a unimolecular rate law and therefore reaction (5.56) is becoming rate determining. The relative rates of decomposition of the protiated (ArCHO) and deuterated (ArCDO) aldehydes are consistent with the proposed mechanism.

The results for the decomposition of 2,4,6-trimethoxybenzaldehyde (Burkett *et al.* 1959) in aqueous HCl, HBr and $HClO_4$ showed rate maxima similar to those for the trialkylbenzaldehydes in aqueous H_2SO_4. General acid catalysis is observed and tends to a better correlation with C_{acid} than with h_0 in dilute solutions. This correlation suggests a bimolecular mechanism is operating and is therefore consistent with reaction (5.57) ($A = H_2O$) as the rate-determining step in the decarbonylation. Bunnett (1961a) was unable to deduce a value for w for this reaction because the data for different catalysing acids required the assumption of different values for pK_{SH^+} (S = ArCHO) in order to obtain satisfactory linear plots. This is obviously unacceptable. Decarbonylation reactions have been reviewed by Schubert and Kintner (1966).

The deacylation reaction of 2,6-dimethylacetophenone correlates with H_0 in $72{\cdot}5\% < {}_{H}C_{2SO_4} < 96\%$ providing the extent of equilibrium protonation of the reacting ketone is taken into account (Schubert and Latourette, 1952). The rate-determining step is therefore a slow unimolecular reaction of the protonated ketone. The interpolated $pK_{SH^+} = -7{\cdot}8$ for 2,6-dimethylacetophenone is reasonable (see Section 3.2.4). The rates of cyclodehydration of 2-*p*-toluidino- and 2-anilino-pent-2-en-4-one in aqueous H_2SO_4 similarly correlate with the Hammett acidity function (Bonner *et al.* 1955). Plots of k_1 against $(C_{H_2SO_4}/C_{HSO_4^-})$ were also linear confirming that equation (2.4) is applicable for $84{\cdot}5\% < C_{H_2SO_4} < 97{\cdot}3\%$. Although cleavage of a C—H bond is not the predominant feature of the slow step an isotope study has shown that some weakening of the C—H bond probably does occur (Bonner and Wilkins, 1955). Because the reacting anils are completely protonated at high H_2SO_4 concentrations the linear correlation between $\log_{10}k_1$ and $-H_0$ indicates that the slowly reacting species must be a diprotonated anil (Bonner and Barnard, 1958; Bonner, 1960). Thus correlation of $\log_{10}k_1$ with H_+ would be more appropriate. The present result is further evidence that H_+ and H_0 are approximately parallel functions of acid concentration.

Noyce and coworkers have made a detailed study of the acid catalysed isomerisation of *cis*-chalcones. The rate of isomerization of *cis*-chalcone (or *cis*-benzalacetophenone, $C_6H_5CH{=}CHCOC_6H_5$) does not follow the Hammett acidity function either for aqueous H_2SO_4 or $HClO_4$ (Noyce *et al.* 1959). The reaction is faster in D_2SO_4/D_2O than in H_2SO_4/H_2O (Noyce *et al.* 1961). Similar results were obtained for the isomerizations of *cis*-4-chlorochalcone and *cis*-4-nitrochalcone in aqueous H_2SO_4 (Noyce and Jorgenson, 1961). The accepted mechanism consistent with the experimental data is represented by equations (5.58) and (5.59) where Ar = C_6H_5, 4-Cl—C_6H_4 or 4-NO_2—C_6H_4.

$$\mathrm{(H)(Ar)C{=}C(H)C({=}O)C_6H_5} + H^+ \rightleftharpoons \mathrm{(H)(Ar)C{=}C(H)C({=}\overset{+}{O}H)C_6H_5} \quad \text{(fast)} \qquad (5.58)$$

$$\mathrm{(H)(Ar)C{=}C(H)C({=}\overset{+}{O}H)C_6H_5} + H_2O \longrightarrow \mathrm{(H)(Ar)C(\overset{+}{O}H_2){-}C(H){=}C(OH)C_6H_5} \quad \text{(slow)} \qquad (5.59)$$

For the isomerization of *cis*-4-methoxychalcone in 5% dioxane/95% water/ H_2SO_4 a graph of $\log_{10}k_1$ against $-H_0$ ($H_0 > -1{\cdot}5$) was accurately linear with slope 1·14 (Noyce and Jorgenson, 1961). The mechanism must be different from that for the other chalcones. The acidity function dependence and the observations that the reaction is faster in D_2O than H_2O and that no deuterium appears in the product *trans*-isomer is consistent with a mechanism in which the slow step is the unimolecular rotation about the C=C bond in the protonated substrate (reaction (5.60) where Ar = 4-MeO—C_6H_4). The dehydration of β-aryl-β-hydroxyketones similarly exhibits the two mechanisms

$$\underset{(cis)}{\text{H(Ar)C=CH–C}(=\overset{+}{\text{O}}\text{H})\text{C}_6\text{H}_5} \longrightarrow \underset{(trans)}{\text{H(Ar)C=CH–C}(=\overset{+}{\text{O}}\text{H})\text{C}_6\text{H}_5} \quad \text{(slow)} \qquad (5.60)$$

(5.31)–(5.33) and (5.34)–(5.36), the former being peculiar to 4-(*p*-methoxyphenyl)-4-hydroxy-2-butanone.

Noyce and Jorgenson (1963a) extended their study of the isomerization of *cis*-chalcones and showed that for some chalcones there was a change in mechanism with increasing acid concentration. The rate for *cis*-4-nitrochalcone is a maximum in 78% H_2SO_4. Independent measurements of the

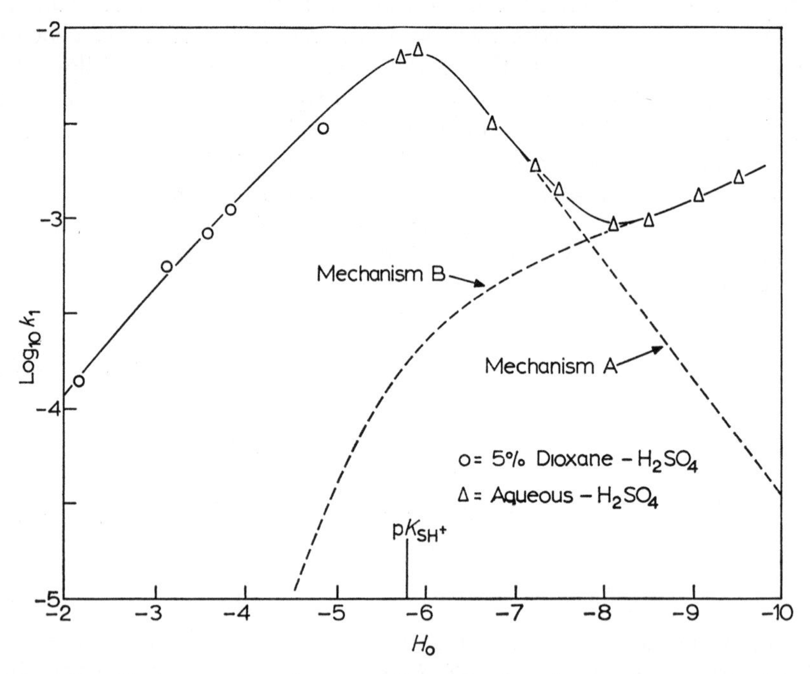

FIG. 5.14. Variation of reaction rate with acidity function for the isomerization of *cis*-chalcone in aqueous H_2SO_4. Reproduced by permission from Noyce and Jorgenson (1963a).

spectrum of *cis*-4-nitrochalcone in aqueous H_2SO_4 showed that in 70–80% H_2SO_4 the chalcone becomes appreciably protonated. Thus $pK_{BH^+} = -6{\cdot}75$ at 5°. The hundredfold decrease in rate between 78% and 98% H_2SO_4 is associated with the effect on the rate of step (5.59) of the decrease in water activity which occurs in this concentration range. The mechanism is therefore reactions (5.58) and (5.59). The variation of reaction rate with H_0 for *cis*-chalcone is shown in Fig. 5.14 in which mechanism A refers to reactions (5.58)

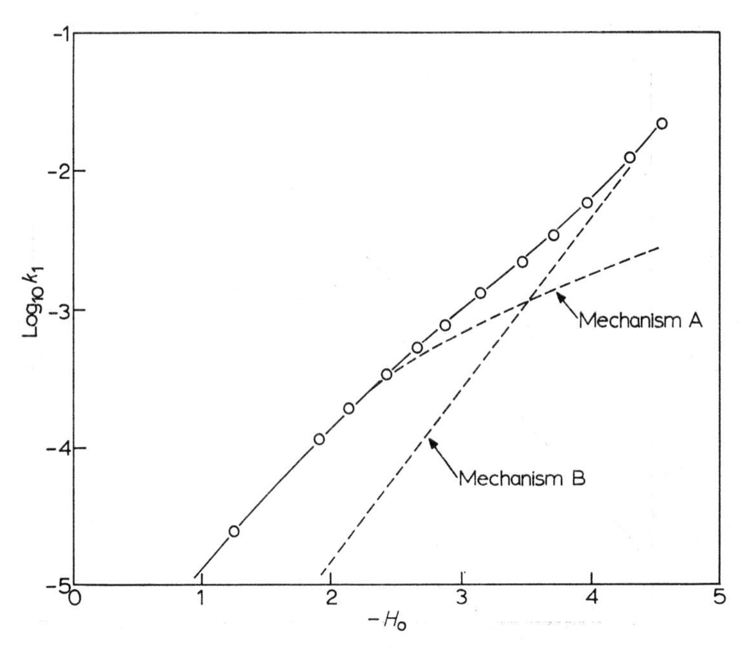

FIG. 5.15. Variation of reaction rate with H_0 for the isomerization of *cis*-2-methoxychalcone in 5% dioxane/95% water/H_2SO_4 at 25°. Reproduced by permission from Noyce and Jorgenson (1963b).

and (5.59) and mechanism B to reactions (5.58) and (5.60). The kinetics are in accord with mechanism A at the lower acid concentrations. This is reflected in the slope ($\ll 1$) of the $\log_{10} k_1$ against $-H_0$ plot, the maximum at 72% H_2SO_4 (cf. $pK_{SH^+} = -5{\cdot}87$) and the decrease in rate up to ca. 88% H_2SO_4. However above 88% H_2SO_4 a further increase in rate occurs and as indicated in the figure this is consistent with the incursion of the unimolecular mechanism B. *cis*-2-Methylchalcone shows a similar change in mechanism in concentrated H_2SO_4, solutions. The rates of isomerization of *cis*-2,4-dimethoxychalcone, *cis*-2,6-dimethoxychalcone, *cis*-2,6-dimethoxychalone and *cis*-2,4,6-tri-methoxychalcone were all in accord with mechanism B, although because of the great electronic accelerating effect of methoxy substituents the reactions could only be investigated for comparatively small acid concentrations.

Noyce and Jorgenson (1963b) have demonstrated the use of Bunnett's w factor as a test for change in mechanism. The variation of rate of isomerization of *cis*-2-methoxychalcone with H_0 is shown in Fig. 5.15. The dotted curves

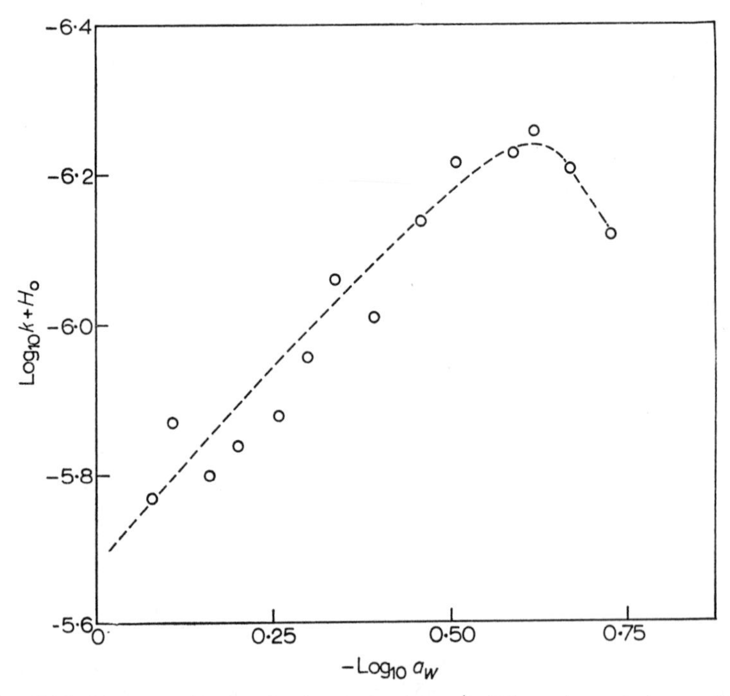

FIG. 5.16. Bunnett plot for the isomerization of *cis*-2-methoxychalcone. Reproduced by permission from Noyce and Jorgenson (1963b).

indicate the expected shape of the experimental curve for the two alternative mechanisms. A change of mechanism is apparently occurring in the range $-2{\cdot}5 > H_0 > 4$. Unfortunately the reaction becomes too fast for study at higher acidities. For *cis*-2-methoxy chalcone $pK_{SH^+} = -5{\cdot}33$ and therefore for the range of acidities investigated a plot of $(\log_{10}k_1 + H_0)$ against $\log_{10}a_w$ should enable the evaluation of w. The graph (Fig. 5.16) clearly demonstrates the mechanistic change. At low acid concentrations $w =$ ca. 1·1 (water acting as a nucleophile) in accord with slow step (5.54) of mechanism B. At the highest concentrations $w < 0$ suggesting (Table 4.2; Bunnett, 1961b) that a unimolecular step such as reaction (5.60) (mechanism A) is becoming predominant.

Some other reactions of aldehydes for which the rates have been correlated with H_0 are the chromic acid oxidation of benzaldehyde (Wiberg and Mill, 1958) and the condensation of formaldehyde with olefins (Hellin and Coussemant, 1957) and with *o*-xylene and 1,2,4-trimethylbenzene (Mironov *et al.* 1966).

5.16. Reactions of Carboxylic Acids

The acid catalysed decarboxylation of trialkylbenzoic acids (Schubert, 1949; Schubert *et al.* 1954) is thought to proceed via a slow bimolecular proton transfer from the protonated acid to a water (or other base) molecule. The reactions are analogous to the decarbonylation of trialkylaldehydes for which mechanism (5.55)–(5.57) has been suggested. An added complication is the formation of acylium ions $ArCO^+$ at high acid concentrations. The concentrations of the protonated acid and the acylium ions can sometimes be independently measured by ultraviolet spectroscopy. Thus, for example, if it is assumed that 2,4,6-trimethoxybenzoic acid is present as unprotonated substrate in solutions for which $H_0 < -3$ then the acidity function dependence of the rate of decarboxylation of this acid suggests a unimolecular mechanism for the reaction (Schubert *et al.* 1955). However, the observed ultraviolet spectra show that $pK_{SH^+} =$ ca. $-0{\cdot}9$ to $-1{\cdot}5$ for 2,4,6-trimethoxybenzoic acid which is therefore nearly completely converted to its conjugated acid when $H_0 = -3$. Taking this into account the kinetic results are more consistent with the bimolecular mechanism.

A plot of $\log_{10} k_1$ against $-H_0$ for the decarboxylation of 2,4,6-trimethylbenzoic acid had a slope of 0·85 at low acid concentrations (Long and Varker, quoted by Long and Paul, 1957). This gives $w =$ ca. $+0{\cdot}2$ which by Bunnett's criteria is intermediate between the requirement for the unimolecular and bimolecular mechanisms. A maximum rate occurs when $H_0 \approx pK_{SH^+} = -7{\cdot}3$ for 2,4,6-trimethylbenzoic acid (Schubert *et al.* 1954). The maximum and subsequent decrease in rate at high acidities is similar to results for the isomerization of chalcones (Noyce and Jorgenson, 1963a) discussed above. The formation of the acylium ion will also contribute to the rate decrease at high acidities. Schubert's explanation of the decrease has been criticized by Long and Paul (1957). The decarboxylation of azulene-1-carboxylic acid is acid catalysed at acid concentrations in the range 10^{-3} mole litre^{-1} $< C_{H^+} < 0{\cdot}03$ mole litre^{-1} (H_2SO_4 or $HClO_4$) the reaction rate is independent of acidity despite showing a small medium effect. Longridge and Long (1968) conclude that the mechanism of the reaction at the high acidities involves a rate-determining unimolecular decomposition of the species $^{+}$H—Ar—COO^{-}. A large decrease in reaction rate occurs at even higher acidities when appreciable equilibrium concentrations of the unreactive conjugate acid formed by 3-protonation of azulene-1-carboxylic acid exist in the solutions.

Anderson and Garbutt (1963) have found that $0{\cdot}60 \leqslant -d\log_{10}k_1/dH_0 \leqslant 0{\cdot}73$ for the decarboxylation of a series of uronic acids in aqueous H_2SO_4. In aqueous HCl the variation of rates gave w factors of ca. 3 although the Bunnett plots were not linear. This value would indicate that the slow step is bimolecular and that water is involved probably as a nucleophile. However Anderson and Garbutt argued that the entropies of activation ($-4{\cdot}6$ e.u. $\geqslant \Delta S^{+} \geqslant -17{\cdot}1$ e.u.; mean $-10{\cdot}6$ e.u.) were not sufficiently negative for an A-2 mechanism to be acceptable. Recent compilations (Schaleger and Long, 1963; Whalley, 1964) suggest that this might be a false assumption. The entropies seemingly eliminate an A-1 mechanism. It is not certain whether the reactions proceed by a slow protonation of the substrate (A-S_E2) as proposed by Anderson and Garbutt or by a rate-determining bimolecular reaction of water and protonated substrate as indicated by the Bunnett w factors.

In the original paper on acidity functions Hammett and Deyrup (1932) showed that $(\log_{10}k_1 + H_0) \approx$ constant for the condensation of 2-benzoylbenzoic acid to anthraquinone. This was emphasized by Deane (1945) who extended the data into oleum solutions up to 28·8% free SO_3. The correlation between $\log_{10}k_1$ and $-H_0$ was not very satisfactory. 2-Benzoylbenzoic acid gives a van't Hoff factor $i = 4$ in anhydrous sulphuric acid (Newman *et al.* 1945). This is in accord with Deane's (1945) proposal than an acylium ion is the reactive species in the cyclization. Newman (1942) has given a mechanism in which the slow step is the unimolecular cyclization of an acylium ion. However Long and Paul (1957) pointed out that a correlation of reaction rate with J_0 rather than H_0 might then be expected and concluded that further investigation was desirable. The rates of cyclization in aqueous P_2O_5 have been studied by Downing (1961).

The cyclization of 3,3-dimethyl-5-ketohexanoic acid to dimedone gives approximately linear plots for $\log_{10}k_1$ plotted against $-H_0$ and $\log_{10}C_{H_2SO_4}$ (Henshall *et al.* 1955). The slopes were ca. 0·4 and 5 respectively and therefore neither extreme of the Zucker–Hammett hypothesis is even approximately obeyed. The w factor is 1·61 (Bunnett and Olsen, 1966b) which would imply a bimolecular rate-determining step involving water as a nucleophile. Similar w values are found for the related cyclization reactions of a series of 6-alkyl substituted 3,3-dimethyl-5-keto-hexanoic acids (Silbermann and Henshall, 1957; Bunnett and Olsen, 1966b). Slopes of ca. 1 for 5,5′-diethyl-2,2′-diphenic acid, 1·34 for 2,2′-diphenic acid and 1·5 for 5,5′-dichloro-2,2′-diphenic acid were obtained from linear plots of $\log_{10}k_1$ against $-H_0$ for the cyclization reactions of these acids in aqueous H_2SO_4 (March and Henshall, 1962). A unimolecular slow step is probable and is consistent with the observed entropies of activation which were small and positive. It is proposed that both the conjugate acid and the acylium ion formed by ionization of the substrate diphenic acid in concentrated H_2SO_4 are intermediates in the reaction sequence.

Several studies of the acid catalysed decomposition of carboxylic acids to carbon monoxide were made before the advent of the acidity function concept. Elliott and Hammick (1951) deduced the slopes of the $\log_{10}k_1$ against $-H_0$ graphs for these earlier data. Although the slopes will be altered slightly owing to the more recent accurate determinations of the H_0 scale for concentrated sulphuric acid solutions (Fig. 2.1) the general conclusions to be drawn from the results are still the same. For formic acid (De Right, 1933) and malic acid (Whitford, 1925) the slopes were ca. 1 which combined with the expected equilibrium protonation (cf. equation (1.31); Table 1.4) of these substrates suggests that the slow step is the unimolecular decomposition of their conjugate acids. For citric acid (Wiig, 1930) the slope was 2 and a unimolecular decomposition of the diprotonated acid probably occurs. Benzoylformic acid (Elliott and Hammick, 1951) is similarly doubly protonated before decomposition. This has been confirmed by Vinnik *et al.* (1957, 1959) who obtained a slope of 1·8 for the $\log_{10}k_1$ against $-H_0$ graph. For both citric acid and benzoylformic acid the rate levels off at the highest acidities ($H_0 <$ ca. $-9{\cdot}8$). The cause of this is uncertain. The rates of decarbonylation of oxalic acid (Lichty, 1907) conform to slopes of 1 and 2 in different ranges of H_2SO_4 concentration. This reaction is apparently less straightforward than the reactions of the other acids which have been studied.

Dittmar (1929) measured the rates of decarbonylation of triphenylacetic acid catalysed by concentrated (> 93 %) aqueous H_2SO_4. The rates correlate satisfactorily with the J_0 acidity function and not with H_0 (Elliott and Hammick, 1951; Deno and Taft, 1954; Deno *et al.* 1955). The following mechanism is therefore plausible.

$$Ph_3CCOOH + H^+ \rightleftharpoons Ph_3CCO^+ + H_2O \quad \text{(fast)}$$

$$Ph_3CCO^+ \rightarrow Ph_3C^+ + CO \quad \text{(slow)}$$

Although several alkyl benzoates hydrolyse in aqueous acid by an A-2 mechanism the hydrolysis of methyl mesitoate is A-1 (Section 5.1). Bunton *et al.* (1960a) have proved the similar difference in mechanism between the oxygen exchange reactions of benzoic acid and mesitoic acid. The rates of reaction for the former (in aqueous H_2SO_4) paralleled the H_2SO_4 concentration (hence $w = 8{\cdot}8$; $\phi = 0{\cdot}94$) whereas for the latter (in 60 % dioxane/40 % water/$HClO_4$) a close correlation with h_0 was obtained. The entropies of activation were ca. -30 e.u. and 9 e.u. respectively. These suggest an A-2 mechanism for benzoic acid and an A-1 mechanism for mesitoic acid, conclusions which are consistent with the implications of the Zucker–Hammett hypothesis when applied to these reactions. A similar result is obtained for the hydration of fumaric acid to malic acid catalysed by HCl (Rozelle and Alberty, 1957). The reaction rate correlates satisfactorily with the acid concentration and the entropy of activation is ca. –24 e.u. A bimolecular slow step (A-2) is implied.

The rate of oxidation of mandelic acid by vanadium(V) is directly proportional to the concentration of catalysing acid (Shanker and Swami, 1963). The entropy of activation is –20 e.u. This indicates a rate-determining step in which the transition state contains a molecule of water. The reactive oxidant in both aqueous $HClO_4$ and H_2SO_4 is probably the $V(OH)_3^{2+}$ cation formed in accord with equilibrium (5.40). Similar results were obtained by Mehrotra (1968) for the oxidation of 2-hydroxy-2-methylpropanoic acid by vanadium(V). For the reaction at 40° catalysed by $HClO_4$ and H_2SO_4 the observed *w* factors were 4·2 and 5·3 respectively. Although Mehrotra concludes that $V(OH)_3^{2+}$ is the reactive species in aqueous $HClO_4$ he suggests that $(VO_2.H_2O.H_2SO_4)^+$ is important in aqueous H_2SO_4, However the entropies of activation were 1·5 e.u. (in $H_2SO_4 = 2$ mole litre^{-1}) and –1·0 e.u. (in $HClO_4 = 2$ mole litre^{-1}), figures which differ appreciably from that of –20 e.u. for the oxidation of mandelic acid.

5.17. Sulphonation Reactions

The rates of sulphonation of seven benzene derivatives in oleum solutions were shown by Brand and Horning (1952; Brand, 1950) to be consistent with equation (5.61) in which p_{SO_3} is the partial pressure of SO_3 (Brand and Rutherford, 1952). The proposed mechanism (5.62)–(5.64) involves the stepwise

$$\log_{10}k_1 = \log_{10}p_{SO_3} - H_0 + \text{constant} \tag{5.61}$$

addition of SO_3 and H^+ to the aromatic substrate (Brand *et al.* 1959).

$$ArH + SO_3 \rightleftharpoons Ar\begin{matrix}\diagup H \\ \diagdown SO_3\end{matrix} \quad \text{(fast)} \tag{5.62}$$

$$Ar\begin{matrix}\diagup H \\ \diagdown SO_3\end{matrix} + H^+ \rightleftharpoons {}^+Ar\begin{matrix}\diagup H \\ \diagdown SO_3H\end{matrix} \tag{5.63}$$

$${}^+Ar\begin{matrix}\diagup H \\ \diagdown SO_3H\end{matrix} \longrightarrow ArSO_3H, H^+ \tag{5.64}$$

An alternative mechanism suggested by Gold and Satchell (1956a) has been criticized (Brand *et al.* 1959; Long and Paul, 1957). Cerfontain (1961) has shown that the slopes of plots of $\log_{10}k_1$ against $\log_{10}p_{SO_3} - H_0$ (equation 5.61) are 1·24 for 3-toluene sulphonic acid, benzenesulphonic acid and 4-bromo phenyltrimethylammonium methyl sulphate and 1·25 for phenyltrimethylammonium methyl sulphate and 4-tolyltrimethylammonium methyl sulphate. For 4-nitrotoluene and nitrobenzene the slopes were 1·07 and 1·04 respectively. The lower values for the nitro compounds are either due to a different activity coefficient behaviour or to a specific interaction between $ArNO_2$ and SO_3 resulting in an unreactive complex. Cerfontain concludes that Brand's mechanism (5.62)–(5.64) is correct. An added complication can arise if the aromatic substrate is appreciably protonated in the acid concentrations necessary to make sulphonation proceed (Cerfontain, 1961; Kaandorp *et al.* 1962; Cerfontain *et al.* 1963). Appropriate correction must be made to the kinetic equations and requires knowledge of pK_{SH^+} where $S = ArH$.

In aqueous sulphuric acid solutions SO_3 is the most probable sulphonating species. Other possibilities have been discussed in some detail (Kaandorp *et al.* 1962). The suggested mechanism for aqueous solution is reaction (5.62) followed by

$$Ar\begin{matrix}H\\SO_3\end{matrix} + B \rightleftharpoons ArSO_3^- + BH^+ \quad (5.65)$$

$$ArSO_3^- + H^+ \rightleftharpoons ArSO_3H \quad (5.66)$$

Here B is a base (HSO_4^- or less likely H_2SO_4). The rate constant for the sulphonation of benzene in 98·8% aqueous H_2SO_4 is inversely proportional to added concentrations of $NaHSO_4$. For this reaction in a series of aqueous H_2SO_4 solutions a plot of $\log_{10}k_1$ against $-H_0$ was linear with a slope of ca. 2 (Kilpatrick *et al.* 1960). The acidity function dependence has played a part in the elucidation of the mechanism of sulphonation. However many other factors must be considered and a detailed knowledge of the species present in concentrated sulphuric acid solutions is desirable. The rate of sulphonation of benzene > ethylbenzene > isopropylbenzene > t-butylbenzene (Kaandorp *et al.* 1963). Clearly steric hindrance to 2-substitution is significant. For t-butylbenzene in $72{\cdot}4\% < C_{H_2SO_4} < 91{\cdot}0$ wt % aqueous H_2SO_4 dealkylation also occurs (Kaandorp *et al.* 1963). The rates of dealkylation approximately follow the J_0 acidity function. This is a case in which carbon protonation occurs and the acidity function behaviour will be significantly different from that for nitrogen or oxygen bases. It would seem from this result that for a reaction involving protonation of a carbon base in which water is not involved in the transition state the Bunnett w factor should be somewhat negative (see Table 4.2). For the chlorosulphonation of 4-phenylurethanesulphonic acid by chlorosulphonic acid in sulphuric acid $\log_{10}k_1$ approximately correlates with

$2H_0$. This has been taken to imply that a double protonation is necessary for the formation of the transition state (Palm, 1956, 1958).

5.18. Nitration and Denitration

The nitration of organic compounds in nitric acid/sulphuric acid solutions involves the nitronium ion NO_2^+ as the nitrating agent (Miller *et al.* 1964). The evidence for this has already been presented in Section 4.2 above. Raman spectra of HNO_3 in aqueous H_2SO_4 confirm that whereas molecular HNO_3 is the predominant species in 85% H_2SO_4, above 92% H_2SO_4 over 90% of the HNO_3 is ionized to NO_2^+ (Deno *et al.* 1961; Bonner and Williams, 1951; Bayliss and Watts, 1963). The mixed anhydride O_2NOSO_3H probably also exists in the most concentrated H_2SO_4 solutions.

The observation (Westheimer and Kharasch, 1946) that $\log_{10}(C_{R^+}/C_{ROH})$ for the ionization of triarylcarbinols and $\log_{10}k_1$ for the nitration of several aromatic nitro compounds were parallel functions of sulphuric acid concentration led to the definition of the J_0 acidity function (Lowen *et al.* 1950; Williams and Lowen, 1950; Gold and Hawes, 1951). The relationship between $\log_{10}k_1$ and J_0 is given by equation (4.34). For example, Lowen *et al.* (1950) obtained a linear graph of slope 1·20 from a plot of $\log_{10}k_1$ against $-J_0$ for the nitration reactions of the ions $4\text{-}MeC_6H_4NMe_3^+$ and $C_6H_5NMe_3^+$.

Bonner and Frizel (1959) suggested that it would be more appropriate to correlate $\log_{10}k_1$ for nitration reactions with $\log_{10}(C_{NO^+}/C_{HNO_2})$ for equilibrium (5.67) in which HA represents the strong acid (usually H_2SO_4).

$$HNO_2 + 2HA \rightleftharpoons NO^+ + H_3O^+ + 2A^- \qquad (5.67)$$

Fig. 5.17 shows the test of this proposal for the nitration of isopentyl alcohol in aqueous $HNO_3/HClO_4$ mixtures. The slopes of the plots are 0·27 for A and C, and 0·31 for B. The nitrous acid ionization (Singer and Vamplew, 1956) gives a slightly better agreement with the rates of nitration. A close parallel between the ionization behaviour of HNO_3 and HNO_2 in aqueous $HClO_4$ is apparent. However the general conclusion that $\log_{10}k_1$ approximately correlates with J_0 is sufficiently well characterized to be acceptable confirmation that NO_2^+ is the nitrating species. More recent measurements of $\log_{10}(C_{NO^+}/C_{HNO_2})$ in aqueous $HClO_4$ has shown that this logarithm in fact exactly parallels the J_0 acidity function (Bayliss *et al.* 1963). The exact parallel is however not applicable for the ionization of HNO_2 in aqueous H_2SO_4 solutions.

Coombes *et al.* (1968) have plotted $\log_{10}k_2$ (where k_2 is the second-order rate constant) against $-(J_0 + \log_{10}a_w)$ and have used the plots (which have ca. unit slope) as the basis of a comparison of the relative reactivity of a series of aromatic compounds. Up to 68% H_2SO_4 in water the relative rates of nitration of toluene and benzene were the same as for aqueous $HClO_4$ solutions and gave a measure of the relative reactivity of toluene and benzene. However for $C_{H_2SO_4} > 68\%$ the relative rates are not a reliable measure of reactivity. Thus both in aqueous H_2SO_4 and $HClO_4$ there is a concentration limit above which the introduction of activating groups into the aromatic species produces no increase in rate of nitration. The reactions studied were very fast and the

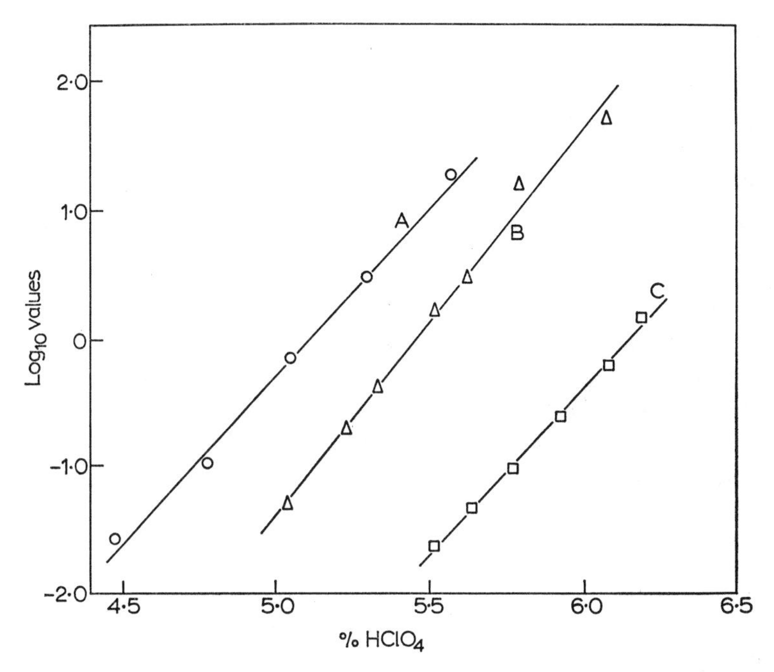

FIG. 5.17. Comparison of nitration rates and ionization ratios in $H_2O/HNO_3/$-$HClO_4$ mixtures. A: $\log_{10}(C_{NO^+}/C_{HNO_2})$; B: $\log_{10}(C_{R^+}/C_{ROH})$ (Indicator methyl 4-trimethylammoniumphenylmethyl ether perchlorate); C: $\log_{10} k_1$ (substrate isopentyl alcohol). (Bonner and Frizel, 1959.)

concentration limit occurs when the reaction rate is governed by the encounter of the aromatic molecules and the nitronium ions. Increasing the viscosity of the solvent magnifies this effect.

The rate constants for the nitration of many benzene and xylene derivatives exhibit a maximum at about 90 wt % H_2SO_4 in water. This has been attributed to a decrease in the activity coefficients y_S (equations 4.29 and 4.34) at high acid concentrations (Vinnik *et al.* 1967). However this cannot be the only explanation (Bonner and Brown, 1966). Maxima are also shown by the rate constants for the nitration of pyridines and pyridine 1-oxides in aqueous H_2SO_4 (Johnson *et al.* 1967a, 1967b, 1967c; Katritzky and Kingsland, 1968). Katritzky and coworkers have given an account of the implications of the "rate profiles" on the mechanism of the nitration reactions. In particular nitration of the substrates and of their conjugate acids is distinguished. In considering the variation of reaction rate with acid concentration or acidity function the extent of conversion of substrate S to SH^+ in a rapid pre-equilibrium must be allowed for. In this context the protonation of the pyridine 1-oxides follow the H_A acidity function (Section 3.2). A convenient test of whether nitration of a base or its conjugate acid is occurring is to use the Moodie–Schofield method in which $\log_{10}k_2$ is plotted against $-(J_0 + \log_{10}a_w)$ (Coombes *et al.* 1968; Gleghorn *et al.* 1963, 1966; Moodie *et al.* 1969). Slopes of about 1 are obtained for nitration of a conjugate acid cation. For nitration of a free base in equilibrium with its conjugate acid the slopes are appreciably less than unity. For the isoquinolinium, quinolinium, *N*-methylquinolinium and *N*-methylisoquinolinium ions (Moodie *et al.* 1963), 2-phenylpyridine 1-oxide and 2-phenylpyridine (Katritzky and Kingsland, 1968) the plots had slopes = ca. 1 showing that nitration of the conjugate acids was occurring. The 6-nitration of 3,5-dimethoxy-2-nitropyridine occurs on the free base (Johnson *et al.* 1967c). For cinnoline 2-oxide both the oxide and its conjugate acid undergo nitration (Gleghorn *et al.* 1968). Examples of Moodie–Schofield plots for a series of quaternary ammonium cations are shown in Fig. 5.18.

The effects of three non-electrolytes (Akand and Wyatt, 1967) and several salts (Surfleet and Wyatt, 1965; Bonner and Brown, 1966) on the rates of some nitration reactions in anhydrous H_2SO_4 have been investigated. Thus when $(NH_4)_2SO_4$ is added the rate of nitration of the trimethylphenylammonium ion is increased by up to ca. 25 times when $C_{(NH_4)_2SO_4} = 4$ mole litre^{-1} (Bonner and Brown, 1966). At higher concentrations of $(NH_4)_2SO_4$ the rate decreases. This is analogous to the increase in the rate of nitration of many aromatic compounds when water is added to anhydrous H_2SO_4 (Vinnik *et al.* 1967). There is a maximum rate at ca. 10 wt % water.

The mechanism of the denitration of nitroguanidines (P) to guanidines (G) in concentrated acid solutions is summarized by equation (5.68) (Simkins and Williams, 1952; Williams and Simkins, 1953; Bonner and Lockhart, 1958).

$$\mathrm{P} \underset{\text{Fast}}{\overset{\mathrm{H^+}}{\rightleftharpoons}} \mathrm{PH^+} \underset{\text{Fast}}{\overset{\mathrm{H^+}}{\rightleftharpoons}} \mathrm{PH_2}^{2+} \underset{\text{Slow}}{\rightleftharpoons} \mathrm{GH^+} + \mathrm{NO_2}^+ \qquad (5.68)$$

The rate of nitration of the guanidinium ion GH^+ in aqueous H_2SO_4 parallels the J_0 acidity function and therefore NO_2^+ is the nitrating species (Simkins and Williams, 1952; Williams and Simkins, 1953). For the reverse denitration reaction in strong acid $C_{PH_2^{2+}}$ will be very small and a nitroguanidine will exist as PH^+ in the solutions. It follows that for equation (5.68) the rates of denitration should parallel the H_+ acidity function (Section 3.3). Bonner and Lockhart (1958; Lockhart, 1966), in the absence of measurements of H_+, used H_0 to test this result. For the denitration in aqueous $HClO_4$ a plot of $\log_{10}k_1$

against $-H_0$ had a slope of 1·03, and the same result was obtained for aqueous methanesulphonic acid. P.m.r. spectra have confirmed the double protonation of nitroguanidines (Birchall and Gillespie, 1963; Lockhart, 1966). Protonation

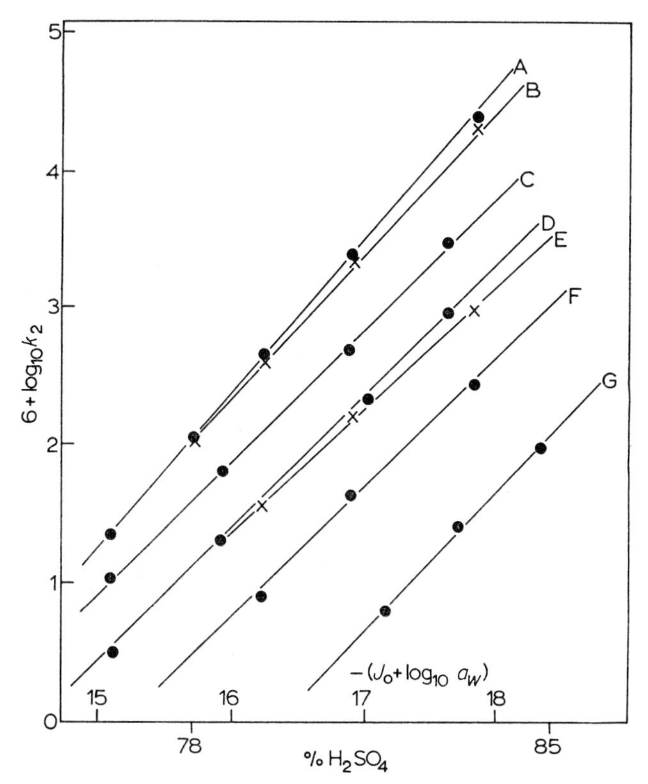

FIG. 5.18. Variation of rates of nitration with acidity (25°). Quaternary cations are A: 2-methylquinolinium; B: isoquinolinium; C: 2-hydroxyisoquinolinium; D: 2-methoxyisoquinolinium; E: quinolinium; F: 1-methylquinolinium; G: 1-hydroxyquinolinium. Reproduced by permission from Gleghorn *et al.* (1968).

occurs on the nitrimino and amino groups. For aqueous H_2SO_4 a slope of −1·3 is obtained for the plot of $\log_{10}k_1$ against H_0. No change in mechanism is envisaged but specific solvation of the transition state by H_3O^+ and HSO_4^- is thought to influence the reaction rates in H_2SO_4. The effect is quantitatively expressed by the equation

$$\log_{10}k_1 + H_0 - \log_{10}a_w - \log_{10}a_{H_2SO_4} = \text{constant}$$

(Hardy-Klein, 1957; Bonner and Lockhart, 1958; Lockhart, 1966).

5.19. Aromatic Hydrogen Exchange

Gold and Satchell (1955a, 1955b, 1955c, 1955d, 1956b; Satchell, 1956, 1958, 1959) studied the rates of isotope exchange of hydrogen on the aromatic nuclei of benzene, and several phenols and phenol ethers. The rates of deuterium or tritium loss from the aromatic nuclei increased more steeply than the concentrations of the catalysing acids. Linear relationships between $\log_{10} k_1$ and $-H_0$ with slopes approximating to 1 were observed for the exchange reactions in aqueous H_2SO_4, $HClO_4$, HCl, H_3PO_4, $KHSO_4$ and $CHCl_2COOH$. Gold and Satchell (1955a) applied the Zucker–Hammett hypothesis to this result and proposed an A-1 mechanism in which the rapid pre-equilibrium addition of a proton to the aromatic substrate was followed by a slow rate-determining rearrangement of the resulting intermediate. If this mechanism is correct aromatic H-exchange should undergo specific hydrogen ion catalysis. However a further mechanistic possibility consistent with the dependence of reaction rate on H_0 is an A-S_E2 mechanism (equations 4.23–4.25) in which the slow step is the electrophilic attack of the aromatic substrate by H^+ (Long and Paul, 1957; Melander and Myhre, 1959). An A-S_E2 reaction should be general acid catalysed.

Kresge and Chiang (1959) showed that the rate of tritium loss from 1,3,5-trimethoxybenzene-2-t in aqueous solution is proportional to the concentration of undissociated acetic acid in the solution. Similarly the rate of H-exchange of azulene-d_2(1,3) in aqueous solutions containing a fixed hydrogen ion concentration but varying formic acid concentrations was proportional to the concentration of undissociated formic acid (Colapietro and Long, 1960). General acid catalysis has therefore been established for H-exchange reactions. Thus the Brønsted relation (Bell, 1941) equation (5.69) is applicable for the detritiation of 1,3,5-trimethoxybenzene-2-t catalysed by the acids water, the

$$k_A = 4{\cdot}72 \times 10^{-2} (K_A)^{0{\cdot}518} \tag{5.69}$$

ammonium cation, the dihydrogen phosphate anion, acetic acid, formic acid, monofluoroacetic acid and the H_3O^+ ion (Kresge and Chiang, 1961; Kresge, 1965). The mechanism involves a rate-determining slow proton transfer and may be represented by equations (5.70) and (5.71) (Kresge and Chiang, 1959). The isotope effects on the acid catalysed hydrogen exchange of 1,3,5-trimethoxybenzene are consistent with this mechanism (Batts and Gold, 1964).

$$\mathrm{H'Ar + HA \rightleftharpoons H'HAr^+ + A^-} \tag{5.70}$$

$$\mathrm{H'HAr^+ + A^- \rightleftharpoons HAr + H'A} \tag{5.71}$$

General acid catalysis has also been established for the H-exchange of 1,3-dimethoxybenzene in moderately concentrated sulphuric acid solutions (Kresge *et al.* 1965a; Kresge *et al.* 1967). Figure 5.19 shows the rates of detritiation of 1,3-dimethoxybenzene-4-t as a function of acidity function (measured by the equilibrium protonation of azulene) for aqueous $HClO_4$ and H_2SO_4 as catalysing acids. The higher rate in H_2SO_4 over that in $HClO_4$ for a particular equilibrium acidity is consistent with general catalysis by HSO_4^- and H_2SO_4 in the former acid. This gives an overall rate which is greater than that for catalysis by H_3O^+ alone. The suggestion (Gold, 1961) that a fundamental difference exists between proton transfer from strong and weak acids is not substantiated by this result.

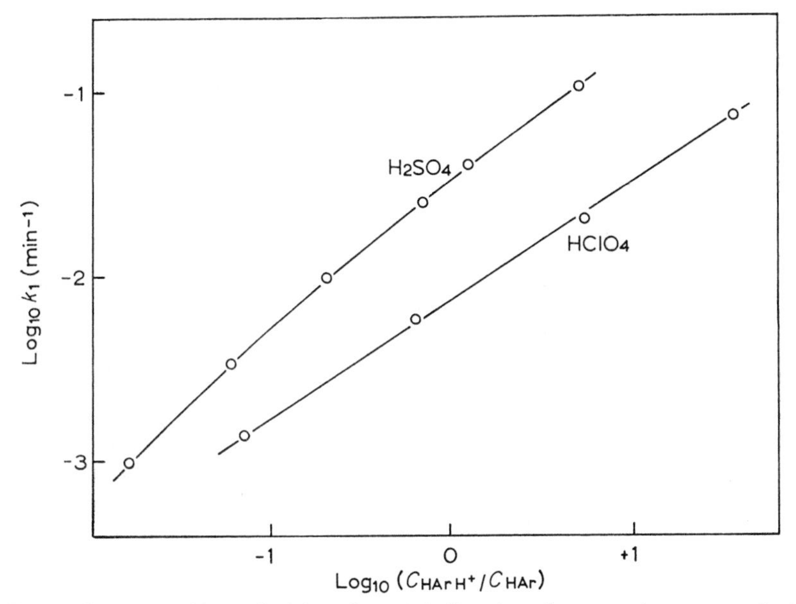

FIG. 5.19. Rates of loss of tritium from 1,3-dimethoxybenzene-4-t compared with the equilibrium protonation of azulene in aqueous $HClO_4$ and H_2SO_4 at 25°. Reproduced by permission from Kresge *et al.* (1965a).

The hydrogen exchange of azulenes is subject to general acid catalysis (Colapietro and Long, 1960; Longridge and Long, 1967; Gruen and Long, 1967). The detritiation of quasiazulene-3-t gave a linear Brønsted plot ($\alpha = 0{\cdot}54 \pm 0{\cdot}03$) for catalysing acids with acidities ranging over 16 pK_a units (H_2O to H_3O^+) (Thomas and Long, 1964a). It is interesting to note here that the linearity of Brønsted plots over wide ranges of acidity is unexpected (Albery, 1967). Thomas and Long (1964) also found that the Brønsted plots for the removal of tritium from azulene-1-t had different slopes α for different classes of catalysing acid. Thus $\alpha = 0{\cdot}61$ for carboxylic acids, $\alpha = 0{\cdot}88$ for dicarboxylic acid monoanions and $\alpha = 0{\cdot}67$ for anilinium ions. Equations (5.70) and (5.71) have been accepted for the mechanism of H-exchange of azulenes and detailed free energy profiles for the two stage A-S_E2 reaction have been considered (Schulze and Long, 1964; Challis and Long, 1965a). The detritiation of cyc[3.2.2]azine exhibits general acid catalysis and has an acidity function dependence similar to that for the corresponding reaction of azulene. A similar mechanism is implied (Thomas and Long, 1964b).

Consideration of equation (5.70) where H_3O^+ is the catalysing acid HA requires that water is involved as a proton transfer agent in the slow step of the A-S_E2 reaction. This mechanism is strongly supported by the observed general acid catalysis for both benzene and azulene nuclei. The approximate parallel between $\log_{10} k_1$ and $-H_0$ for both benzene and azulene derivatives leads to the conclusion that the Zucker–Hammett hypothesis is not applicable for these reactions. The H_0 correlations are further emphasized by the w factors which are usually negative for hydrogen isotope exhange (Bunnett, 1961a). On the Bunnett (1961b) criteria for the protonation of oxygen and nitrogen bases this would imply that no water molecule is involved in the slow step of the acid catalysed reaction. However this is clearly wrong and a distinction must be drawn between substrates which are protonated on carbon and those which are protonated on nitrogen or oxygen. Hence Bunnett's proposal (Table 4.2) that when $w = $ ca. 0 water is involved as a proton transfer agent in the rate-determining step for substrates which are carbon bases. This conclusion is supported by the results for the acid catalysed hydration of olefins (Section 5.13).

The difference in kinetic behaviour between carbon and nitrogen or oxygen bases is akin to the difference between the equilibrium protonation behaviour of olefines (Section 3.2.5) and amines or ketones. The latter is embodied in the deviations between the H_R' and the H_0, H_0''', H_B and H_A acidity functions (Fig. 3.2). Bunnett's approach is to deduce a w (or w^*) factor by comparison of reaction rates with the Hammett H_0 acidity function. It would be more appropriate to compare the rates for a particular substrate with the acidity function applicable to the equilibrium protonation of that substrate (or structurally similar molecules). Thus the relationship between the rates of reactions involving carbon protonation and the H_R' or azulene (Long and Schulze, 1964) acidity function should be similar to the relationship between the rates of reactions involving nitrogen protonation and the H_0 acidity function. This idea is embodied in Kresge's α-coefficient concept (Kresge *et al.* 1965b) which has already been discussed in Section 4.4 above. The w values (+3 and −2 respectively) for the H-exchange reactions of 1,3,5-trihydroxybenzene and 1,3,5-trimethoxybenzene does not indicate a different mechanism for the two reactions but merely reflects the different variations of (y_{HAr}/y^+) (equation 4.59) with increasing acid concentration.

The rates of protonation of azulene and deprotonation of the conjugate acid of azulene in aqueous $HClO_4$ have been measured by Challis and Long

(1965a, 1965b). Plots of $\log_{10}k_1$ (at 7·3°) against $-H_0$ had slopes of 1·26 for the protonation and −0·68 for the deprotonation reactions. Bunnett and Olsen (1966b) have used these results to illustrate the additive properties of their parameter ϕ (equation 3.22 and Section 4.3.2). The following reaction scheme (5.72) represents the reversible protonation of azulene (AzH) in aqueous solution. The ϕ_1 parameter is deduced using equation (3.22) and the experimental results for the equilibrium protonation of azulene. Use of equation

$$\begin{array}{ccc} \text{AzH} & \underset{}{\overset{\phi_1}{\rightleftharpoons}} & \text{AzH}_2^+ \\ \phi_2 \searrow & & \swarrow \phi_3 \\ & \text{transition state} & \end{array} \qquad (5.72)$$

(4.53) enables calculation of ϕ_2 and ϕ_3 from the rates of the protonation and deprotonation reactions respectively. Hence $\phi_1(5{\cdot}7°) = -1{\cdot}54$, and $\phi_2 = -0{\cdot}42$ and $\phi_3 = 1{\cdot}11$ both at 7·3°. The additivity relationship (5.73) is therefore obeyed.

$$\phi_1 = \phi_2 - \phi_3 \qquad (5.73)$$

The rates of H-exchange of 2,6-lutidine and 2,4,6-collidine in aqueous H_2SO_4 give w factors of $-0{\cdot}55 \leqslant w \leqslant -0{\cdot}21$ depending on the temperature at which the kinetic measurements were carried out (Katritzky and Ridgewell, 1963). These reactions involve A-S_E2 substitution of the conjugate acids of the parent bases and correlation of reaction rate with the H_+ acidity function might be more appropriate. However insufficient information about the protonation of cations in concentrated acid solutions is available. The similarity between the w values for the H-exchange of pyridinium cations and the neutral benzenes and azulenes is further evidence for the conclusion that the acidity functions for the protonation of neutral and singly positively charged species are probably parallel functions of acid concentration. The present comparison is for the protonation of carbon bases.

Katritzky and coworkers have also studied the dependence on acidity function of the hydrogen exchange reactions of substituted pyridines (Bean *et al.* 1967b), pyridine-1-oxides (Bean *et al.* 1967a), pyridones, pyrones and quinolone (Bellingham *et al.* 1967, 1968), anilines (Bean and Katritzky, 1968) and pyridazines (Katritzky and Pojarlieff, 1968). Complicated rate profiles (plots of $\log_{10}k_1$ against $-H_0$) were often obtained and were interpreted in terms of the relative reactivities of the neutral, singly protonated and doubly protonated substrate species formed in rapid pre-equilibrium in the reaction solutions. Thus, for example, for 4-aminopyridine $\log_{10}k_1$ is a linear function of $-H_0$ when $0 > H_0 > -6$. However at higher acidities than $H_0 = -6{\cdot}3$ the rate falls off (at least as far as $H_0 = -8$). This is because $pK_a =$ ca. $-6{\cdot}3$ for the second protonation of 4-aminopyridine. The linear section of the rate profile corresponds to the H-exchange reaction of the monoprotonated pyridine. Above $H_0 = -6{\cdot}3$ protonation to the comparatively unreactive diprotonated species becomes predominant and the observed reaction rate decreases. A similar rate profile was obtained for 4-aminopyridine-1-oxide.

5.20. Miscellaneous Reactions

Albery and Bell (1961) have found that the rate of decomposition of ethyl diazoacetate in aqueous $HClO_4$ at 0° is proportional to h_0. The Zucker–Hammett hypothesis would require that water is not involved in the rate-determining step in the reaction. This conclusion is incorrect and the Zucker–Hammett idea is not applicable in this case. Thus the rates of reaction in aqueous hydrochloric acid are up to $2\frac{1}{2}$ times greater than in $HClO_4$ of the same concentration although the differences in acidity are less than 10% in the concentration range studied. This suggests that catalysis by chloride ions is occurring in aqueous HCl and causes an increase in the observed rate of reaction. It is logical to suppose that the weaker nucleophile water reacts in a similar way to the chloride ion. The proposed mechanism is summarized in equations (5.74)–(5.76). The rapid pre-equilibrium is in accord with the observed isotope effect $(k_{D_2O}/k_{H_2O}) = 2{\cdot}7$ (Gross *et al.* 1936). Product analysis

$$\overset{-}{N}{=}\overset{+}{N}{=}CHCO_2Et + H^+ \rightleftharpoons N{\equiv}\overset{+}{N}CH_2CO_2Et \quad \text{(fast)} \qquad (5.74)$$

$$N{\equiv}\overset{+}{N}CH_2CO_2Et + H_2O \rightarrow N_2 + CH_2OHCO_2Et + H^+ \quad \text{(slow)} \qquad (5.75)$$

$$N{\equiv}\overset{+}{N}CH_2CO_2Et + Cl^- \rightarrow N_2 + CH_2ClCO_2Et \quad \text{(slow)} \qquad (5.76)$$

confirms reaction (5.76) (Albery and Bell, 1961) and further evidence for reaction (5.75) has been reviewed by More O'Ferrall (1967). Step (5.74) involves carbon protonation and therefore by analogy with the hydration of olefines and hydrogen exchange reactions the correlation between $\log_{10}k_1$ and $-H_0$ suggests that water is involved in the rate-determining step in the reaction. The acidity function behaviour is not anomalous but further confirms the distinction which must be drawn between reactions involving carbon bases and those involving oxygen or nitrogen bases. In the present example the Bunnett w factor is ca. 0 for a reaction of a carbon base involving water as a nucleophile in the rate-determining step (reaction 5.75). For the H-exchange reactions $w =$ ca. 0 with water acting as a proton transfer agent in the slow step (equation 5.70 with HA as H_2O).

The rates of diazotization of benzamide in aqueous H_2SO_4 correlate with the J_0 acidity function. The mechanism therefore involves a rapid pre-equilibrium ionization of nitrous acid to the nitrosonium ion NO^+ followed by a rate determining attack of NO^+ on unprotonated benzamide (Ladenheim and Bender, 1960).

$$HONO + H^+ \rightleftharpoons NO^+ + H_2O \quad \text{(fast)} \tag{5.77}$$

$$C_6H_5CONH_2 + NO^+ \rightarrow \text{products} \quad \text{(slow)} \tag{5.78}$$

In concentrated acid solution the variation of rates of diazotization with acid concentration is often complex. However this is due to a combination of different kinetic terms plus a pronounced medium effect. The rate of diazotization of 4-nitroaniline in aqueous $HClO_4/NaClO_4$ mixtures of constant ionic strength (3 M) is given by equation (5.79) whereas for aniline and 4-toluidine equation (5.80) is applicable (Challis and Ridd, 1960, 1961). The rate

$$\text{Rate} = k_2\, C_{ArNH_3^+}\, C_{HNO_2} \tag{5.79}$$

$$\text{Rate} = k_3\, C_{ArNH_3^+}\, C_{HNO_2}\, h_0 \tag{5.80}$$

of diazotization of 2-chloroaniline is explicable in terms of the sum of two kinetic terms and is of the form equation (5.79) plus equation (5.80) (Challis and Ridd, 1961). Equation (5.79) corresponds to the reaction of the free amine with the nitrous acidium ion, and equation (5.80) to a mechanism which involves nitrosation of the protonated amine such that the departing proton is still present in the transition state (Ridd, 1961). The diazotization of aniline, 4-toluidine and 4-nitroaniline in 57–61% aqueous $HClO_4$ conforms to the rate equation (5.81) (Challis and Ridd, 1960). At these acid concentrations nitrous acid is completely converted to ionized nitrosonium salts. The proposed mechanism is summarized in equations (5.82) and (5.83).

$$\text{Rate} = kC_{ArNH_3^+}\, C_{NO^+}\, h_0^{-2} \tag{5.81}$$

$$ArNH_3^+ + NO^+ \rightleftharpoons ArNH_2NO^+ + H^+ \quad \text{(fast)} \tag{5.82}$$

$$ArNH_2NO^+ \rightarrow ArNHNO + H^+ \quad \text{(slow)} \tag{5.83}$$

In 66–74% aqueous H_2SO_4 the rates are faster than in aqueous $HClO_4$ of similar acidity because of base catalysis by HSO_4^- ions in the slow reaction (5.83) (Challis and Ridd, 1960). The reaction of nitrous acid with alkyl hydroxylamines in aqueous acid gives results essentially similar to those obtained by Challis and Ridd (1960, 1961; Ridd, 1961) for the diazotization of amines (Morgan *et al.* 1969).

The decomposition of hyponitrous acid in aqueous $HClO_4$ occurs by two mechanisms one being independent of acidity and the rate of the other being proportional to $h_0^{0\cdot8}$ (Hughes and Stedman, 1964). The latter is probably a fast equilibrium protonation of $H_2N_2O_2$ followed by the slow fission (to $N_2O + H_3O^+$) of the protonated acid. Buchholz and Powell (1963) suggested the transfer of the proton in the slow step occurs either to two water molecules or, at higher acid (H_2SO_4) concentrations, to an H_2SO_4 molecule. This arose because a Bunnett plot gave $w = 2$ at low acidities but tended towards $w = 0$ at high H_2SO_4 concentrations. This is a further example (cf. Fig. 5.16) where a Bunnett plot indicates a change in mechanism as the acid concentration is increased.

The rate of racemization of D-sec-butyl hydrogen sulphate in 45–65% aqueous H_2SO_4 correlates very closely with h_0 (Deno and Newman, 1951). Hence $w = -0{\cdot}02$ (Bunnett, 1961a) and $\phi = -0{\cdot}006$ (Bunnett and Olsen, 1966b). The mechanism probably involves a rapid pre-equilibrium protonation of sec-$BuHSO_4$ followed by the unimolecular slow formation of a sec-butyl carbonium ion from the protonated alkyl sulphate (Long and Paul, 1957).

Plots of $\log k$ against H_0 are linear for the chromic acid oxidations of substituted benzaldehydes (Wiberg and Mill, 1958), iso-propanol (Rocek and Krupicka, 1958), diphenylmethane (Wiberg and Evans, 1960) and alkylbenzenes (Brandenberger *et al.* 1961). The mechanism of these reactions involves an interaction between the organic substrate and a protonated species of chromium(VI). Wiberg *et al.* write the protonation equilibrium as equation (5.84) and note that an H_- acidity function would be more appropriate to

$$HCrO_4^- + H^+ \rightleftharpoons H_2CrO_4 \tag{5.84}$$

describe this equilibrium. Alternatively CrO_3H^+ (rather than H_2CrO_4) has been proposed as the reactive entity in the oxidation of alkylbenzenes. However the acidity function dependence does not enable these alternatives to be satisfactorily distinguished. H_- acidity functions for concentrated acid solutions have been less well studied than acidity functions for the protonation of electrically neutral bases (Section 3.3).

In aqueous $HClO_4$ (1–5 M) the rate of iodination of 4-chloroaniline with iodine monochloride is inversely proportional to h_0 (Berliner, 1956). Thus $\log_{10} k_{obs}$ gave a slope of 1·17 when plotted against H_0. The reactive iodine species is probably the hypoiodous acidium ion formed via equation (5.85). I^+ cations are an alternative possibility. The inverse acidity function dependence follows since it is the free amine (equation 5.86) which is iodinated. Since pK_a(4-chloroanilinium ion) = ca. 4 in water the equilibrium (5.86) lies pre-

$$ICl + H_2O \rightleftharpoons H_2OI^+ + Cl^- \tag{5.85}$$

$$4\text{-}Cl\text{—}C_6H_4NH_3^+ \rightleftharpoons 4\text{-}Cl\text{—}C_6H_4NH_2 + H^+ \tag{5.86}$$

dominantly on the left-hand side when $C_{HClO_4} > 1$ mole litre^{-1}. The concentration of the reactive neutral amine is therefore inversely proportional to h_0.

The experimental rate constants ($k_1 = -d\log_{10} C_{R_2SO}/dt$) for the reduction and racemization of 4-tolyl methyl sulphoxide by halide ions in aqueous $HClO_4$ vary with H_0 as shown in Fig. 5.20 (Landini *et al.* 1968). The changes in slope from ca. 2·2 at low acidities to ca. 1·3 (0·9 for the racemization by Cl^-) at high acidities has been interpreted in terms of the following mechanism (X^- represents a halide anion).

$$R_1R_2SO + H^+ \rightleftharpoons R_1R_2\overset{+}{S}OH \qquad \text{(fast)} \tag{5.87}$$

$$R_1R_2\overset{+}{S}OH + X^- + H^+ \rightarrow R_1R_2\overset{+}{S}X + H_2O \quad \text{(slow)} \tag{5.88}$$

Reaction (5.88) controls the overall rate and is followed by fast interaction of $R_1R_2SX^+$ with X^- (reduction or racemization) or with H_2O (racemization).

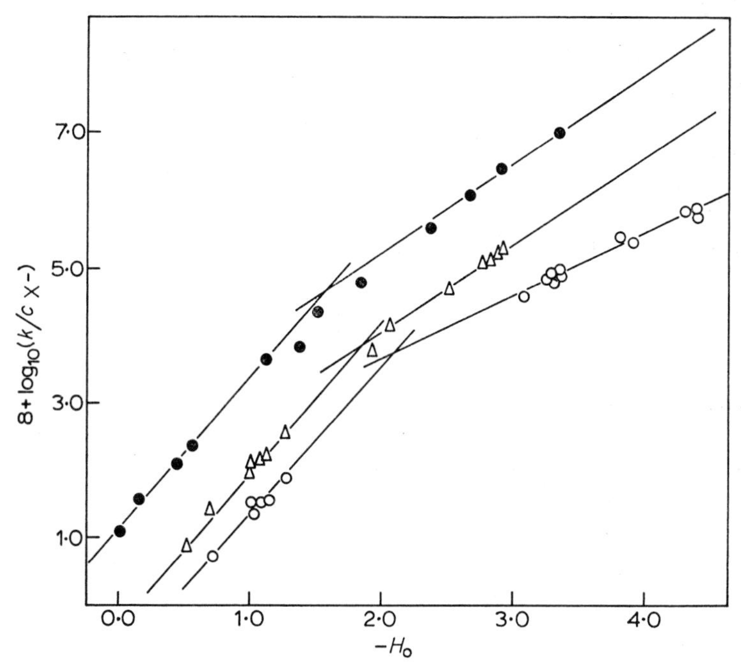

FIG. 5.20. Rates of reduction of 4-tolyl methyl sulphoxide by I^- (●) and of racemization of (+)-4-tolyl methyl sulphoxide by Br^-(△) and by Cl^-(○) in aqueous $HClO_4$ at 25°. Reproduced by permission from Landini *et al.* 1968.

Equation (5.89) is the general kinetic equation which becomes (5.90) at high acidities when protonation of the sulphoxide (reaction 5.87) is virtually complete. At low acidities ($C_S \gg C_{SH^+}$ where S is the unprotonated sulphoxide) equation (5.91) is applicable. K_a is the acid ionization constant of the protonated sulphoxide. As the acid concentration is increased above $C_{HClO_4} \sim 1$

$$\text{Rate} = k_3\, C_{SH^+}\, C_{X^-}\, C_{H^+}\, y_{SH^+}\, y_{X^-}\, y_{H^+}/y^{\ddagger} \tag{5.89}$$

$$\text{Rate} = k_3\, C_S\, C_{X^-}\, C_{H^+}\, y_{SH^+}\, y_{X^-}\, y_{H^+}/y^{\ddagger} \tag{5.90}$$

$$\text{Rate} = k_3\, C_S\, C_{X^-}\, C_{H^+}^2\, y_S\, y_{X^-}\, y_{H^+}^2/K_a\, y^{\ddagger} \tag{5.91}$$

mole litre^{-1} the correlations of rate with acidity function are consistent with a trend from equation (5.91) to equation (5.90).

Hydrogen peroxide and benzeneboronic acid react to form phenol and boric acid in aqueous acid solutions. The rate of reaction in aqueous H_3PO_4 (up to 9·67 mole litre^{-1}) paralleled the Hammett acidity function (Kuivila, 1955). Bunnett factors w of 1·91 and 2·73 (Bunnett, 1961a) were obtained for the reaction in concentrated aqueous H_2SO_4 and $HClO_4$ respectively. Bunnett's (1961b) criterion for mechanism would require that water is involved as a nucleophile in the rate determining step of the reaction. The reaction is first order in both hydrogen peroxide and benzeneboronic acid. Other reactions for which a water molecule is involved in the transition state include the hydrogen–deuterium exchange reactions of NH_3D^+ ($w = 3{\cdot}09$), $CH_3NH_2D^+$ ($w = 4{\cdot}70$), and $(CH_3)_3ND^+$ ($w = 6{\cdot}83$) (Emerson *et al.* 1960) and the acid cleavage of methylmercuric iodide (Kreevoy, 1957). For the latter in aqueous H_2SO_4 $w = 6{\cdot}59$, whereas in aqueous $HClO_4$ $w = 3{\cdot}38$ (Bunnett, 1961a). The rate determining step is the bimolecular electrophilic attack of H_3O^+ on CH_3HgI (equation 5.92).

$$H_3O^+ + CH_3HgI \rightarrow CH_4 + HgI^+ + H_2O \tag{5.92}$$

Water is involved as a proton transfer agent (for which Bunnett, 1961b, gives $w > 3{\cdot}3$) in the transition state. This is not strictly an acid *catalysed* reaction as H^+ is consumed in forming the products. The rates of the acid catalysed aquations of *cis*- and *trans*-dinitrobisethylenediaminecobalt(III) salts in concentrated ($< 3{\cdot}5$ mole litre^{-1}) aqueous $HClO_4$ and HNO_3 (solutions made to constant ionic strength by adding $NaClO_4$ or $NaNO_3$) were directly proportional to the acid concentrations (Staples, 1964). The S_N2CA mechanism (5.93)–(5.94) is in accord with this result.

$$[\text{Co en}_2(NO_2)_2]^+ + H_3O^+ \rightleftharpoons [\text{Co en}_2(NO_2)(NO_2H)]^{2+} + H_2O \quad \text{(fast)} \tag{5.93}$$

$$[\text{Co en}_2(NO_2)(NO_2H)]^{2+} + H_2O \rightarrow [\text{Co en}_2(NO_2)(H_2O)]^{2+} + HNO_2 \quad \text{(slow)} \tag{5.94}$$

5.21. Conclusion

In concentrated acid solutions the variation of the rate of an acid catalysed reaction with increasing acid concentration depends on the reaction mechanism and the relative values of the rate constants in each step of the mechanism. However the activity coefficient behaviour of the reactants and the transition state also makes a significant contribution to the variation of rate with acid concentration. This activity coefficient behaviour in concentrated electrolyte solutions is intimately connected with the solvation requirements (or solute–solvent interactions) of the reactants and transition state. In order to assess the mechanism of a reaction from rate measurements for a series of catalysing acid concentrations the effect of activity coefficients on the experimental rates must be estimated and eliminated.

The Zucker–Hammett hypothesis required that $\log_{10}k_1$ for an acid catalysed reaction should be a linear function with unit slope of either the stoichiometric acid concentration or the Hammett acidity function. There is a wealth of evidence which proves that these two alternatives are not adhered to. Indeed the exact fitting of either of them for a given reaction may be regarded as fortuitous. Thus the correlation of $\log_{10}k_1$ with $-H_0$ aims at eliminating the activity coefficient term $(y_S y_{H^+}/y^{\ddagger})$ in the rate equation by equating it with $(y_B y_{H^+}/y_{BH^+})$ in the definition of H_0. In the same way as the protonation of different classes of base (e.g. amines, amides, olefins) give rise to different acidity functions (H_0, H_A, H_R') so the variation of $(y_S y_{H^+}/y^{\ddagger})$ with acid concentration depends upon the structure and solvation requirements of S and the transition state $\ddagger$. The exact requirement of the Zucker–Hammett hypothesis is therefore rarely obeyed. However for reacting substrates which are nitrogen or oxygen bases $\log_{10}k_1$ is often linear in either $-H_0$ or $\log_{10}C_{acid}$ (but not both). The slope of the graph which is linear for a particular reaction may differ from one but the linearity may be taken as a rough indication of an A-1 ($-H_0$) or an A-2 ($\log_{10}C_{acid}$) reaction. If $\log_{10}k_1$ against $-H_0$ is linear and has a slope above ca. 0·8 then this suggests an A-1 mechanism. If the slope is < ca. 0·8 or the graph is curved and the $\log_{10}k_1$ against $\log_{10}C_{acid}$ plot is linear this suggests an A-2 mechanism. In the past this has been the most used treatment for the diagnosis of mechanism from an acidity function correlation. However it is very dangerous to base the determination of mechanism on this evidence alone. Rather the acidity function dependence merely gives an indication of mechanism and further confirmatory evidence is desirable. Entropies and volumes of activation, isotope effects, and tests for specific hydrogen ion or general acid catalysis are useful further guides for the validity of a possible mechanism.

An extreme case where the Zucker–Hammett approach fails completely is for the reactions of carbon bases. This is exemplified by the acid catalysed hydration of olefins. The failure is analogous to the large deviation between the H_0 and H_R' acidity functions. The variation $(y_S/y^{\ddagger})$, where S is a carbon base, with acid concentration differs enormously from that for (y_B/y_{BH^+}) where B is a primary amine but is similar to (y_B/y_{BH^+}) if B is another carbon base. Thus it would be better to correlate the rates of the reactions of carbon bases with H_R' rather than with H_0. However a linear plot of $\log_{10}k_1$ against $-H_0$ with slope ca. 1 for the reaction of a substrate which is a carbon base apparently implies an A-2 mechanism (at least for the cases far so studied).

Bunnett's approach bases the determination of mechanism from H_0 correlations on the observed values of the w and w^* factors for a large number of reactions of known mechanism. His criteria for mechanism (Table 4.2) are reasonably easy to test and give a useful means of representing the acidity function behaviour for a reaction in terms of w and or w^*. The clear distinction between oxygen and nitrogen bases, which behave similarly, and carbon bases must be maintained. However it is unwise to take the Bunnett criteria alone as the only evidence for the mechanism of a particular reaction. Other independent confirmatory evidence is desirable. The connection between w or w^* and the solvation requirements of the reactants and transition state (equations 4.52 and 4.49 respectively) is well worth remembering. Thus two substrate molecules which react by the same mechanism may give quite different acidity function behaviour because of a difference between the solute–solvent interactions of the substrate and the transition state in the two cases. The hydrogen exchange reactions of 1,3,5-trimethoxybenzene and 1,3,5-trihydroxybenzene are examples. Bunnett's criteria not only give an indication of reaction mechanism but may also give information about the changes in solvation requirements of the transition states for the reactions of series of structurally related molecules. In general the Bunnett approach is probably the most useful method at present available for the interpretation in terms of mechanism of the response to changes in acid concentration of the rates of acid catalysed reactions.

REFERENCES

Akand, M. A., and Wyatt, P. A. H. (1967). *J. Chem. Soc. B*, 1326.
Albery, W. J. (1967). *Progr. Reaction Kinetics*, **4**, 353.
Albery, W. J., and Bell, R. P. (1961). *Trans. Faraday Soc.* **57**, 1942.
Alexander, E. R., Busch, H. M., and Webster, G. L. (1952). *J. Amer. Chem. Soc.* **74**, 3173.
Anderson, D. M. W., and Garbutt, S. (1963). *J. Chem. Soc.* 3204.
Archer, G., and Bell, R. P. (1959). *J. Chem. Soc.* 3228.
Armour, C., Bunton, C. A., Patai, S., Selman, L. H., and Vernon, C. A. (1961). *J. Chem. Soc.* 412.
Armstrong, V. C., Farlow, D. W., and Moodie, R. B. (1968). *J. Chem. Soc. B.* 1099.
Bakule, R., and Long, F. A. (1963). *J. Amer. Chem. Soc.* **85**, 2309.
Baliga, B. T., and Whalley, E. (1964). *Canad. J. Chem.* **42**, 1019.
Baliga, B. T., and Whalley, E. (1965). *Canad. J. Chem.* **43**, 2453.
Barnard, P. W. C., Bunton, C. A., Kellerman, D., Mhala, M. M., Silver, B., Vernon, C. A., and Welch, V. A. (1966). *J. Chem. Soc. B*, 227.
Barnard, P. W. C., Bunton, C. A., Llewellyn, D. R., Oldham, K. G., Silver, B. L., and Vernon, C. A. (1955). *Chem. and Ind.* 760.
Batts, B. D., and Gold, V. (1964). *J. Chem. Soc.* 4284.
Bayliss, N. S., and Watts, D. W. (1963). *Aust. J. Chem.* **16**, 943.
Bayliss, N. S., Dingle, R., Watts, D. W., and Wilkie, R. J. (1963). *Aust. J. Chem.* **16**, 933.
Bean, G. P., and Katritzky, A. R. (1968). *J. Chem. Soc. B*, 864.
Bean, G. P., Brignell, P. J., Johnson, C. D., Katritzky, A. R., Ridgewell, B. J., Tarhan, H. O., and White, A. M. (1967a). *J. Chem. Soc. B*, 1222.
Bean, G. P., Johnson, C. D., Katritzky, A. R., Ridgewell, B. J., and White, A. M. (1967b). *J. Chem. Soc. B*, 1219.
Bell, R. P. (1941). "Acid-Base Catalysis." Oxford U.P. London.
Bell, R. P., and Brown, A. H. (1954). *J. Chem. Soc.* 774.

Bell, R. P., and Lukianenko, B. (1957). *J. Chem. Soc.* 1686.
Bell, R. P., and Rawlinson, D. J. (1958). *J. Chem. Soc.* 4387.
Bell, R. P., Dowding, A. L., and Noble, J. A. (1955). *J. Chem. Soc.* 3106.
Bell, R. P., Bascombe, K. N., and McCoubrey, J. C. (1956). *J. Chem. Soc.* 1286.
Bell, R. P., Preston, J., and Whitney, R. B. (1962). *J. Chem. Soc.* 1166.
Bellingham, P., Johnson, C. D., and Katritzky, A. R. (1967). *J. Chem. Soc. B*, 1226.
Bellingham, P., Johnson, C. D., and Katritzky, A. R. (1968). *J. Chem. Soc. B*, 866.
BeMiller, J. N. (1967). *Adv. Carbohydrate Chem.* **22**, 25.
Berliner, E. (1956). *J. Amer. Chem. Soc.* **78**, 3632.
Birchall, T., and Gillespie, R. J. (1963). *Canad. J. Chem.* **41**, 2642.
Bonner, T. G. (1960). *J. Chem. Soc.* 3493.
Bonner, T. G., and Barnard, M. (1958). *J. Chem. Soc.* 4176.
Bonner, T. G., and Brown, F. (1966). *J. Chem. Soc. B*, 658.
Bonner, T. G., and Frizel, D. E. (1959a). *J. Chem. Soc.* 3894.
Bonner, T. G., and Frizel, D. E. (1959b). *J. Chem. Soc.* 3903.
Bonner, T. G., and Lockhart, J. C. (1958). *J. Chem. Soc.* 3852.
Bonner, T. G., and Wilkins, J. M. (1955). *J. Chem. Soc.* 2358.
Bonner, T. G., and Williams, G. (1951). *Chem. and Ind.* **70**, 820.
Bonner, T. G., Thorne, M. P., and Wilkins, J. M. (1955). *J. Chem. Soc.* 2351.
Brand, J. C. D. (1950). *J. Chem. Soc.* 1004.
Brand, J. C. D., and Horning, W. C. (1952). *J. Chem. Soc.* 3922.
Brand, J. C. D., and Rutherford, A. (1952). *J. Chem. Soc.* 3916.
Brand, J. C. D., Jarvie, A. W. P., and Horning, W. C. (1959). *J. Chem. Soc.* 3844.
Brandenberger, S. G., Maas, L. W., and Dvoretzky, I. (1961). *J. Amer. Chem. Soc.* **83**, 2146.
Brice, L. K., and Lindsay, L. P. (1960). *J. Amer. Chem. Soc.* **82**, 3538.
Brønsted, J. N., and Grove, C. (1930). *J. Amer. Chem. Soc.* **52**, 1394.
Brønsted, J. N., and Wynne-Jones, W. F. K. (1929). *Trans. Faraday Soc.* **25**, 59.
Brønsted, J. N., Kilpatrick, M., and Kilpatrick, M. L. (1929). *J. Amer. Chem. Soc.* **51**, 428.
Brown, H. C., and Okamoto, Y. (1958). *J. Amer. Chem. Soc.* **80**, 4979.
Buchholz, J. R., and Powell, R. E. (1963). *J. Amer. Chem. Soc.* **85**, 509.
Bunnett, J. F. (1961a). *J. Amer. Chem. Soc.* **83**, 4956.
Bunnett, J. F. (1961b). *J. Amer. Chem. Soc.* **83**, 4968.
Bunnett, J. F. (1961c). *J. Amer. Chem. Soc.* **83**, 4973.
Bunnett, J. F. (1961d). *J. Amer. Chem. Soc.* **83**, 4978.
Bunnett, J. F., and Buncel, E. (1961). *J. Amer. Chem. Soc.* **83**, 1117.
Bunnett, J. F., and Olsen, F. P. (1966a). *Canad. J. Chem.* **44**, 1899.
Bunnett, J. F., and Olsen, F. P. (1966b). *Canad. J. Chem.* **44**, 1917.
Bunnett, J. F., Buncel, E., and Nahabedian, K. V. (1962). *J. Amer. Chem. Soc.* **84**, 4136.
Bunton, C. A., and Llewellyn, D. R. (1957). *J. Chem. Soc.* 3402.
Bunton, C. A., and Wood, J. L. (1955). *J. Chem. Soc.* 1522.
Bunton, C. A., Konasiewicz, A., and Llewellyn, D. R. (1955a). *J. Chem. Soc.* 604.
Bunton, C. A., Lewis, T. A., Llewellyn, D. R., and Vernon, C. A. (1955b). *J. Chem. Soc.* 4419.
Bunton, C. A., Hadwick, T., Llewellyn, D. R., and Pocker, Y. (1956). *Chem. and Ind.* **75**, 547.
Bunton, C. A., O'Connor, C., and Turney, T.A. (1967). *Chem. and Ind.* 1835.

Bunton, C. A., de la Mare, P. B. D., Greaseley, P. M., Llewellyn, D. R., Pratt, N. H., and Tillett, J. G.(1958a). *J. Chem. Soc.* 4751.
Bunton, C. A., de la Mare, P. B. D., Leonard, A., Llewellyn, D. R., Pearson, R. B., Pritchard, J. G., and Tillett, J. G. (1958b). *J. Chem. Soc.* 4761.
Bunton, C. A., de la Mare, P. B. D., and Tillett, J. G. (1958c). *J. Chem. Soc.* 4754.
Bunton, C. A., Hadwick, T., Llewellyn, D. R., and Pocker, Y. (1958d). *J. Chem. Soc.* 403.
Bunton, C. A., Llewellyn, D. R., Oldham, K. G., and Vernon, C. A. (1958e). *J. Chem. Soc.* 3574.
Bunton, C. A., Llewellyn, D. R., Oldham, K. G., and Vernon, C. A. (1958f). *J. Chem. Soc.* 3588.
Bunton, C. A., de la Mare, P. B. D., and Tillett, J. G. (1959). *J. Chem. Soc.* 1766.
Bunton, C. A., James, D. H., and Senior, J. B. (1960a). *J. Chem. Soc.* 3364.
Bunton, C. A., Mhala, M. M., Oldham, K. G., and Vernon, C. A. (1960b). *J. Chem. Soc.* 3293.
Bunton, C. A., Kellerman, D., Oldham, K. G., and Vernon, C. A. (1966). *J. Chem. Soc. B*, 292.
Burkett, H., Schubert, W. M., Schultz, F., Murphy, R. B., and Talbott, R. (1959) *J. Amer. Chem. Soc.* **81**, 3923.
Campbell, H. J., and Edward, J. T. (1960). *Canad. J. Chem.* **38**, 2109.
Candlin, J. P., and Wilkins, R. G. (1961). *J. Chem. Soc.* 3625.
Candlin, J. P., and Wilkins, R. G. (1965). *J. Amer. Chem. Soc.* **87**, 1490.
Cerfontain, H. (1961). *Rec. Trav. Chim.* **80**, 296.
Cerfontain, H., Kaandorp, A. W., and Sixma, F. L. J. (1963). *Rec. Trav. Chim.* **82**, 565.
Challis, B. C., and Long, F. A. (1965a). *Discuss. Faraday Soc.* **39**, 67.
Challis, B. C., and Long, F. A. (1965b). *J. Amer. Chem. Soc.* **87**, 1196.
Challis, B. C., and Ridd, J. H. (1960). *Proc. Chem. Soc.* 245.
Challis, B. C., and Ridd, J. H. (1961). *Proc. Chem. Soc.* 173.
Chmiel, C. T., and Long, F. A. (1956). *J. Amer. Chem. Soc.* **78**, 3326.
Ciapetta, F. G., and Kilpatrick, M. (1948). *J. Amer. Chem. Soc.* **70**, 639.
Colapietro, J., and Long, F. A. (1960). *Chem. and Ind.* 1056.
Collins, C. J. (1960). *Quart. Rev.* **14**, 357.
Coombes, R. G., Moodie, R. B., and Schofield, K. (1968). *J. Chem. Soc. B*, 800.
Coombes, R. G., Moodie, R. B., and Schofield, K. (1969). *J. Chem. Soc. B*, 52.
Cordes, E. H. (1967). *Progr. Phys. Org. Chem.* **4**, 1.
Datta, S. C., Day, J. N. E., and Ingold, C. K. (1939). *J. Chem. Soc.* 838.
Dawson, H. M., and Powis, F. (1913). *J. Chem. Soc.* 2135.
Day, J. N. E., and Ingold, C. K. (1941). *Trans. Faraday Soc.* **37**, 686.
Dayagi, S. (1961). *Bull. Res. Council of Israel*, **10A**, 152.
Deane, C. W. (1945). *J. Amer. Chem. Soc.* **67**, 329.
de la Mare, P. B. D., Hughes, E. D., Ingold, C. K., and Pocker, Y. (1954). *J. Chem. Soc.* 2930.
Deno, N. C., and Newman, M. S. (1950). *J. Amer. Chem. Soc.* **72**, 3852.
Deno, N. C., and Newman, M. S. (1951). *J. Amer. Chem. Soc.* **73**, 1920.
Deno, N. C., and Perizzolo, C. (1957a). *J. Org. Chem.* **22**, 836.
Deno, N. C., and Perizzolo, C. (1957b). *J. Amer. Chem. Soc.* **79**, 1345.
Deno, N. C., and Taft, R. W. (1954). *J. Amer. Chem. Soc.* **76**, 244.
Deno, N. C., Jaruzelski, J. J., and Schriesheim, A. (1955). *J. Amer. Chem. Soc.* **77**, 3044.
Deno, N. C., Edwards, T., and Perizzolo, C. (1957). *J. Amer. Chem. Soc.* **79**, 2110.

Deno, N. C., Peterson, H. J., and Sacher, E. (1961). *J. Phys. Chem.* **65**, 199.
De Right, R. E. (1933). *J. Amer. Chem. Soc.* **55**, 4761.
Dittmar, H. R. (1929). *J. Phys. Chem.* **33**, 533.
Downing, R. G. (1961). *Diss. Abs.* **22**, 1825.
Drenth, W., and Hogeveen, H. (1960). *Rec. Trav. Chim.* **79**, 1002.
Duboux, M., and de Sousa, A. (1940). *Helv. Chim. Acta*, **23**, 1381.
Duncan, J. F., and Lynn, K. R. (1956a). *J. Chem. Soc.* 3512.
Duncan, J. F., and Lynn, K. R. (1956b). *J. Chem. Soc.* 3519.
Duncan, J. F., and Lynn, K. R. (1956c). *J. Chem. Soc.* 3674.
Duncan, J. F., and Lynn, K. R. (1957a). *Aust. J. Chem.* **10**, 1.
Duncan, J. F., and Lynn, K. R. (1957b). *Aust. J. Chem.* **10**, 7.
Edward, J. T. (1955). *Chem. and Ind.* 1102.
Edward, J. T., and Meacock, S. C. R. (1957a). *J. Chem. Soc.* 2000.
Edward, J. T., and Meacock, S. C. R. (1957b). *J. Chem. Soc.* 2007.
Edward, J. T., Hutchison, H. P., and Meacock, S. C. R. (1955). *J. Chem. Soc.* 2520.
Elliott, W. W., and Hammick, D. Ll. (1951). *J. Chem. Soc.* 3402.
Emerson, M. T., Grunwald, E., Kaplan, M. L., and Kromhout, R. A. (1960). *J. Amer. Chem. Soc.* **82**, 6307.
Fraenkel, G., and Franconi, C. (1960). *J. Amer. Chem. Soc.* **82**, 4478.
Friedman, H. B., and Elmore, G. V. (1941). *J. Amer. Chem. Soc.* **63**, 864.
Fuller, M. W., and Schubert, W. M. (1963). *J. Amer. Chem. Soc.* **85**, 108.
Gleghorn, J., Moodie, R. B., Schofield, K., and Williamson, M. J. (1963). *Chem. and Ind.* 1283.
Gleghorn, J., Moodie, R. B., Schofield, K., and Williamson, M. J. (1966). *J. Chem. Soc. B*, 870.
Gleghorn, J., Moodie, R. B., Quereshi, E. A., and Schofield, K. (1968). *J. Chem. Soc. B*, 316.
Gold, V. (1961). *Proc. Chem. Soc.* 453.
Gold, V., and Hawes, B. W. V. (1951). *J. Chem. Soc.* 2102.
Gold, V., and Hilton, J. (1955a). *J. Chem. Soc.* 838.
Gold, V., and Hilton, J. (1955b). *J. Chem. Soc.* 843.
Gold, V., and Riley, T. (1962). *J. Chem. Soc.* 4183.
Gold, V., and Satchell, D. P. N. (1955a). *Nature*, **176**, 602.
Gold, V., and Satchell, D. P. N. (1955b). *J. Chem. Soc.* 3609.
Gold, V., and Satchell, D. P. N. (1955c). *J. Chem. Soc.* 3619.
Gold, V., and Satchell, D. P. N. (1955d). *J. Chem. Soc.* 3622.
Gold, V., and Satchell, D. P. N. (1956a). *J. Chem. Soc.* 1635.
Gold, V., and Satchell, D. P. N. (1956b). *J. Chem. Soc.* 2743.
Gross, P., Steiner, H., and Krauss, F. (1936). *Trans. Faraday Soc.* **32**, 877.
Gruen, L. C., and Long, F. A. (1967). *J. Amer. Chem. Soc.* **89**, 1287.
Grunwald, E., Heller, A., and Klein, F. S. (1957). *J. Chem. Soc.* 2604.
Hamer, D., and Leslie, J. (1960). *J. Chem. Soc.* 4198.
Hammett, L. P. (1935). *Chem. Rev.* **16**, 67.
Hammett, L. P., and Deyrup, A. J. (1932). *J. Amer. Chem. Soc.* **54**, 2721.
Hammett, L. P., and Paul, M. A. (1934). *J. Amer. Chem. Soc.* **56**, 830.
Hardy-Klein, M. L. (1957). *J. Chem. Soc.* **70**, 3312.
Harris, R. T., and Weale, K. E. (1956). *J. Chem. Soc.* 953.
Hellin, M., and Coussemant, F. (1957). *Compt. rend.* **245**, 2504.
Henshall, T., Silbermann, W. E., and Webster, J. G. (1955). *J. Amer. Chem. Soc.* **77**, 6656.

Hogeveen, H., and Drenth, W. (1963a). *Rec. Trav. Chim.* **82**, 375.
Hogeveen, H., and Drenth, W. (1963b). *Rec. Trav. Chim.* **82**, 410.
Huckings, B. J., and Johnson, M. D. (1964). *J. Chem. Soc.* 5371.
Hughes, M. N., and Stedman, G. (1964). *J. Chem. Soc.* 163.
Huisgen, R., and Brade, H. (1957). *Chem. Ber.* **90**, 1432.
Ingold, C. K. (1953). "Structure and Mechanism in Organic Chemistry", Ch. 14. Bell, London.
Jaques, D. (1965). *J. Chem. Soc.* 3874.
Jaques, D., and Leisten, J. A. (1964). *J. Chem. Soc.* 2683.
Jellinek, H. H. G., and Gordon, A. (1949). *J. Phys. Chem.* **53**, 996.
Jellinek, H. H. G., and Urwin, J. R. (1953). *J. Phys. Chem.* **57**, 900.
Johnson, C. D., Katritzky, A. R., Ridgewell, B. J., and Viney, M. (1967a). *J. Chem. Soc. B*, 1204.
Johnson, C. D., Katritzky, A. R., Shakir, N., and Viney, M. (1967b). *J. Chem. Soc. B*, 1213.
Johnson, C. D., Katritzky, A. R., and Viney, M. (1967c). *J. Chem. Soc. B*, 1211.
Julian, K., and Walters, W. A. (1962). *J. Chem. Soc.* 819.
Kaandorp, A. W., Cerfontain, H., and Sixma, F. L. J. (1962). *Rec. Trav. Chim.* **81**, 969.
Kaandorp, A. W., Cerfontain, H., and Sixma, F. L. J. (1963). *Rec. Trav. Chim.* **82**, 113.
Katritzky, A. R., and Kingsland, M. (1968). *J. Chem. Soc. B*, 862.
Katritzky, A. R., and Pojarlieff, I. (1968). *J. Chem. Soc. B*, 873.
Katritzky, A. R., and Ridgewell, B. J. (1963). *J. Chem. Soc.* 3753.
Kent, P. W., and Barnett, J. E. G. (1964). *J. Chem. Soc.* 6196.
Kilpatrick, M. (1963). *J. Amer. Chem. Soc.* **85**, 1036.
Kilpatrick, M., Meyer, M. W., and Kilpatrick, M. L. (1960). *J. Phys. Chem.* **64**, 1433.
Kilpatrick, M. L. (1947). *J. Amer. Chem. Soc.* **69**, 40.
Koskikallio, J., and Whalley, E. (1959a). *Trans. Faraday Soc.* **55**, 815.
Koskikallio, J., and Whalley, E. (1959b). *Canad. J. Chem.* **37**, 788.
Koskikallio, J., Pouli, D., and Whalley, E. (1959). *Canad. J. Chem.* **37**, 1360.
Kreevoy, M. M. (1957). *J. Amer. Chem. Soc.* **79**, 5927.
Kreevoy, M. M., and Taft, R. W. (1955a). *J. Amer. Chem. Soc.* **77**, 3146.
Kreevoy, M. M., and Taft, R. W. (1955b). *J. Amer. Chem. Soc.* **77**, 5590.
Kresge, A. J. (1965). *Discuss. Faraday Soc.* **39**, 46.
Kresge, A. J., and Chiang, Y. (1959). *J. Amer. Chem. Soc.* **81**, 5509.
Kresge, A. J., and Chiang, Y. (1961). *J. Amer. Chem. Soc.* **83**, 2877.
Kresge, A. J., Hakka, L. E., Mylonakis, S., and Sato, Y. (1965a). *Discuss. Faraday Soc.* **39**, 75.
Kresge, A. J., More O'Ferrall, R. A., Hakka, L. E., and Vitullo, V. P. (1965b). *Chem. Comm.* 46.
Kresge, A. J., Chiang, Y., and Sato, Y. (1967). *J. Amer. Chem. Soc.* **89**, 4418.
Krieble, V. K., and Holst, K. A. (1938). *J. Amer. Chem. Soc.* **60**, 2977.
Kuivila, H. G. (1955). *J. Am. Chem. Soc.* **77**, 4014.
Kwart, H., and Herbig, J. (1963). *J. Amer. Chem. Soc.* **85**, 226.
Kwart, H., and Weisfeld, L. B. (1958). *J. Amer. Chem. Soc.* **80**, 4670.
Lacey, R. N. (1960). *J. Chem. Soc.* 1633.
Ladenheim, H., and Bender, M. L. (1960). *J. Amer. Chem. Soc.* **82**, 1895.
Landini, D., Montanari, F., Modena, G., and Scorrano, G. (1968). *Chem. Comm.* 86.
Lane, C. A. (1964). *J. Amer. Chem. Soc.* **86**, 2521.

Lapidot, A., Samuel, D., and Weiss-Broday, M. (1964). *J. Chem. Soc.* 637.
Lapworth, A. (1904). *J. Chem. Soc.* 30.
Leininger, P. M., and Kilpatrick, M. (1939). *J. Amer. Chem. Soc.* **61**, 2510.
Leisten, J. A. (1956). *J. Chem. Soc.* 1572.
Leisten, J. A. (1959). *J. Chem. Soc.* 765.
Levy, J. B., Taft, R. W., and Hammett, L. P. (1953). *J. Amer. Chem. Soc.* **75**, 1253.
Ley, J. B., and Vernon, C. A. (1957a). *J. Chem. Soc.* 2987.
Ley, J. B., and Vernon, C. A. (1957b). *J. Chem. Soc.* 3256.
Liang, H. T., and Bartlett, P. D. (1958). *J. Amer. Chem. Soc.* **80**, 5585.
Librovich, N. B., and Vinnik, M. I. (1968). *Russ. J. Phys. Chem.* **42**, 1337.
Lichty, D. M. (1907). *J. Phys. Chem.* **11**, 225.
Lockhart, J. C. (1966). *J. Chem. Soc. B*, 1174.
Long, F. A., and Bakule, R. (1963). *J. Amer. Chem. Soc.* **85**, 2313.
Long, F. A., and Friedman, L. (1950). *J. Amer. Chem. Soc.* **72**, 3692.
Long, F. A., and McIntyre, D. (1954). *J. Amer. Chem. Soc.* **76**, 3243.
Long, F. A., and Paul, M. A. (1957). *Chem. Rev.* **57**, 935.
Long, F. A., and Pritchard, J. G. (1956). *J. Amer. Chem. Soc.* **78**, 2663.
Long, F. A., and Purchase, M. (1950). *J. Amer. Chem. Soc.* **72**, 3267.
Long, F. A., and Schulze, J. (1964). *J. Amer. Chem. Soc.* **86**, 327.
Long, F. A., Dunkle, F. B., and McDevit, W. F. (1951a). *J. Phys. Chem.* **55**, 829.
Long, F. A., McDevit, W. F., and Dunkle, F. B. (1951b). *J. Phys. Chem.* **55**, 813.
Long, F. A., Pritchard, J. G., and Stafford, F. E. (1957). *J. Amer. Chem. Soc.* **79**, 2362.
Longridge, J. L., and Long, F. A. (1968). *J. Amer. Chem. Soc.* **90**, 3088, 3092.
Longridge, J. L., and Long, F. A. (1967). *J. Amer. Chem. Soc.* **89**, 1292.
Lowen, A. M., Murray, M. A., and Williams, G. (1950). *J. Chem. Soc.* 3318.
Lucas, H. J., and Eberz, W. F. (1934). *J. Amer. Chem. Soc.* **56**, 460.
Lucas, H. J., and Liu, Y. (1934). *J. Amer. Chem. Soc.* **56**, 2138.
Lucas, H. J., Stewart, W. T., and Pressman, D. (1944). *J. Amer. Chem. Soc.* **66**, 1818.
March, D. M., and Henshall, T. (1962). *J. Phys. Chem.* **66**, 840.
McIntyre, D., and Long, F. A. (1954). *J. Amer. Chem. Soc.* **76**, 3240.
Mehrotra, R. N. (1968). *J. Chem. Soc. B*, 642.
Melander, L., and Myhre, P. C. (1959). *Arkiv. Kemi.* **13**, 507.
Miller, R. C., Noyce, D. S., and Vermeulen, T. (1964). *Ind. Eng. Chem.* **56**, 43.
Mironov, G. S., Vetrova, V. V., Kozlova, I. P., and Farberov, M. I. (1966). *Russ. J. Appl. Chem.* **39**, 1502.
Moodie, R. B., Renton, J. R., and Schofield, K. (1969). *J. Chem. Soc. B*, 578.
Moodie, R. B., Wale, P. D., and Whaite, T. J. (1963). *J. Chem. Soc.* 4273.
More O'Ferrall, R. A. (1967). *Adv. Phys. Org. Chem.* **5**, 331.
Morgan, T. D. B., Stedman, G., and Hughes, M. N. (1968). *J. Chem. Soc. B*, 344.
Nagakura, S., Minegishi, A., and Stanfield, K. (1957). *J. Amer. Chem. Soc.* **79**, 1033.
Newman, M. (1941). *J. Amer. Chem. Soc.* **63**, 2431.
Newman, M. S. (1942). *J. Amer. Chem. Soc.* **64**, 2324.
Newman, M. S., Kuivila, H. G., and Garrett, A. B. (1945). *J. Amer. Chem. Soc.* **67**, 704.
Novikova, L. I., and Konkin, A. A. (1959). *Russ. J. Appl. Chem.* **32**, 1104.
Noyce, D. S., and Avarbock, H. S. (1962). *J. Amer. Chem. Soc.* **84**, 1644.
Noyce, D. S., and Jorgenson, M. J. (1961). *J. Amer. Chem. Soc.* **83**, 2525.
Noyce, D. S., and Jorgenson, M. J. (1963a). *J. Amer. Chem. Soc.* **85**, 2420.
Noyce, D. S., and Jorgenson, M. J. (1963b). *J. Amer. Chem. Soc.* **85**, 2427.
Noyce, D. S., and Lane, C. A. (1962a). *J. Amer. Chem. Soc.* **84**, 1635.
Noyce, D. S., and Lane, C. A. (1962b). *J. Amer. Chem. Soc.* **84**, 1641.

Noyce, D. S., and Reed, W. L. (1958). *J. Amer. Chem. Soc.* **80**, 5539.
Noyce, D. S., and Snyder, L. R. (1958). *J. Amer. Chem. Soc.* **80**, 4324.
Noyce, D. S., Pryor, W. A., and King, P. A. (1959). *J. Amer. Chem. Soc.* **81**, 5423.
Noyce, D. S., Woo, G. L., and Jorgenson, M. J. (1961). *J. Amer. Chem. Soc.* **83**, 1160.
Noyce, D. S., Avarbock, H. S., and Reed, W. L. (1962a). *J. Amer. Chem. Soc.* **84**, 1647.
Noyce, D. S., King, P. A., Kirby, F. B., and Reed, W. L. (1962b). *J. Amer. Chem. Soc.* **84**, 1632.
Noyce, D. S., King, P. A., Lane, C. A., and Reed, W. L. (1962c). *J. Amer. Chem. Soc.* **84**, 1638.
Noyce, D. S., Hartter, D. R., and Miles, F. B. (1968a). *J. Amer. Chem. Soc.* **90**, 3794.
Noyce, D. S., Hartter, D. R., and Pollack, R. M. (1968b). *J. Amer. Chem. Soc.* **90**, 3791.
Ogata, Y., Okano, M., and Ganke, T. (1956). *J. Amer. Chem. Soc.* **78**, 2962.
O'Gorman, J. M., and Lucas, H. J. (1950). *J. Amer. Chem. Soc.* **72**, 5489.
Olson, A. R., and Hyde, J. L. (1941). *J. Amer. Chem. Soc.* **63**, 2459.
Olson, A. R., and Miller, R. J. (1938). *J. Amer. Chem. Soc.* **60**, 2687.
Orr, W. J. C., and Butler, J. A. V. (1937). *J. Chem. Soc.* 330.
Osborn, A. R., and Whalley, E. (1961). *Canad. J. Chem.* **39**, 597.
Overend, W. G., Rees, C. W., and Sequeira, J. S. (1962). *J. Chem. Soc.* 3429.
Palm, V. A. (1956). *Proc. Acad. Sci. U.S.S.R. Sect. Chem.* **108**, 289.
Palm, V. A. (1958). *Zhur. fiz. Khim.* **32**, 620.
Paul, M. A. (1950). *J. Amer. Chem. Soc.* **72**, 3813.
Paul, M. A. (1954). *J. Amer. Chem. Soc.* **76**, 3236.
Polanyi, M., and Szabo, A. L. (1934). *Trans. Faraday Soc.* **30**, 508.
Pritchard, J. G., and Long, F. A. (1956a). *J. Amer. Chem. Soc.* **78**, 2667.
Pritchard, J. G., and Long, F. A. (1956b). *J. Amer. Chem. Soc.* **78**, 6008.
Pritchard, J. G., and Long, F. A. (1958). *J. Amer. Chem. Soc.* **80**, 4162.
Purlee, E. L., and Taft, R. W. (1956). *J. Amer. Chem. Soc.* **78**, 5807.
Rabinovitch, B. S., and Winkler, C. A. (1964). *Canad. J. Res.* **20B**, 73.
Reid, E. E. (1899). *Amer. Chem. J.* **21**, 284.
Reid, E. E. (1900). *Amer. Chem. J.* **24**, 397.
Ridd, J. H. (1961). *Quart. Rev.* **15**, 418.
Roberts, I., and Urey, H. C. (1938). *J. Amer. Chem. Soc.* **60**, 2391.
Rocek, J., and Krupicka, J. (1958). *Coll. Czech. Chem. Comm.* **23**, 2068.
Rosenthal, D., and Taylor, T. I. (1957). *J. Amer. Chem. Soc.* **79**, 2684.
Rozelle, L. T., and Alberty, R. A. (1957). *J. Phys. Chem.* **61**, 1637.
Salomaa, P. (1957a). *Acta Chem. Scand.* **11**, 132.
Salomaa, P. (1957b). *Acta Chem. Scand.* **11**, 141.
Salomaa, P. (1957c). *Acta Chem. Scand.* **11**, 235.
Salomaa, P. (1957d). *Acta Chem. Scand.* **11**, 239.
Salomaa, P. (1957e). *Acta Chem. Scand.* **11**, 247.
Salomaa, P. (1959). *Suomen Kemi*, **B32**, 145.
Salomaa, P., and Kankaanperä, A. (1961). *Acta Chem. Scand.* **15**, 871.
Salomaa, P., and Linnantie, R. (1958). *Acta Chem. Scand.* **12**, 2051.
Satchell, D. P. N. (1956). *J. Chem. Soc.* 3911.
Satchell, D. P. N. (1957). *J. Chem. Soc.* 2878.
Satchell, D. P. N. (1958). *J. Chem. Soc.* 3904.
Satchell, D. P. N. (1959). *J. Chem. Soc.* 463.
Schaleger, L. L., and Long, F. A. (1963). *Adv. Phys. Org. Chem.* **1**, 1.
Schubert, W. M. (1949). *J. Amer. Chem. Soc.* **71**, 2639.

Schubert, W. M., and Burkett, H. (1956). *J. Amer. Chem. Soc.* **78**, 64.
Schubert, W. M., and Kintner, R. R. (1966). *In* "The Chemistry of the Carbonyl Group" (S. Patai, ed.), p. 695. Interscience, London.
Schubert, W. M., and Latourette, H. K. (1952). *J. Amer. Chem. Soc.* **74**, 1829.
Schubert, W. M., and Quacchia, R. H. (1963). *J. Amer. Chem. Soc.* **85**, 1284.
Schubert, W. M., and Zahler, R. E. (1954). *J. Amer. Chem. Soc.* **76**, 1.
Schubert, W. M., Donohue, J., and Gardner, J. D. (1954). *J. Amer. Chem. Soc.* **76**, 9.
Schubert, W. M., Zahler, R. E., and Robins, J. (1955). *J. Amer. Chem. Soc.* **77**, 2293.
Schubert, W. M., Lamm, B., and Keeffe, J. R. (1964). *J. Amer. Chem. Soc.* **86**, 4727.
Schulze, J., and Long, F. A. (1964). *J. Amer. Chem. Soc.* **86**, 331.
Semke, L. K., Thompson, N. S., and Williams, D. G. (1964). *J. Org. Chem.* **29**, 1041.
Shanker, R., and Swami, S. N. (1963). *J. Indian Chem. Soc.* **40**, 105.
Silbermann, W. E., and Henshall, T. (1957). *J. Amer. Chem. Soc.* **79**, 4107.
Simkins, R. J. J., and Williams, G. (1952). *J. Chem. Soc.* 3086.
Singer, K., and Vamplew, P. A. (1956). *J. Chem. Soc.* 3971.
Skrabal, A., Stockmair, W., and Schreiner, H. (1934). *Z. phys. Chem.* **A169**, 177.
Smith, W. B., Bowman, R. E., and Kmet, T. J. (1959). *J. Amer. Chem. Soc.* **81**, 997.
Stamhuis, E. J., and Drenth, W. (1963a). *Rec. Trav. Chim.* **82**, 385.
Stamhuis, E. J., and Drenth, W. (1963b). *Rec. Trav. Chim.* **82**, 394.
Staples, P. J. (1964). *J. Chem. Soc.* 2534.
Stasiuk, F., Sheppard, W. A., and Bourns, A. N. (1956). *Canad. J. Chem.* **34**, 123.
Sullivan, M. J., and Kilpatrick, M. L. (1945). *J. Amer. Chem. Soc.* **67**, 1815.
Surfleet, B., and Wyatt, P. A. H. (1965). *J. Chem. Soc.* 6524.
Swain, C. G., and Rosenberg, A. S. (1961). *J. Amer. Chem. Soc.* **83**, 2154.
Taft, R. W. (1952). *J. Amer. Chem. Soc.* **74**, 5372.
Taft, R. W., Purlee, E. L., Riesz, P., and De Fazio, C. A. (1955). *J. Amer. Chem. Soc.* **77**, 837.
Taft, R. W., Deno, N. C., and Skell, P. S. (1958). *Ann. Rev. Phys. Chem.* **9**, 287.
Taylor, T. W. J. (1930). *J. Chem. Soc.* 2741.
Thomas, R. J., and Long, F. A. (1964a). *J. Amer. Chem. Soc.* **86**, 4770.
Thomas, R. J., and Long, F. A. (1964b). *J. Org. Chem.* **29**, 3411.
Timell, T. E. (1964). *Canad. J. Chem.* **42**, 1456.
Treffers, H. P., and Hammett, L. P. (1937). *J. Amer. Chem. Soc.* **59**, 1708.
Vinnik, M. I., and Librovich, N. B. (1967). *Russ. J. Phys. Chem.* **41**, 1075.
Vinnik, M. I., and Medvetskaya, I. M. (1967). *Russ. J. Phys. Chem.* **41**, 947.
Vinnik, M. I., Ryabova, R. S., and Chirkov, N. M. (1957). *Proc. Acad. Sci. U.S.S.R.* **117**, 793.
Vinnik, M. I., Ryabova, R. S., and Chirkov, N. M. (1959). *Russ. J. Phys. Chem.* **33**, 575.
Vinnik, M. I., Grabovskaya, Z. E., and Arzamaskova, L. N. (1967). *Russ. J. Phys. Chem.* **41**, 580.
Westheimer, F. H., and Kharasch, M. S. (1946). *J. Amer. Chem. Soc.* **68**, 1871.
Whalley, E. (1959). *Trans. Faraday Soc.* **55**, 798.
Whalley, E. (1964). *Adv. Phys. Org. Chem.* **2**, 93.
Whalley, E. (1966). *Ber. Phys. Chem.* **70**, 958.
Whitford, E. L. (1925). *J. Amer. Chem. Soc.* **47**, 953.
Whitmore, F. C., and Rothrock, H. S. (1932). *J. Amer. Chem. Soc.* **54**, 3431.
Wiberg, K. B. (1955). *Chem. Rev.* **55**, 713.
Wiberg, K. B., and Evans, R. J. (1960). *Tetrahedron*, **8**, 313.
Wiberg, K. B., and Mill, T. (1958). *J. Amer. Chem. Soc.* **80**, 3022.

Wiig, E. O. (1930). *J. Amer. Chem. Soc.* **52**, 4737.
Williams, G., and Clark, D. J. (1956). *J. Chem. Soc.* 1304.
Williams, G., and Lowen, A. M., (1950). *J. Chem. Soc.* 3312.
Williams, G., and Simkins, R. J. J. (1953). *J. Chem. Soc.* 1386.
Winstein, S., and Lucas, H. J. (1937). *J. Amer. Chem. Soc.* **59**, 1461.
Withey, R. J., and Whalley, E. (1963a). *Trans. Faraday Soc.* **59**, 901.
Withey, R. J., and Whalley, E. (1963b). *Canad. J. Chem.* **41**, 546.
Yates, K., and McClelland, R. A. (1967). *J. Amer. Chem. Soc.* **89**, 2686.
Yates, K., and Riordan, J. C. (1965). *Canad. J. Chem.* **43**, 2328.
Yates, K., and Stevens, J. B. (1965). *Canad. J. Chem.* **43**, 529.
Zucker, L., and Hammett, L. P. (1939). *J. Amer. Chem. Soc.* **61**, 2791.

CHAPTER 6

Non-aqueous and Mixed Aqueous Solvents

Many studies have been made of chemical equilibria and reaction rates in non-aqueous and mixed aqueous solutions of strong acids and attempts have been made to correlate the results using acidity function concepts. As with aqueous solutions equilibrium protonation measurements lead to the definition of appropriate acidity functions. Correlation of reaction rates with the acidity functions may then lead to useful information about the mechanisms of the reactions. The complications and limitations of the acidity function approach have already been emphasized in the preceeding chapters for aqueous acid solutions. In general the situation becomes more complex for mixed aqueous acid solutions. This is in part due to the addition of another component to the solvent system. Also for solvents of low dielectric constant there is the added possibility of appreciable ion association. In this chapter the usefulness of the acidity function concept when applied to non-aqueous and mixed aqueous acid solutions will be examined.

6.1. Anhydrous Formic Acid

The value, referred to water as standard state, of H_0 is –2·22 for anhydrous formic acid (Paul and Long, 1957; Stewart and Mathews, 1960). Sodium formate and aniline behave as completely ionized strong bases in formic acid (Hammett and Deyrup, 1932). Both solutes depress the acidity of formic acid. Thus $H_0 = -0{\cdot}22$ for 0·10 mole litre^{-1} sodium formate and $H_0 = -0{\cdot}24$ for 0·10 mole litre^{-1} aniline solution. Measurements for sulphuric acid in the concentration range 0·0266 mole litre$^{-1} \leqslant C_{H_2SO_4} \leqslant$ 0·1064 mole litre^{-1} using 4-nitrodiphenylamine and 2,4-dichloro-6-nitroaniline as indicators show that H_2SO_4 is a strong acid in anhydrous formic acid. Thus $H_0 = -3{\cdot}60$ when $C_{H_2SO_4} = 0{\cdot}0266$ mole litre^{-1} and $H_0 = -4{\cdot}32$ when $C_{H_2SO_4} = 0{\cdot}1064$ mole litre^{-1}. Benzene sulphonic acid is not such a strong acid in HCOOH as is H_2SO_4.

Taking $pK_{BH^+} = -0{\cdot}29$ for 2-nitroaniline the pK_{BH^+} values for the other indicators studied by Hammett and Deyrup (1932) are by stepwise comparison: –1·06 for 4-chloro-2-nitroaniline, –2·63 for 4-nitrodiphenylamine and –3·43 for 2,4-dichloro-6-nitroaniline. These figures are deduced from measurements for anhydrous formic acid but refer to water as the standard reference state. They compare favourably with the values (Table 2.28) of –1·03, –2·49 and –3·29 respectively deduced from results for aqueous acid solutions. Thus equation (6.1) is applicable (Paul and Long, 1957) where $K^{S}_{BH^+}$ is the ionization constant of BH^+ in solvent S and K_{BH^+} is for water solvent. The activity

$$pK^{S}_{BH^+} - pK_{BH^+} = \log_{10}\left(\frac{y_B{}^0 y^0_{H^+}}{y^0_{BH^+}}\right) \approx \text{constant} \tag{6.1}$$

coefficients y^0 are the limiting values at infinite dilution in S relative to the limiting infinite dilution values in water. The present results give the constant in equation (6.1) as 5·10 when S is HCOOH (Paul and Long, 1957). Thus anhydrous HCOOH is a much more acidic solvent than water. The validity of pK_{BH^+} data deduced from indicator measurements using anhydrous solvents depends upon there being a constant difference between $pK^{S}_{BH^+}$ and pK_{BH^+} for the series of successively more weakly basic indicators used in the stepwise comparison of ionization ratios. This condition is not always fulfilled, even by indicators with the same functional basic groups (Bell, 1959).

Figures of –6·85 at 20° (Grinter and Mason, 1964) and –6·49 at 25° (Stewart and Mathews, 1960) have been quoted for J_0 for formic acid containing 2% of water. Both values are based on $pK_{ROH} = 6{\cdot}63$ (equation 3.3; Table 3.3) for triphenylcarbinol (Deno *et al.* 1955). In accord with the corresponding solute effects on H_0 of anhydrous formic acid sodium formate increases J_0 (J_0 becomes less negative—a decrease in acidity) and H_2SO_4 decreases J_0 (an increase in acidity) for 98% aqueous formic acid (Grinter and Mason, 1964). For anhydrous HCOOH J_0 is probably of the order of –8 (Stewart and Mathews, 1960).

6.2. Acetic Acid and Acetic Acid/Water Mixtures

6.2.1. *Equilibrium Measurements*

The dielectric constant of acetic acid is 6·13 at 20°C (Smyth and Rogers, 1930). The formation of ion pairs and higher ionic aggregates therefore has a major influence on acid–base equilibria in acetic acid solvent (Kolthoff and Bruckenstein, 1956; Bruckenstein and Kolthoff, 1956). Thus the protonation of a weakly basic indicator B in a solution of an acid HX in acetic acid may be written as equation (6.2). The protonated base exists almost entirely as the ion

$$B + HX \rightleftharpoons BH^+X^- \tag{6.2}$$

pair species BH^+X^- and therefore the experimental H_0 acidity function defined from indicator measurements for acetic acid solutions is given by equation

$$H_0 = pK_{BH^+} - \log_{10}(C_{BH^+X^-}/C_B) \tag{6.3}$$

(6.3) in which K_{BH^+} is the acid ionization constant of BH^+ in water (H_0 therefore is referred to standard state water). This leads to equation (6.4) in which K_{BHX} is the equilibrium constant for reaction (6.2) and C_{HX} is the sum of the equilibrium concentrations of ionized HX and ion pairs H^+X^-.

$$H_0 = pK_{BH^+} - \log_{10} K_{BHX} - \log_{10} C_{HX} \quad (6.4)$$

Equation (6.4) predicts that a linear graph with slope of -1 should be obtained when H_0 is plotted against $-\log_{10} c_{HX}$. This is in accord with experiment providing $c_{HX} <$ ca. 0·01 mole litre^{-1} for perchloric acid, sulphuric acid, 4-toluene sulphonic acid and hydrochloric acid in anhydrous acetic acid (Ludwig and Adams, 1954; Kolthoff and Bruckenstein, 1956; Torck *et al.* 1962a). Smith and Elliott (1953) have also proved this relationship for $HClO_4$, HBr, H_2SO_4, HCl and seven sulphonic acids in acetic acid containing 0·12% water. However, for a given concentration of each acid in the series the values of H_0 ranged over ca. 2 H_0 units. Thus, for example, with $C_{HX} = 0{\cdot}001$ mole litre^{-1}, $H_0 = -0{\cdot}56$ for $HClO_4$, $H_0 = +0{\cdot}26$ for H_2SO_4 and $H_0 = +1{\cdot}01$ for HCl in acetic acid containing 0·12% water (Smith and Elliott, 1953). These figures are corrected to $pK_{BH^+} = -0{\cdot}29$ for 2-nitroaniline (Table 2.28). Similar results were obtained for anhydrous acetic acid (Torck *et al.* 1962a). For a series of acids HX a series of parallel lines result when H_0 is plotted against $\log_{10} C_{HX}$. This clearly reflects the influence of ion association on the measured H_0 scales. Gramstad's (1959) results for $HClO_4$ and trifluoromethanesulphonic acid in acetic acid containing 0·195% water and some data for phosphoric acid in CH_3COOH (Knessl *et al.* 1955) are consistent with this conclusion. Also conductivity measurements have shown that the specific conductivities of a series of acids at the same concentration in acetic acid are in the order $CF_3SO_3H > HClO_4 > HBr > H_2SO_4 > CH_3SO_3H > HCl$ (Kolthoff and Willman, 1934; Gramstad, 1959). On the assumption that the specific conductivities reflect the extent of dissociation of the acids, there is a direct correlation between this sequence and the same order of decreasing acidity of the acid solutions as indicated by the experimental H_0 values.

Homer and Moodie (1965) determined the ionization ratios for several amides in solutions of H_2SO_4 in acetic acid. Plots of $\log_{10}(C_{BH^+}/C_B)$ against $\log_{10} C_{H_2SO_4}$ had unit slope for three primary amides and one tertiary amide but slopes of 1·2 for three secondary amides.

Hall and Spengeman (1940) have measured H_0 for H_2SO_4 in acetic acid up to $C_{H_2SO_4} = 8$ mole litre^{-1}. At these high concentrations the linear relationship between H_0 and $\log_{10} C_{H_2SO_4}$ is no longer valid. Hall and Spengeman used as indicators the primary amines 2-nitroaniline, 2,4-dichloro-6-nitroaniline, 4-chloro-2-nitroaniline and 2,6 -dinitro-4-methylaniline and the tertiary amine *N,N*-dimethyl-2,4,6-trinitroaniline. For the latter the experimental acidity function decreased more rapidly with increasing H_2SO_4 concentration than the

H_0 scale defined by the primary amines. This result is analogous to the deviation between the protonation behaviour of primary and tertiary amines in aqueous solutions of H_2SO_4 (Section 3.2.1; Fig. 3.2). For 2-nitroaniline, 4-chloro-2-nitroaniline and 2,6-dinitro-4-methylaniline the observed ionization ratios led to the definition of an H_0 scale which was consistent with earlier measurements by Paul and Hammett (1936) using 2-nitroaniline indicator. The H_0 scales for aqueous H_2SO_4 and CH_3COOH/H_2SO_4 are approximately parallel functions of H_2SO_4 concentration, although for the latter the H_0 values are ca. 2·75 units more negative.

The indicator 2,4-dichloro-6-nitroaniline in CH_3COOH/H_2SO_4 gives an H_0 scale which parallels that defined by the other three primary amines used by Hall and Spengeman (1940), but which is consistently ca. 0·4 H_0 units less negative (Paul and Long, 1957). Comparison of the H_0 scales defined by the protonation behaviour of two different weakly basic indicators B and C may be made via equation (6.4). Thus equation (6.5) is applicable (Bruckenstein, 1960). It follows that differences in H_0 for two indicators arise either because

$$H_0(\mathrm{B}) - H_0(\mathrm{C}) = \mathrm{p}K_{\mathrm{BH^+}} - \mathrm{p}K_{\mathrm{CH^+}} - \log_{10}(K_{\mathrm{BHX}}/K_{\mathrm{CHX}}) \tag{6.5}$$

$\Delta \mathrm{p}K_a$ in going from water to anhydrous acetic acid are not equal for the two indicators or because K_{BHX} does not equal K_{CHX} in acetic acid. For structurally similar indicators K_{BHX} and K_{CHX} would be of the same order of magnitude and $\Delta \mathrm{p}K_a$ would also be approximately constant. It follows that a common H_0 scale would be expected. However, other examples in which differences exist for structurally similar indicators have become common place (Roček, 1957; Satchell, 1958a, 1958c; Zajac and Nowicki, 1965). The indicator *p*-naphtholbenzein which was used by Smith and Elliott (1953) and Kolthoff and Bruckenstein (1956) is unsuitable for use at high (ca. 100°C) temperatures as it then becomes unstable (Martin and Reece, 1959). Its use at ordinary temperatures has also been questioned (Bethell *et al.* 1958).

One further consequence of extensive ion pairing occurring in acetic acid is that comparison of the H_0 values $H_0(\mathrm{HX})$ and $H_0(\mathrm{HY})$ defined by the same indicator B for two acids HX and HY in CH_3COOH does not give a straightforward indication of the relative strengths of the acids. Equation (6.6) can be derived (Bruckenstein, 1956) in which K_{HX} is the dissociation constant of HX and $K_{\mathrm{BH^+X^-}}$ is the dissociation constant of the ion pair $\mathrm{BH^+X^-}$ in acetic acid (similarly for K_{HY} and $K_{\mathrm{BH^+Y^-}}$).

$$H_0(\mathrm{HX}) - H_0(\mathrm{HY}) = \log_{10}(K_{\mathrm{HY}}\,K_{\mathrm{BH^+X^-}}/K_{\mathrm{HX}}\,K_{\mathrm{BH^+Y^-}}) \tag{6.6}$$

The relative values of H_0 therefore not only reflect the relative strengths of the acids HX and HY but are also influenced by the relative ability of the acid anions X^- and Y^- to form ion pairs with BH^+ the protonated indicator.

The addition of water to an H_2SO_4/CH_3COOH mixture decreases the acidity of the solution (Gold and Hawes, 1951). Figure 6.1. shows how H_0 for some $H_2O/CH_3COOH/H_2SO_4$ mixtures varies with the water and sulphuric acid concentrations. Noyce and Castelfranco (1951) studied the entire range

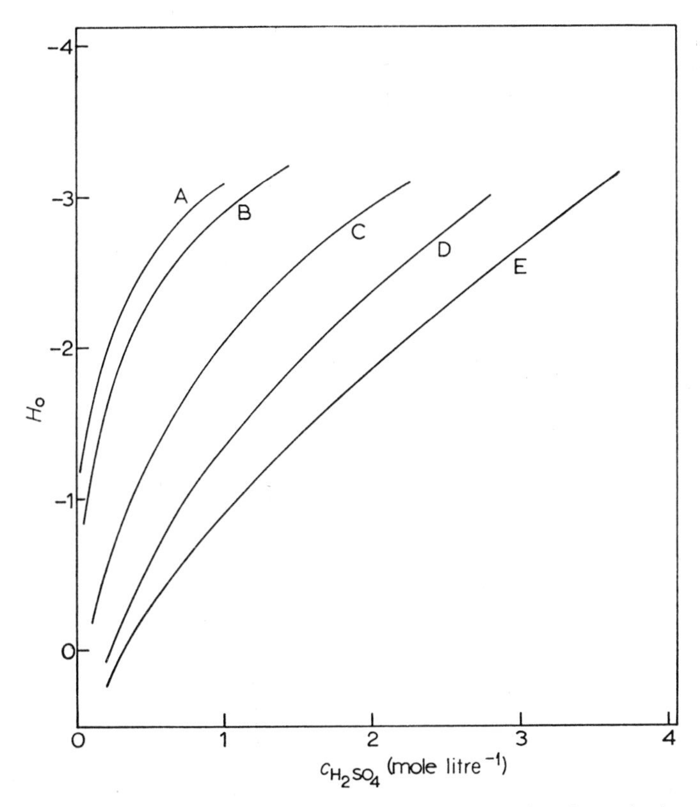

FIG. 6.1. Variation of H_0 with sulphuric acid concentration for anhydrous acetic acid: curve A (Hall and Spengeman, 1940), and four CH_3COOH/H_2O mixtures (Roček, 1957). % CH_3COOH are: curve B 99%; C 95%; D 90%; E 85%. Indicator was 4-chloro-2-nitroaniline pK_{BH^+}(water) = −1·03 (Table 2.28).

0–100% of acetic acid concentrations for 1 mole litre^{-1} and 0·005 mole litre^{-1} sulphuric acid. For the former a maximum in H_0 (minimum acidity) occurred at ca. 37% acetic acid. Results for these solutions have also been reported by Ogata and Okano (1956), Hamer and Leslie (1960), Torck *et al.* (1962a, 1960) and Valenta (1964).

The acidity of the solution of several other acids in CH_3COOH is also decreased by the addition of water. The acids for which quantitative H_0 results are available include perchloric acid (Ludwig and Adams, 1954;

Torck *et al.* 1960, 1962a; Valenta, 1964; Roček, 1957; Wiberg and Evans, 1958), hydrochloric acid (Schwarzenbach and Stensby, 1959; Ogata and Okano, 1956), hydrobromic acid (Zajac and Nowicki, 1965; Schwarzenbach and Stensby, 1959), phosphoric acid (Knessl *et al.* 1955; Roček, 1957) and toluene sulphonic acid (Valenta, 1964). The H_0 values for HCl, HBr and $HClO_4$ each at 0·01 mole litre^{-1} in the full 0–100% range of acetic acid–water mixtures all show a maximum (minimum acidity) at ca. 55% CH_3COOH (Schwarzenbach and Stensby, 1959). Furthermore, for a particular H_2O/CH_3COOH ratio the H_0 values for the three acids are never different by more than ca. 0·1 H_0 units except for $>99\%$ CH_3COOH. Here there was a large deviation in accord with Smith and Elliott's (1953) results for several acids in acetic acid containing 0·12% water. A linear relationship has been established between H_0 and the logarithm of the water activity for 0·3 mole litre^{-1} solutions of HBr, $HClO_4$ and H_2SO_4 in 85%, 91% and 95% aqueous acetic acid (Zajac and Nowicki, 1965). Analogous correlations between H_0 and $\log_{10} a_w$ for aqueous acids are discussed in Section 2.3.1.

Bruckenstein (1960) has derived equation (6.7) for the effect of an equilibrium concentration C_w of water on H_0 for an acid HX in acetic acid. This follows from equation (6.4). The effect of the water is expressed by reaction (6.8)

$$H_0 = pK_{BH^+} - \log_{10} K_{BHX} - \log_{10} C_{HX^+} \log_{10}(1 + K_f\, C_w) \qquad (6.7)$$

$$H_2O + HX \rightleftharpoons H_3O^+X^- \qquad (6.8)$$

for which the equilibrium constant is K_f. Equation (6.7) is applicable for water concentrations less than about 0·4 mole litre^{-1}. Thus Ludwig and Adams (1954) observed that H_0 plotted aginst $-\log_{10} C_{HClO_4}$ gave a series of parallel lines with unit slope, one for each of a series of different water concentrations in acetic acid. The displacement of the lines is consistent with the value of 34 (Kolthoff and Bruckenstein, 1956) for K_f when X^- is ClO_4^-. When $C_w >$ ca. 0·4 mole litre^{-1} solvent structure, solvation effects and the dielectric constant of the solvent all begin to alter appreciably and the simple correlation is no longer valid.

The results discussed above have largely been based on the use of primary amines as weakly basic indicators. The acidity functions are all referred to standard state water via pK_{BH^+} for the indicators in water (equation 6.3). "Best" values of pK_{BH^+} are given in Table 2.28 and many of the earlier results require adjusting in accord with these more recent pK_{BH^+} data. Because indicators of the same structural class can lead to different H_0 scales (equation 6.5) it is necessary to emphasize the particular indicators used to define any scale.

The acidity of solutions of $SnCl_4$ in acetic acid indicates that the species $H_2SnCl_4(OCOCH_3)_2$ is of comparable strength to H_2SO_4. However, Satchell (1958c) stresses that the apparent relative acidity is to some extent a function of the indicator used to make the measurement. For $ZnCl_2$ in acetic acid the species $H_2ZnCl_2(OCOCH_3)_2$ is responsible for the acidity of the solutions. The acidity functions for $ZnCl_2/HCl/CH_3COOH$ mixtures exhibit maxima when the $ZnCl_2$ concentration is varied with the HCl concentration constant (Bethell *et al.* 1958). The three species $H_2ZnCl_2(OCOCH_3)_2$, H_2ZnCl_4 and $H_2ZnCl_3(OCOCH_3)$ are present in the solutions. As the $ZnCl_2$ concentration is increased H_2ZnCl_4 which is formed at low concentrations is replaced by the less acidic $H_2ZnCl_3(OCOCH_3)$ thus giving rise to the observed maxima in the H_0 curves. No maxima were observed for the $SnCl_4/HCl/CH_3COOH$ system (Satchell, 1958c). The effect of lithium chloride on the ionization ratios of primary amines in acetic acid is consistent with a linear relationship between $\log_{10}(C_{BH^+}/C_B)$ and $\log_{10} C_{LiCl}$ (Gaboriaud, 1967). An H_0 acidity function for 0·005 to 32 wt % BF_3 in acetic acid has been determined by Bel'skii and Vinnik (1964). $H_0 = -5{\cdot}41$ for 32 % BF_3. At low concentrations (< 8 % BF_3) H_0 varies linearly with the logarithm of the BF_3 concentration. Bel'skii and Vinnik deduced pK_{BH^+} values for a series of amines by the stepwise procedure based on $pK_{BH^+} = 0{\cdot}99$ for 4-nitroaniline in water. Their results again show that the deduced pK_{BH^+} values sometimes differ significantly from the correct values (Table 2.28) for water. Thus pK_{BH^+} was –0·20 for 2-nitroaniline (cf. –0·29 in water), –1·75 for 2-nitro-5-chloroaniline (–1·54), –3·90 for 2,4-dichloro-6-nitroaniline (–3·29) and –4·63 for 2,4-dinitroaniline (–4·48). Bel'skii and Vinnik intimate that it would be preferable to define the acidity function for acetic acid solvent in terms of the ionization constants of the indicators in acetic acid. This would merely shift their scale numerically by a constant amount equal to the difference between pK_{BH^+} for 4-nitroaniline in acetic acid and in water. The scale would then refer to anhydrous acetic acid as standard state. This would seem a better approach than the more common one in which acidity function scales for anhydrous solvents are referred to water as standard state.

The ionization of triphenylcarbinol (Gold and Hawes, 1951) in H_2SO_4/-CH_3COOH mixtures has been studied by visible spectrometry (Gold and Hawes, 1951) and by p.m.r. spectroscopy (O'Reilly and Leftin, 1960). A set of J_0 values can thence be deduced via equation (3.2) with $pK_{ROH} = -6{\cdot}6$ (Table 3.3). The relationship between J_0 and H_0 is similar to that for aqueous H_2SO_4 solutions. Addition of small concentrations of water to acetic acid containing H_2SO_4 represses the ionization of carbinol indicators (Gold and Hawes, 1951; Bethell and Gold, 1958a). The relationship between H_0 and J_0 is maintained for water concentrations at least up to 0·555 mole litre^{-1}. White and Stout (1962) have determined J_0 for the range of H_2SO_4 concentrations $4\% \leqslant C_{H_2SO_4} \leqslant 86\%$ in an aqueous acetic acid solvent containing 80 wt % acetic acid. Values range from $J_0 = -1{\cdot}80$ when $C_{H_2SO_4} = 4\%$ to $J_0 = -15{\cdot}94$ when $C_{H_2SO_4} = 86\%$. The corresponding figures for aqueous H_2SO_4 are –0·11 and –15·7 respectively (Table 3.1).

6.2.2. *Kinetic Measurements*

Correlation of reaction rates with acidity functions for acetic acid solvent has relied on the Zucker–Hammett approach in which the linearity of $\log_{10} k_1$ against H_0 plots is tested. Satchell's (1956) results for the rates of hydrogen exchange of anisole in H_2SO_4/CH_3COOH furnish an example (Fig. 6.2). The

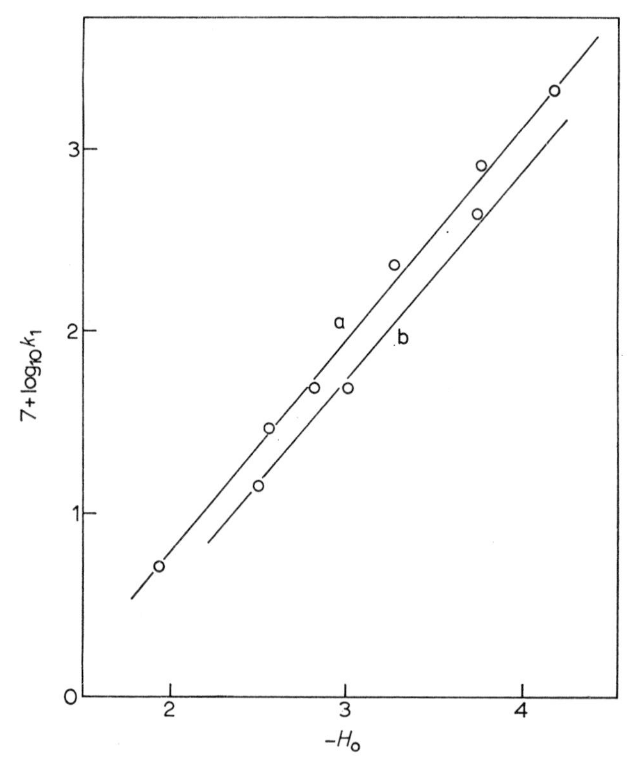

FIG. 6.2. Variation of rates of exchange (25°) of (a) 4-deutero anisole and (b) 2-deutero anisole in CH_3COOH/H_2SO_4 solutions with the H_0 acidity function. Reproduced by permission from Satchell (1956).

plots are straight with −1·18 slope. The same mechanism as that for aqueous solutions is implied (Section 5.19). Linear plots of $\log_{10} k_1$ for the exchange reaction of 4-deutero anisole against H_0 were also found for solutions of $SnCl_4$ in acetic acid, for the $SnCl_4/HCl/CH_3COOH$ solvent system (Satchell, 1958c) and for $ZnCl_2$ in acetic acid (Satchell 1958b).

The rates of aralkylation of anisole by diarylmethanols (XOH) in acetic acid/sulphuric acid solutions correlate with the equilibrium ionization ratios of two carbinols in the same solutions (Bethell and Gold, 1958a). The reactions therefore involve a rate-determining bimolecular attack of anisole by the carbonium ion X^+ formed via the rapid pre-equilibrium (6.9).

$$XOH + H^+ \rightleftharpoons X^+ + H_2O \qquad (6.9)$$

The appropriate correlation for the $ZnCl_2$ catalysed reaction is identical to that for catalysis by H_2SO_4 showing that the $ZnCl_2$ generates Brønsted acidity in the solutions by interaction with the co-catalyst acetic acid (Bethell and Gold 1958b). The rates of diarylmethylation of anisole in $ZnCl_2/CH_3COOH$ containing a constant added HCl concentration exhibit a maximum and a minimum as the $ZnCl_2$ concentration is increased. However, these effects are also reflected in the variation with $ZnCl_2$ concentration of the ionization ratios of indicators with similar structures to the aralkylation reagents (Bethell *et al.* 1958). The rate of chloromethylation of mesitylene by formaldehyde and HCl in 90 v/v % aqueous acetic acid may be written in the form k_2[mesitylene] [formaldehyde] (Ogata and Okano, 1956). A slope of $-0{\cdot}96$ was obtained from the linear plot of $\log_{10} k_2$ against H_0. The mechanism probably involves a rate determining attack of anisole by the protonated formaldehyde cation $^+CH_2OH$.

The rate constants k for the sulphuric acid catalysed dehydration of 1-methylcyclohexanol are porportional to h_0 (Roček, 1960). The $\log_{10} k$ against H_0 plots for 100 %, 99 %, 95 %, 90 % and 85 % aqueous acetic acid all fell on the same straight line. The fast pre-equilibrium protonation of the alcohol is followed by a slow rate-determining loss of water to form a methylcycylohexyl carbonium ion which rapidly loses a proton to give the product olefin. The loss of acetic acid from 1-methylcyclohexyl acetate procedes by a similar mechanism and gives a linear correlation of $\log_{10} k$ with H_0 for H_2SO_4 catalyst in 100 %, 99 %, and 90 % aqueous acetic acid. The rates of dehydration of 1-(4-anisyl)-1,2,2-triphenyl ethanol in acetic acid and aqueous acetic acid also correlate with H_0 (Valenta, 1964). A study of the HBr and H_2SO_4 catalysed removal of the *N*-benzyloxycarbonyl group from *N*-benzyloxycarbonylglycine ethyl ester in acetic acid led Homer *et al.* (1965) to emphasize the importance of ion pair formation on the rates of reactions in media of low dielectric constant. The acidity dependence of the rates was steeper for the H_2SO_4 than for the HBr catalysed reaction. This was ascribed to the involvement of the ion pair SH^+Br^- (where S is the substrate ester) in the slow step in solutions containing the more nucleophilic (than HSO_4^-) bromide ion.

The first correlation of reaction rate with H_0 for anhydrous acetic acid was made by Paul (1950) for the H_2SO_4 catalysed depolymerization of trioxane. The rate of decomposition is decreased by the addition of water (Hamer and Leslie, 1960). This is consistent with the decrease in acidity produced when water is added to acetic acid although a reduction might also arise because of the possibility of catalysis by undissociated acetic acid molecules (Section 5.9).

Plots of $\log_{10} k_1$ against $-H_0$ were linear with ca. unit slope for the depolymerization in 80 vol. % aqueous acetic acid catalysed by $HClO_4$ and H_2SO_4 (Torck *et al.* 1960, 1962b). However, for a particular H_0 value the plots were staggered by ca. 0·4 $\log_{10} k_1$ units for the two catalysing acids. The ionization of an H_0 indicator B in an acid HX is given by equation (6.2). For two different acids HX and HY the respective equilibrium constants K_{BHX} and K_{BHY} will not be equal. The rate-determining step in the depolymerization will involve a slow unimolecular reaction of the ion pair SH^+X^- (catalyst HX) or SH^+Y^- (catalyst HY) where S is trioxane. The equilibrium constants K_{SHX} and K_{SHY} for the rapid pre-equilibrium formation of the ion pairs from S + HX and S + HY respectively will also not be the same. The relative rates of depolymerization catalysed by solutions of two different acids with the same H_0 values will be given by equation (6.10). The results for the $HClO_4$ and H_2SO_4 catalysed reactions of trioxane are in accord with this equation. A similar deviation has

$$\log_{10} k_{HX} - \log_{10} k_{HY} = \log_{10} (K_{SHX} K_{BHY}/K_{SHY} K_{BHX}) \qquad (6.10)$$

been observed between the rates of the HCl and $HClO_4$ catalysed decomposition of paraldehyde in anhydrous acetic acid (Torck *et al.* 1960, 1962b). For the same reaction in water, where ion association is absent, catalysis by HNO_3, $HClO_4$, HCl or H_2SO_4 leads to a common H_0 against $\log_{10} k_1$ plot (Bell and Brown, 1954). The rates of the H_2SO_4 catalysed decomposition of trimeric acetone peroxide in acetic acid also correlate with H_0 (Furuya and Ogata, 1963).

The $HClO_4$ catalysed rearrangement of benzopinacol to the corresponding pinacolone occurs by two paths, one direct and one via the intermediate tetraphenylethylene oxide (Gebhart and Adams, 1954). The rates of pinacolone formation either direct from the pinacol or from the tetraphenylethylene oxide parallelled h_0. An A-1 slow rate-determining reaction of the protonated substrate is indicated in either case. The rates of the H_2SO_4 catalysed rearrangements of several benzhydryl azides and 1,1-diarylethyl azides in acetic acid are also directly proportional to h_0 (Gudmundsen and McEwen, 1957). Similar results were obtained by Patai and Dayagi (1962) for the formation of the triphenylmethyl carbonium ion from compounds of the type $Ph_3C.CHX_2$ where X is an amide, ester or cyanide group. The entropies of activation $-4{\cdot}3$ e.u. $\leqslant \Delta S^{\neq} \leqslant 2{\cdot}9$ e.u. were taken as supporting the evidence for a unimolecular slow step implied by the acidity function correlation. However, these $\Delta S^{\neq}$ values are intermediate between typical results for proven A-1 and A-2 reactions (Whalley, 1964).

The condensation of benzaldehyde with methyl ethyl ketone (Noyce and Snyder, 1959) or acetophone (Noyce and Pryor, 1955) in acetic acid solvent involves a bimolecular interaction between the ketone enol and the protonated aldehyde in the slow condensation step. Proportionality of reaction rate with

h_0 is observed. The acid catalysed condensation reaction of anisaldehyde and methyl ethyl ketone behaves similarly (Noyce and Snyder, 1958) although the reaction of 4-nitrobenzaldehyde with methyl ethyl ketone does not fit the acidity function correlation (Noyce and Snyder, 1959).

Examples for which $\log_{10} k_1$ does not parallel H_0 are the acid catalysed bromination reactions of 3-nitroacetophenone (Paul and Hammett, 1936) and deoxybenzoin (Cieciuch and Westheimer, 1963). However, in both these cases there is evidence that the acid anions (HSO_4^- for the former reaction and Cl^- or Br^- for the latter reaction) catalyse the slow α-proton transfer in the rate-determining step of the enolization process. Thus for catalysis by perchloric acid there is no such effect. A quantitative description of the reaction rates is complicated by ion pairing in solutions with high acetic acid content.

The rates of loss of hydrogen peroxide in the H_2SO_4 catalysed peroxide oxidation of 4-hydroxybenzaldehyde in aqueous acetic acid are proportional to the product of the concentrations of the aldehyde and H_2O_2 (Ogata and Tabushi, 1959). The second-order rate constants are dependent on k_p the rate constant for the reaction of peracetic acid which is the reactive intermediate in the oxidation. Although k_p was proportional to h_0, $\log_{10} k_p$ did not parallel H_0. This is consistent with equation (6.11) in which k_p is split into an acid dependent

$$k_p = k + k_A h_0 \tag{6.11}$$

and an acid independent term. Ogata and Tabushi proposed that the mechanism involves rate determining formation both of the protonated complex $ArCHO.CH_3CO_3H.H^+$ and of the unprotonated intermediate $ArCHO.CH_3CO_3H$. Ayagi (1961) proposed that the formation of acetamide and triphenylmethyl carbonium ions from *N*-(triphenylmethyl)-acetamide in anhydrous H_2SO_4/CH_3COOH mixtures is an $A_{AL}1$ reaction. The correlation of rate with H_0 required used of equation (4.13) because *N*-(triphenylmethyl)-acetamide was appreciably protonated in the strongly acidic solutions.

The perchloric acid catalysed Thiele acetylations of benzoquinone and toluquinone in acetic acid/acetic anhydride mixtures were studied by Mackenzie and Winter (1944). They correlated the rates with the ionization ratios of 2,2-dimethoxyquinone in the same solutions. However, Long and Paul (1957) consider that the results do not give an unambiguous indication of a particular mechanism for the reactions.

A graph of $\log_{10} k_1$ against $-H_0$ with approximately unit slope for an acid catalysed raction in aqueous or anhydrous acetic acid solutions has generally been taken as indicative of an A-1 mechanism. However, in drawing such a conclusion the many assumptions inherent in the interpretation of acidity function behaviour should not be lost sight of. It is worth noting that for some of the correlations described in this section the H_0 values are in the region where $-H_0$ is a linear function with unit slope of the logarithm of the catalysing acid concentration. Thus $\log_{10} k_1$ would also show a linear dependence (slope ca. 1) on $\log_{10} C_{acid}$. The Zucker–Hammett approach would require a different interpretation to be put on this result! Considerable caution should be exercised in the interpretation of rate data for reactions in these solvent systems.

6.3. Anhydrous Trifluoroacetic Acid

The original measurements of H_0 for solutions of H_2SO_4 in anhydrous trifluoroacetic acid (Mackor *et al.* 1957) have been corrected in accord with the Paul and Long (1957) revised indicator pK_{BH^+} values by Dallinga and ter Maten (1960). These authors have also listed H_0 scales for the following acids in CF_3COOH: perchloric acid (0·0068 to 0·080 molal), HBF_3OH (0·142 to 0·616 molal), $(CF_3)_2PO_2H$ (0·043 to 0·386 molal) and F_2PO_2H (0·099 to 1·175 molal). For a particular acid concentration the $-H_0$ values for the series of acids studied were in the order $HClO_4 > HBF_3OH > H_2SO_4 > (CF_3)_2PO_2H > F_2PO_2H$. Solutions of H_2SO_4 in CF_3COOH are much more acidic than the corresponding H_2SO_4 solutions in either formic acid or acetic acid solvents. D_2SO_4 in CF_3COOD is slightly less acidic than H_2SO_4 in CF_3COOH (Dallinga and ter Maten, 1960). An H_0 scale for H_2SO_4 up to 18·70 mole litre^{-1} ($H_0 = -11{\cdot}10$) in CF_3COOH has been determined by Hyman and Garber (1959) who have also studied HF/CF_3COOH mixtures over the full composition range. They quote $H_0 = -3{\cdot}03$ for anhydrous trifluoroacetic acid and $H_0 = -9{\cdot}97$ for 100% HF. The latter value is less negative than the figures deduced by measurements on aqueous HF solutions (Table 2.12) although the discrepancy is consistent with the difference between the pK_{BH^+} for 2,4,6-trinitroaniline used by Hyman and Garber (1959) and the most probable value given in Table 2.28.

Plots of $\log_{10} k_1$ against $-H_0$ for the H_2SO_4 catalysed hydrogen exchange reactions of benzene and alkylsubstituted benzenes in trifluoroacetic acid were linear with unit slope (Mackor *et al.* 1957). The same mechanism as for aqueous solution is likely (Section 5.19). Acid catalysed hydrogen exchange of hydrogen atoms in the t-butyl groups of t-butylbenzene and t-butyltoluenes occurs in trifluoroacetic acid (Dallinga and ter Maten, 1960). The proposed mechanism involves a pre-equilibrium protonation of the aromatic molecule followed by a rate-determining reaction between the protonated substrate and a molecule of the catalysing acid. Exchange takes place rapidly in the resulting cation. This mechanism is in accord with the observed linear relationship between $\log_{10} k_1$ and $-H_0 + \log_{10} C_{acid}$. The experimental points for $HClO_4$, HBF_3OH, H_2SO_4, $(CF_3)_2PO_2H$ and F_2PO_2H as catalysing acids all fell on the same straight line.

6.4. Alcohols and Alcohol/Water Mixtures

6.4.1. *Equilibrium Measurements*

An acidity function for hydrochloric acid in ethanol has been measured by Braude (1948) who used 4-nitroaniline as indicator. Nahlovsky and Chvalovsky (1968) have extended the H_0 scale up to $C_{HCl} = 8{\cdot}27$ mole litre^{-1} by the stepwise comparison of $\log_{10}(C_{BH^+}/C_B)$ for benzeneazodiphenylamine, 4-nitroaniline, 2-nitroaniline, 4-chloro-2-nitroaniline and 4-nitrodiphenylamine. With $pK_{BH^+} = 0{\cdot}99$ for 4-nitroaniline values of pK for the other indicators were 1·45, –0·54, –1·51 and –2·92 respectively. These do not agree with pK_{BH^+} for these indicators in water (see Table 2.28). The difference between $\log_{10}(C_{BH^+}/C_B)$ for two indicators in the same HCl/EtOH solution is related to the difference between pK_{BH^+} *in ethanol* for the two indicators by equation (1.43). Thus if pK_{BH^+} (ethanol) for one indicator is known pK_{BH^+} for the others can be deduced which leads to the definition of an acidity function for HCl in EtOH which is referred to ethanol standard state. The Nahlovsky and Chvalovsky H_0 scale refers to water standard state and is compared in Fig. 6.3 with the corresponding scales for HCl in 5% v/v water in ethanol (Nahlovsky and Chvalovsky, 1968), HCl in 48% v/v water in ethanol (Satchell, 1957b), and aqueous HCl (Table 2.9). Baines and Eaborn (1956) measured ionization ratios of 4-nitroaniline for 0·109 mole litre^{-1} $\leqslant C_{HCl} \leqslant$ 1·954 mole litre^{-1} in 5% v/v water in ethanol. Their results are consistent with the appropriate H_0 scale in Fig. 6.3. Bunton *et al.* (1957) have quoted an H_0 scale for methanesulphonic acid in methanol.

Figure 6.3 shows that for a particular HCl concentration over the full range of aqueous ethanol mixtures a minimum acidity (maximum in H_0) occurs at a particular solvent composition. This was first clearly established by Braude and Stern (1948a) for 0·1 mole litre^{-1} and 1 mole litre^{-1} HCl. The maximum in H_0 occurred around 60 mole % ethanol. Similar results have been obtained by Salomaa (1957a) for HCl (0·1, 0·2, 0·4 and 0·8 mole litre^{-1}) in methanol–water mixtures and by Nachbaur (1960) for HF (2 mole litre^{-1}) in ethanol–water solutions. For low concentrations of HCl in a particular solvent composition $(\log_{10} C_{HCl} + H_0)$ is ca. constant (Braude, 1948; Salomaa, 1957a). The conclusion (Paul and Long, 1957) that this implies that HCl is a strong acid in these solutions is reasonable. However, by analogy to the results for acetic acid solvent (Bruckenstein, 1960) the possibility of other effects, in particular ion association, should also be considered. Other measurements for HCl in aqueous alcohols include those of D'Cruz (1964) for methanol and ethanol, of Kalidas and Palit (1961) for glycol and of Eaborn (1953) for methanol. Rosenthal *et al.* (1964) have shown that the addition of sodium chloride to HCl solutions in 33·4% methanol decreases H_0 by an amount which is directly proportional to the salt concentration. This is analogous to the effect of NaCl on H_0 for aqueous HCl (Paul, 1954).

The most detailed investigation of acidity scales for ethanol–water mixtures over the full composition range has been made by Bates and Schwarzenbach

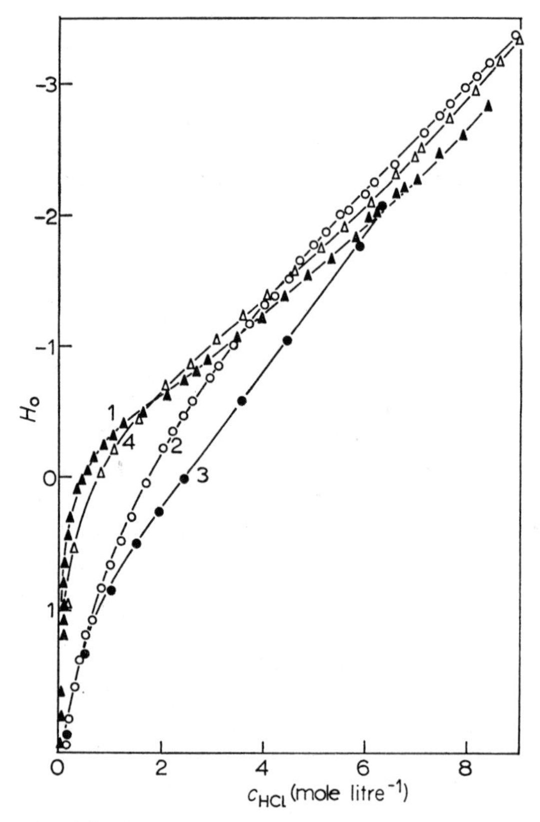

FIG. 6.3. The H_0 acidity functions for solutions of HCl in (1) EtOH, (2) 5% v/v water in EtOH, (3) 48% v/v water in EtOH, (4) water. Reproduced by permission from Nahlovsky and Chvalovsky (1968).

(1955; Bates 1964a). They measured H_0 using an amine indicator, H_- using a phenol indicator, pH by measurement of the e.m.f. of a cell with a hydrogen and a calomel electrode (Section 1.2), and $p(a_{H^+}\gamma_{Cl^-})$ using cell (1.9) with the buffer solution replaced by the solution under investigation (see equation 1.10). The solutes studied were (0·002 M HCl, 0·008 M NaCl), (0·02 M CH_3COOH, 0·01 M CH_3COONa, 0·005 M NaCl), and (0·004 M triethanolammonium chloride, 0·002 M triethanolamine). For the HCl/NaCl system H_0 (3-nitroaniline indicator) had a maximum value at ca. 70 wt % EtOH. This corresponds to a minimum in pK_{BH^+} for 3-nitroaniline at this solvent composition. Explanations of the variations of H_0 with solvent composition have invoked the breakdown of water structure as the alcohol content is increased (Franks and Ives, 1966), the relative basicities of the alcohol and water molecules in a particular solvent mixture, and the competition for a neutral indicator molecule between different forms of solvated protons, for example $(H_2O)_4H^+$ and $ROH(H_2O)_3H^+$ (Braude and Stern, 1948a; Salomaa, 1957a; Wells, 1965, 1966, 1968). Paul and Long (1957) point out that the marked deviations between H_0, H_-, pH and $p(a_{H^+}\gamma_{Cl^-})$ found by Bates and Schwarzenbach (1955) clearly show that the acidity of a solution is not a unique property but is dependent on the method of measurement. Use of the acidity scales must be made in the correct context. Thus for example it is logical to attempt to correlate the rates of acid catalysed reactions of neutral substrates with H_0 and of mono-negatively charged substrates with H_-.

There is evidence (Gutbezahl and Grunwald, 1953; de Ligny *et al.* 1961; Rosenthal *et al.* 1964; Juillard and Simonot, 1968; Wynne-Jones, 1968) that the H_0 and H_- scales for alcohol–water mixtures are not independent of the indicators used to measure the scales sometimes even for structurally similar bases. Some of the recorded deviations arise because of a structural dissimilarity of the bases concerned. Thus the ionization behaviour of 4-aminoazobenzene-4′-sulphonic acid and its *N*-methyl and *N,N*-dimethyl derivatives (de Ligny *et al.* 1961) is in part consistent with the observed deviations between the acidity function scales defined by the protonation behaviour of primary, secondary and tertiary amines in aqueous solutions of strong acids. Deviations for structurally similar bases are less serious but nevertheless lead to the conclusion that acidity function scales lose at least some of their significance for these solutions. The definition by Grunwald and Berkowitz (1951) of an "activity function" Y_0 which for alcohol–water mixtures is a function of solvent only has been criticized by Wynne-Jones (1968). Y_0 is also a function of the structure of the weakly acidic solutes (Reynaud, 1968, 1969).

The pH and other acidity function scales for aqueous acid solutions are referred to water standard state. Similarly it is preferable to refer pH* and acidity function scales for a particular alcohol–water solvent composition to a standard state in the mixed solvent (de Ligny and Alfenaar, 1967). Standard buffer solutions for methanol–water and ethanol–water mixtures have been tabulated (Bates, 1964; Alfenaar and de Ligny, 1967). Acidity function scales for changing acid concentration in a fixed solvent composition will then have similar significance and usefulness as the corresponding scales for aqueous solutions providing ion association is negligible. Correlations of acidity functions with reaction rates or protonation equilibria (and hence the evaluation of ionization constants) must then refer to the particular solvent system as standard state. Acidity functions for a fixed acid concentration in a series of aqueous solutions with increasing alcohol content apparently have less generality and therefore less significance. Their usefulness is more limited.

Acidity functions for sulphuric acid in 20% ethanol–80% aqueous H_2SO_4 mixtures have been measured using azobenzenes (Jaffé and Gardner, 1958; Yeh and Jaffé, 1959) and diphenylamines (Dolman and Stewart, 1967) as bases. Plots of $\log_{10}(C_{BH^+}/C_B)$ against $C_{H_2SO_4}$ were accurately parallel for overlapping indicators in each series. The two scales are not the same in accord with the deviation between acidity functions for aqueous acids defined by structurally dissimilar bases. Less detailed measurements using primary amine indicators for H_2SO_4 in ethanol–water (Kwart and Weisfeld, 1958; Kwart and Herbig, 1963) and isopropanol–water (Bartlett and McCollum, 1956) and for $HClO_4$ in ethanol–water (de Ruyter de Steveninck, 1958) and methanol–water (Eaborn, 1953) have also been reported. Tillett and Young (1968) used 4-nitroaniline, 2-nitroaniline and 4-chloro-2-nitroaniline to measure H_0 for concentrated solutions of $HClO_4$ (up to 4 mole litre^{-1}) in 2-butoxyethanol–water mixtures containing 40, 60 and 80 v/v % of the alcohol. H_0 varied with increasing acid concentration more steeply in the order of alcohol content $40 < 60 < 80$ v/v %. H_0 values for 0·2 mole litre^{-1} and 0·853 mole litre^{-1} solutions of eleven organophosphorous acids (general formula (X)(Y)PO_2H with X and/or Y as phenoxyl, hydrogen, cyclohexoyl, hydroxyl, phenyl or cyclohexyl) in ethylene glycol were determined using 4-nitroaniline as indicator (Cook and Mason, 1966).

The J_0 acidity function for aqueous H_2SO_4 is increased (a decrease in acidity) on the addition of isopropanol (Entelis *et al.* 1960a). The increase in J_0 is a linear function of the propanol concentration. Triphenylcarbinol was used as indicator.

6.4.2. *Kinetic Measurements*

The $HClO_4$ hydrolysis of sucrose and the HCl catalysed depolymerization of paraldehyde both in aqueous ethylene glycol give linear plots of $\log_{10}k_1$ against $-H_0$ with slopes of 0·98 and 1·22 respectively (Kalidas and Palit, 1961). The corresponding ϕ values (Section 4.3.2) are 0·15 and –0·75 (Bunnett and Olsen, 1966). A plot of $\log_{10}k_1$ against $-H_0$ for the depolymerization of paraldehyde catalysed by HCl in 24·4 mole % ethanol in water was 1·09 (Satchell, 1957b). These results are consistent with the established A-1 mechanisms for the reactions. On the other hand the iodination of acetone which is an A-2 reaction is catalysed by $HClO_4$ in 10% and 50% v/v aqueous ethylene glycol at rates which are directly proportional to the acid concentration (Kalidas and Palit, 1961). The ϕ parameters are 0·99 and 1·03 respectively (Bunnett and Olsen, 1966). Using the linearity of $\log_{10}k_1$ against H_0 or $\log_{10}C_{acid}$ as a diagnosis for an A-1 or A-2 mechanism is apparently acceptable for these solutions. However, Satchell (1957a) has shown that the proportionality between k_1 and acid concentration for the iodination of acetone in ethanol–water mixtures becomes unsatisfactory as the ethanol content of the solvent is increased. Thus for 44·4 mole % ethanol the correlation between $\log_{10} k_1$ and H_0 is better than that between $\log_{10} k_1$ and $\log_{10} C_{acid}$. It is unlikely that the mechanism is altered by the changes in solvent composition and therefore it must be concluded that the usefulness of acidity function correlations as criteria for mechanism are potentially ambiguous for alcohol–water solvent systems. Archer and Bell's (1959) surmise that Satchell's results arose because of appreciable equilibrium protonation of the substrate acetone is incorrect (Campbell and Edward, 1960). An apparent anomaly exists for the acid catalysed decomposition of cumene hydroperoxide for which $\log_{10} k_1$ is linear in $-H_0$ (slope 1·45) in 50 w/w % ethanol–water (de Ruyter van Steveninck, 1958) but is linear in $\log_{10} C_{acid}$ for H_2SO_4 in 50 w/w % aqueous acetic acid (Wichterle and Cefelin, 1957). However, in the latter case the acid concentrations studied were such that H_0 depended linearly on $\log_{10} C_{H_2SO_4}$ and therefore $\log_{10} k_1$ also depended linearly on H_0. An A-1 mechanism is probable and the apparent anomaly between the results for the two solvents ceases to exist. The hydrolyses of t-butyl benzoate and t-butyl 2,4,6-trimethylbenzoate in ethanol/water occur via alkyl–oxygen fission (Stimson and Watson, 1954) at rates which parallel the H_0 acidity function (Hawke and Stimson, 1956). This acidity function dependence is in accord with the proposed $A_{AL}1$ mechanism (Long and Paul, 1957).

An A-1 mechanism is also proposed to account for the acidity function dependence of the rates of the acid catalysed decomposition of several, substituted silanes in alcohol–water mixtures (Eaborn, 1953; Baines and Eaborn, 1956). The linear correlation (slope 1·1) between $\log_{10} k_1$ and $-H_0$ for the HCl and H_2SO_4 catalysed rearrangement of propenylethynyl carbinol in 20% v/v ethanol in water is also consistent with an A-1 mechanism. The observed absence of general acid catalysis is in accord with this conclusion. Braude and Stern (1948b) also studied this reaction and the rearrangement of phenylpropenyl carbinol in ethanol–water mixtures over the complete composition range. The rates of reaction exhibited a minimum at a particular ethanol–water composition. This is analogous to the corresponding minimum in the plot of $-H_0$ against solvent composition for a fixed concentration of a strong acid in alcohol–water mixtures.

Kaeding and Andrews (1952; Andrews and Kaeding, 1951) showed that $\log_{10}(k_1/a_w)$ was a linear function of H_0 for the hydrolysis of 4-nitrobenzophenone diethylketal in aqueous ethanolic HCl solutions. Their suggestion that this implies an A-2 mechanism involving a molecule of water in the rate determining step is improbable in view of the accepted A-1 mechanism for acetal or ketal hydrolyses in aqueous acid solutions (Section 5.5). Furthermore the rates of the HCl catalysed hydrolysis of ethylal in $EtOH/H_2O$ are in accord with the A-1 mechanism (Salomaa 1957e). Salomaa drew attention to the deviation between the measured rates of solvolysis and the true rates of hydrolysis of ethylal. The deviation arises because of the competition between water (reaction leading to hydrolysis products) and ethanol (reaction leading back to ethylal) for reaction with the intermediate $\mathrm{Et\overset{+}{O}{=}CH_2}$ formed in the slow step of the overall mechanism. Salomaa (1957b, 1957c, 1957d) has also studied the acid catalysed solvolyses of some alkoxymethyl esters in methanol–water and ethanol–water mixtures. The reactions proceed by $A_{AL}1$ and $A_{AC}2$ mechanisms concurrently and Salomaa (1957b) has devised a method for deducing the separate rates appropriate to each mechanism. The rates for the unimolecular reaction correlate with the H_0 acidity function.

Tests of reaction mechanism based on linear correlations of $\log_{10} k_1$ with H_0 or $\log_{10} C_{acid}$ depend on the assumptions of the Zucker–Hammett hypothesis (Chapter 4) and have even more severe limitations for acids in aqueous–alcohol mixtures than in water alone. Despite this only a few attempts have been made to investigate other approaches to the problem of using acidity function correlations as diagnoses of reaction mechanism. The values of w and w^* for the HCl catalysed hydrolyses of 2- and 4-nitrophenyl acetates in 20 v/v % methanol in water are consistent with the A-2 mechanism for the hydrolysis of simple esters (Rangaiah and Brahmaji Rao, 1967) although the a_w values used in the plots do not refer strictly to the aqueous methanol solvent system.

Gutbezahl and Grunwald (1953) have defined an "activity function" Y_0 for ethanol–water mixtures by equation (6.12) in which y_B and y_{BH^+} are the activity coefficients of a base B and its conjugate acid BH^+ in infinitely dilute non-aqueous solution referred to infinite dilution in water. The quantity m_B is

$$\log_{10}(y_B/y_{BH^+}) = m_B Y_0 \tag{6.12}$$

characteristic of B and Y_0 is considered to depend only on the solvent. Y_0 is zero for water and is taken as $-1{\cdot}000$ for 100% ethanol. Values of Y_0 for aqueous ethanol were hence deduced via equation (6.13) in which $pK_a(BH^+)$ refers to the particular solvent mixture for which Y_0 is applicable and $pK_a^w(BH^+)$ is the corresponding pK_a in water. $pK_a(C_6H_5NH_3^+)$ and $pK_a^w(C_6H_5NH_3^+)$ are for the anilinium ion in the solvent mixture and water respectively. The acids BH^+ chosen for the evaluation of Y_0 were a series of substituted ammonium

$$[pK_a(BH^+) - pK_a^w(BH^+)] - [pK_a(C_6H_5NH_3^+) - pK_a^w(C_6H_5NH_3^+)] = (m_B - m_{C_6H_5NH_2})\, Y_0 \tag{6.13}$$

ions. Correlation of Y_0 with reaction rate for acid catalysed reactions in aqueous ethanol depends on the proposal that a plot of $(\log_{10} k_1 - \log_{10} y_{H^+})$ against Y_0 may either be linear or markedly curved (Gutbezahl and Grunwald, 1953).

For the former it has been suggested that water is not involved in the slow step of the reaction, whereas if the plot is curved the involvement of water in the slow step is implied. Linear plots have been observed for the rearrangement of phenylpropenyl carbinol (Gutbezahl and Grunwald, 1953), the hydration of 3-*p*-menthene (Kwart and Weisfeld, 1958), the oxotropic rearrangement of α-phenylallyl alcohols (Kwart and Herbig, 1963), the hydration of 1-ethoxy-1-propyne (Stamhuis and Drenth, 1963), and the hydration of 1-t-butylthio-1-propyne (Hogeveen and Drenth, 1963). On the other hand the solvolysis of phenyl acetate (an established A-2 mechanism) gives a strongly curved graph (Kwart and Weisfeld, 1958). Despite these apparently satisfactory correlations the concept of a generally applicable Y_0 scale for ethanol–water mixtures is not acceptable (Wynne-Jones, 1968; Reynaud, 1968, 1969). The Y_0 scale has not a general significance and therefore its usefulness must be limited.

The problems associated with the correlation of reaction rates with some acidity scale for fixed concentrations of strong acids in aqueous solutions of alcohols are seemingly unsurmountable at the present time. The equilibrium and kinetic behaviour of chemical reactions in these solutions is intimately dependent on the structures of the solvent mixtures which themselves are incompletely understood (Franks and Ives, 1966). The determination of the effect of solvent composition on the enthalpy and entropy changes which accompany chemical reactions may be much more informative in this context. Thus for example values of the enthalpies and entropies of activation for the solvolysis of t-butyl chloride in aqueous ethanol, aqueous dioxane and aqueous acetic acid, over the complete solvent composition ranges have been measured by Winstein and Fainberg (1957). The importance of these and related results has been stressed by Arnett (1967). Further work in this important field is essential if chemical processes in mixed aqueous solvent systems are to be completely understood.

6.5. Dioxane/Water Mixtures

6.5.1. *Equilibrium Measurements*

Braude and Stern (1948a; Braude, 1948), Eaborn (1953), and Sadek *et al.* (1966) have measured an H_0 scale for solutions of HCl in aqueous dioxane. Braude and Stern's results showed that for a fixed concentration of HCl over the full aqueous dioxane composition range there was a minimum acidity (maximum in H_0) at a particular solvent composition. This is analogous to the corresponding result for aqueous ethanol solutions although because the dielectric constant of aqueous dioxane falls rapidly with increasing dioxane content HCl becomes less completely dissociated as the concentration of dioxane becomes greater. However, this does not account for the maximum in the H_0 against solvent composition curve which apparently arises because of the anomalous structural properties of water which is present as a component of the solvent system. Thus the H_0 scales for 0·1 molar HCl in ethanol–acetone and ethanol–dioxane mixtures fail to exhibit a minimum acidity at a particular solvent composition (Braude and Stern, 1948a).

H_0 scales have been recorded for $HClO_4$ (Bunton *et al.* 1957; Banthorpe *et al.* 1962) and HCl and H_2SO_4 (Torck *et al.* 1960, 1962a, 1962b) in 40–60 and 60–40 water–dioxane mixtures. For all three acids $-dH_0/dC_{acid}$ is greater for 40–60 water–dioxane than for 60–40 water–dioxane. Also for a particular molar concentration of the three acids $H_0(HClO_4) < H_0(H_2SO_4) < H_0(HCl)$. Thus, for example, for $HClO_4$, H_2SO_4 and HCl in 40–60 water–dioxane the H_0 values (based on $pK_{BH^+} = 0{\cdot}99$ for 4-nitroaniline) were 0·75, 1·15 and 1·47 respectively for $C_{acid} = 1$ mole litre^{-1} and –1·84, –1·41 and –0·88 for $C_{acid} = 4$ mole litre^{-1}. Bunton *et al.* (1957) concluded that for the ionization of structurally similar bases in aqueous dioxane solutions equation (1.42) and equation (1.43) were applicable and therefore that the measured H_0 scale has the same validity and significance as the H_0 acidity functions for purely aqueous solutions of strong acids. Torck *et al.* (1962a) concur with this conclusion. Banthorpe *et al.* (1962) measured H_0 for $HClO_4$ in 40–60 aqueous dioxane containing $LiClO_4$ added to constant total ionic strength (0·5 mole litre^{-1} and 1·0 mole litre^{-1}). Addition of $LiClO_4$ increased the acidity of the solutions.

An H_0 scale for H_2SO_4 (up to 74·88%; $H_0 = -6{\cdot}71$) in 5–95% dioxane–water solvent has been determined by Noyce and Jorgenson (1961, 1962). The ionization constants of some substituted chalcones were deduced by correlating their protonation behaviour with the measured H_0 scale. The electronic spectra of the protonated chalcones show pronounced medium effects and methods of correcting for these were compared by Noyce and Jorgenson (see also Section 1.4). Comparison of H_0 for H_2SO_4 in 5–95% dioxane–water with H_0 for H_2SO_4 in water shows that the former is greater (solutions less acidic) than the latter for $C_{H_2SO_4} <$ ca. 6 mole litre^{-1}, whereas the reverse is true for $C_{H_2SO_4} >$ ca. 6 mole litre^{-1}. This reversal may be associated with the fact that H_0 for ca. 6 mole litre^{-1} H_2SO_4 is numerically similar to the value of $pK_{BH^+} = -2{\cdot}92$ (Arnett and Wu, 1962) for dioxane in water. Thus appreciable protonation of dioxane occurs above this acid concentration.

The relationship between the J_0 and H_0 acidity functions for aqueous $HClO_4$ is maintained for $HClO_4$ in 60–40 and 40–60 water–dioxane mixtures (Dahn *et al.* 1960). For the aqueous dioxane solutions J_0 scales are tabulated for C_{HClO_4} up to 4·14 mole litre^{-1} (40% dioxane) and 3·89 mole litre^{-1} (60% dioxane).

6.5.2. *Kinetic Measurements*

There are many examples in which the rates of catalysis of an organic reaction by varying concentrations of a strong acid in a particular aqueous dioxane solvent composition correlate with the H_0 acidity function. Thus

$\log_{10}k_1$ is a linear function of H_0 for the $HClO_4$, HCl and H_2SO_4 catalysed decomposition of trioxane in 40–60 and 60–40 water–dioxane (Torck *et al.* 1960, 1962b), the HCl catalysed dehydration of 1-(4-anisyl)-1,2,2-triphenyl-ethanol (D'Cruz, 1964) the $HClO_4$ catalysed oxygen exchange reaction of 2,4,6-trimethylbenzoic acid in 40–60 water–dioxane (Bunton *et al.* 1960), the H_2SO_4 catalysed dehydration of β,β-diphenyl-β-hydroxypropiophenone in 95–5% water–dioxane (Noyce and Jorgenson, 1963), the $HClO_4$ and H_2SO_4 catalysed hydrolysis of 4-methoxydiphenylmethyl acetate in 40–60 water–dioxane (Bunton and Hadwick, 1957), and the $HClO_4$ catalysed hydrolyses of diazoketones (Dahn and Gold, 1963), acetals and ketals (Kreevoy, 1956), trimethylene oxide (Pritchard and Long, 1958), and sucrose (Bunton *et al.* 1957) in 40–60 and 60–40, 50–50, 60–40, and 40–60 and 60–40 water–dioxane respectively. In several of these examples there is independent evidence from entropies of activation or deuterium isotope effects that the A-1 mechanisms implied by the acidity function correlations are applicable. The linearity of plots of $\log_{10}k_1$ against $-H_0$ for reactions in aqueous dioxane mixtures apparently has similar implications (and limitations) to the corresponding result for purely aqueous solutions of strong acid. Graphs of unit slope will be fortuitous rather than the rule.

The rates of the $HClO_4$ catalysed hydrolysis of ethyl acetate in 40–60 and 60–40 water–dioxane are proportional to the perchloric acid concentration in accord with the expected A-2 mechanism for the reaction (Bunton *et al.* 1957). Similarly the H_2SO_4 catalysed *cis* to *trans* isomerizations of *cis*-chalcone and its 4-nitro and 4-chloro derivatives in 5–95% dioxane–water give curved plots of $\log_{10}k_1$ against $-H_0$ for which the slopes at high acid concentrations were 0·45, 0·50 and 0·55 respectively (Noyce and Jorgenson, 1961). The corresponding entropies of activation are −25·5 cal deg^{-1} mole^{-1} for *cis*-4-nitrochalcone and −23·7 cal deg^{-1} mole^{-1} for *cis*-chalcone. An A-2 mechanism involving bimolecular attack of water on the protonated chalcone is implied (equations 5.58 and 5.59). This contrasts with the proposed unimolecular rate-determining step (5.60) for the isomerization of *cis*-4-methoxychalcone. In this case the $\log_{10}k_1$ against $-H_0$ plot was linear with slope 1·14 (Noyce and Jorgenson, 1961). The slope 2·35 of a $\log_{10}k_1$ against $-H_0$ graph for the $HClO_4$ catalysed rearrangement of hydrozobenzene in 40–60 water–dioxane in part arises from an ionic strength effect (Banthorpe *et al.* 1962). Thus the slope became 2·1 when $LiClO_4$ was added to constant ionic strength for a series of $HClO_4$ solutions. This result conforms with the mechanism proposed by Carlin *et al.* (1951) in which the slow step is the decomposition of the doubly protonated hydrazobenzene molecule.

Very few studies have been reported in which reaction rates are correlated with H_0 for a series of acid solutions with varying dioxane–water composition. $\text{Log}_{10}k_1$ is approximately a linear function of H_0 for the HCl catalysed rearrangements of phenylpropenylcarbinol and ethynylpropenylcarbinol (Braude and Stern, 1948b). The rate-determining step in the reaction is the slow unimolecular rearrangement of the oxonium ion formed by a rapid pre-equilibrium protonation of the substrate carbinol. The HCl catalysed cleavage of the Si—C bond of 4-methoxyphenyltrimethylsilane (Eaborn, 1953) and hydrolysis of ethyl acetate (Corsaro, 1964) both give rise to linear plots of $\log_{10}k_1$ against $(H_0 - \log_{10}a_w)$. In agreement with the accepted (at least for aqueous solution) A-2 mechanism for ethyl acetate hydrolysis this result is interpreted as implying that a molecule of water is involved in bimolecular attack of the protonated substrate in the slow step of both reactions. The rates of the acid catalysed hydrolysis of methyl toluene-4-sulphinate in aqueous dioxane increase faster than C_{acid} but slower than h_0 with changing acid concentration or solvent composition (Bunton and Hendy, 1962). In accord with Bunnett's (1961) proposals this probably suggests an A-2 mechanism, a conclusion which is consistent with the entropy of activation (–12 cal deg^{-1} $mole^{-1}$) for the reaction.

6.6. Mixtures of Water with Aldehydes or Ketones

The H_0 scale measured with 4-nitroaniline and 3-nitroaniline indicators for 0·1 mole $litre^{-1}$ HCl in acetone–water mixtures showed a minimum acidity at ca. 50 mole % acetone (Braude and Stern, 1948a). There is no minimum for ethanol–acetone mixtures, a result which strengthens the conclusion that the observed acidity minima for mixtures of alcohols, dioxane or acetone with water in some way arises because of the anomalous structural properties of water. An acidity function scale for 0·01 mole $litre^{-1}$ HCl in acetaldehyde–water mixtures measured with 4-dimethylaminoazobenzene indicator also exhibits a maximum value of H_0 at ca. 50 mole % acetaldehyde (Ahrens and Strehlow, 1965). An H_0 scale for HCl in acetone gives rise at low HCl concentrations to a linear plot with unit slope of $-H_0$ against $\log_{10}C_{acid}$ (Braude, 1948). Thus HCl is a strong acid in acetone. An H_0 scale has also been measured for 0·0004 mole $litre^{-1} < C_{H_2SO_4} <$ 0·6 mole $litre^{-1}$ in methyl isopropyl ketone containing 0·5% water (Mörikofer *et al.* 1959). The weak base phenylazodiphenylamine gives an acidity function behaviour parallel to that of true Hammett primary amine bases in this solvent.

The rates of hydrolysis of benzyl fluoride catalysed by $HClO_4$ or H_2SO_4 in 10% aqueous acetone are proportional to the corresponding h_0 values for aqueous solutions of the catalysing acids (Swain and Spalding, 1960). A unimolecular decomposition of the protonated fluoride is implied for the rate-determining step. This is consistent with the proposed mechanism for the hydrolyses of some other fluorides in aqueous solutions of strong acids (Section 5.10.3). The approximate constancy of $(\log_{10}k_1 + H_0)$ for the rearrangements of phenylpropenylcarbinol (catalysed by 0·1 mole litre^{-1} HCl) and of ethynylpropenylcarbinol (catalysed by 1 mole litre^{-1} HCl) in acetone–water mixtures is also in accord with an A-1 mechanism for these reactions (Braude and Stern, 1948b).

Corsaro (1964) has shown for the HCl catalysed hydrolysis of ethyl acetate in acetone–water mixtures that $\log_{10} k_1$ is a linear function of $(H_0 - \log_{10} a_w)$. A molecule of water is involved in the rate-determining step and the A-2 mechanism for simple ester hydrolysis is applicable. A similar mechanism holds for the hydrolyses of 2- and 4-nitrophenyl acetates in 20 v/v % aqueous acetone. For these reactions $w =$ ca. 5 and $w^* =$ ca. –2 in accord (Table 4.2) with the involvement of water in the slow steps (Rangaiah and Brahmaji Rao, 1967). The rates of hydrolysis of t-butyl formate in aqueous acetone are proportional to the concentration of acid catalyst (Stimson, 1955c). An $A_{AC}2$ mechanism seems likely and is supported by the similarity between the energies of activation for the reaction and for the established $A_{AC}2$ hydrolyses of methyl and ethyl formate. The hydrolyses of t-butyl 2,4,6-trimethylbenzoate (Stimson, 1955a) and t-butyl benzoate (Stimson, 1955b) are probably $A_{AL}1$ reactions over the whole composition range of acetone–water mixtures. However, for high (ca. 90 v/v %) concentrations of acetone $\log_{10} k_1$ for the hydrolyses does not vary as expected for the linear correlation of $\log_{10} k_1$ with H_0 characteristic of A-1 reactions. The Zucker–Hammett hypothesis gives an incorrect indication of mechanism for these reactions.

6.7. Miscellaneous Solvent Systems

Acidity function scales for 0–100 % chlorosulphonic acid in sulphuric acid at 0, 20, 40, and 60°C have been deduced from measurements with 4-nitrotoluene as indicator (Palm, 1958). The acidity of the solutions decreases with increasing temperature. An H_0 scale for 10–85 wt % HNO_3 in H_2SO_4 has been determined by Novatskii *et al.* (1968) by p.m.r. measurements with 4-nitrobenzaldehyde as indicator. Addition of HCl to aqueous H_3PO_4 or P_2O_5 solutions increases their acidity as indicated by the protonation equilibrium of 2,4-dichloro-6-nitroaniline in these systems (Entelis and Chirkov, 1957). The addition of propylene decreases the acidity which is consistent with the analogous result for the addition of propylene to aqueous H_2SO_4 (Entelis *et al.* 1960b, 1960c).

The effect on H_0 of adding BF_3 to 98·5 % aqueous H_3PO_4 has been studied by Vinnik *et al.* (1957) using 2,4-dinitroaniline, 6-bromo-2,4-dinitroaniline and the non-Hammett base anthraquinone as indicators. The acidity of the solutions increased to a value of $H_0 = -8{\cdot}70$ when the mole ratio $BF_3/(H_3PO_4 + H_2O)$ equalled unity. H_0 scales for BF_3 in aqueous sulphuric acid have also been investigated (Wichterle *et al.* 1955). The acidity of solutions of boron trifluoride etherate in diethyl ether increases with increasing concentration of the etherate (Manelis *et al.* 1959). The increase in acidity is particularly marked as the 1:1 mole ratio of BF_3 to ether (for which $H_0 = -6{\cdot}22$) in the system is reached. The acidity function has been correlated with the rates of the BF_3 catalysed polymerization of isobutene and decomposition of formic acid (Vinnik *et al.* 1960). However, a different interpretation must be put on the H_0 values deduced for BF_3–ether mixtures because the colour change of the indicators arises via Lewis addition of BF_3 rather than Brønsted protonation (Utyanskaya and Vinnik, 1968).

Arnett and Douty (1964) have investigated the suitability of sulpholane as a solvent for acid–base reactions and have compared its behaviour with that of nitromethane which was previously studied by Hammett (Smith and Hammett, 1945; Van Looy and Hammett, 1959). The H_0 scales for HCl ($< 0{\cdot}15$ mole litre^{-1}) and H_2SO_4 ($< 0{\cdot}03$ mole litre^{-1}) in nitromethane were linear functions of $\log_{10} C_{acid}$ with 1·76 slope but displaced by ca. 4 H_0 units from each other. Nitromethane is a very weakly basic, aprotic solvent with a fairly high dielectric constant (38) and a poor capacity for the solvation of ions. Solutions of H_2SO_4 in nitromethane are considerably more acidic than the same concentrations of H_2SO_4 in water because in the former the acidity is due to H_2SO_4 molecules whereas for the latter the acidity is associated, in the usual way, with the ionic species arising from protonation of the solvent (i.e. H_3O^+ in water). The definition of an H_0 scale for H_2SO_4 in sulpholane was satisfactorily accomplished using nine indicators of decreasing basicity. Equations (1.42) and (1.43) were obeyed in the regions of overlap for successive indicators except for 6-bromo-2,4-dinitroaniline which was anomalous. Excellent agreement in behaviour was observed for the Hammett base 2,4,6-trinitroaniline and the non-Hammett base anthraquinone. The pK_{BH^+} values for the indicators used (deduced on the basis of $pK_{BH^+} = -1{\cdot}03$ for 4-chloro-2-nitroaniline) were in reasonable agreement with the values deduced from measurements for aqueous systems. Thus the relative basic strengths of the indicators is not much altered by the change from water to sulpholane solvent. The acidity of solutions of low concentrations of H_2SO_4 in sulpholane are somewhat higher than the corresponding acidity of aqueous solutions. This arises because the weak solvating ability of sulpholane represses the ionization of sulphuric acid, a conclusion which is borne out by the unit cryoscopic i factor for H_2SO_4 in this solvent. Sulpholane is therefore very similar to nitromethane in its behaviour as a medium for acid–base reactions. Measurements by Alder *et al.* (1966) of H_0 scales for HPF_6, HBF_4, $HClO_4$, HSO_3F, HBr, H_2SO_4 and HCl in sulpholane have emphasized in more detail the general findings of Arnett and Douty (1964).

Plots of $\log_{10}(C_{BH^+}/C_B)$ against $C_{H_2SO_4}$ for six primary amine indicators in solutions of H_2SO_4 in dimethyl sulphoxide were accurately parallel where the protonation of different indicators were measurable in overlapping ranges of H_2SO_4 concentration (Cook and Mason, 1966). An acidity function, referred to dimethyl sulphoxide as standard state was deduced from the ionization ratios. Cox and McTigue (1967) have shown that acid catalysed A-1 hydrolysis of acetal and the A-2 hydrolysis of ethyl acetate show distinctly different medium effects in water–dimethyl sulphoxide mixtures. They have discussed the results in terms of the effect of solvent composition on the activity coefficient ratios $(y_{H^+}y_S/y^{\ddagger})$ and $(y_{H^+}y_S a_w/y^{\ddagger})$ in the appropriate rate equations (4.6) and (4.18) respectively.

H_0 scales for some acids in toluene solvent have been discussed and measurements attempted using several weakly basic indicators (Sanders and Berger, 1967). However, the basic assumptions inherent in the acidity function approach are not applicable mainly because of solute aggregation processes (ion association, hydrogen bonding) which readily occur in toluene solvent.

6.8. Conclusions

The acidity function approach is generally as satisfactory for increasing concentrations of strong acids in a particular mixed solvent composition as for solutions of strong acids in water. However, in some cases this is not true, particularly for solvents of low dielectric constant where ion association becomes significant. pH scales for mixed aqueous or non-aqueous solvents are best referred to standard states in the particular solvent systems to which the scales refer (Bates and Robinson, 1966; Bates, 1964b). In many ways it would be preferable to maintain a similar approach for acidity functions rather than the present reference of most scales to water standard state via pK_{BH^+} in water for at least one of the indicators used to define each scale. However, the latter approach does give a more direct indication of the relative acidity of solutions of strong acids in different solvent systems.

The acidity function concept is less useful when applied to fixed concentrations of strong acids in series of solutions with varying solvent composition. The ionization behaviour of weak bases in such systems tends to be more specific. Parallel protonation behaviour is not maintained in many cases even for structurally similar bases. Correlations of reaction rates with acidity function scales are also not as satisfactory as for solutions in which the catalysing acid concentration is varied in a fixed solvent composition. It is these latter systems which are the most informative to study if it is required to determine pK_{BH^+} for a weak base or investigate the mechanism of a reaction in a particular solvent. The effect of solvent composition on chemical equilibria and reaction rates is largely dependent on solvent structure and solute–solvent interactions. The heat and entropy changes accompanying chemical processes give more insight into these effects than linear free energy correlations. It would therefore seem more profitable to tackle the problem of changing solvent composition by detailed study of heat and entropy effects.

REFERENCES

Ahrens, M-L., and Strehlow, H. (1965). *Discuss. Faraday Soc.* **39**, 112.
Alder, R. W., Chalkley, G. R., and Whiting, M. C. (1966). *Chem. Comm.* 405.
Alfenaar, M., and de Ligny, C. L. (1967). *Rec. Trav. Chim.* **86**, 1185.
Andrews, L. J., and Kaeding, W. W. (1951). *J. Amer. Chem. Soc.* **73**, 1007.
Archer, G., and Bell, R. P. (1959). *J. Chem. Soc.* 3228.
Arnett, E. M. (1967). *In* "Physico-Chemical Processes in Mixed Aqueous Solvents" (F. Franks, ed.), p. 105. Heinemann, London.
Arnett, E. M., and Douty, C. F. (1964). *J. Amer. Chem. Soc.* **86**, 409.
Arnett, E. M., and Wu. C. Y. (1962). *J. Amer. Chem. Soc.* **84**, 1684.
Baines, J. E., and Eaborn, C. (1956). *J. Chem. Soc.* 1436.
Banthorpe, D. V., Hughes, E. D., Ingold, C. K., and Roy, J. (1962). *J. Chem. Soc.* 3294.
Bartlett, P. D., and McCollum, J. D. (1956). *J. Amer. Chem. Soc.* **78**, 1441.
Bates, R. G. (1964a). "Determination of pH, Theory and Practice", Ch. 8. Wiley, New York.
Bates, R. G. (1964b). "Determination of pH, Theory and Practice", p. 222. Wiley, New York.
Bates, R. G., and Robinson, R. A. (1966). *In* "Chemical Physics of Ionic Solutions" (B. E. Conway, and R. G. Barradas, eds.), p. 211. Wiley, New York.
Bates, R. G., and Schwarzenbach, G. (1955). *Helv. Chim. Acta*, **38**, 699.
Bell, R. P. (1959). "The Proton in Chemistry," Ch. 4. Methuen, London.
Bell, R. P., and Brown, A. H. (1954). *J. Chem. Soc.* 774.
Bel'skii, V. E., and Vinnik, M. I. (1964). *Russ. J. Phys. Chem.* **38**, 1061.
Bethell, D., and Gold, V. (1958a). *J. Chem. Soc.* 1905.
Bethell, D., and Gold, V. (1958b). *J. Chem. Soc.* 1930.
Bethell, D., Gold, V., and Satchell, D. P. N. (1958). *J. Chem. Soc.* 1918.
Braude, E. A. (1948). *J. Chem. Soc.* 1971.
Braude, E. A., and Stern, E. S. (1948a). *J. Chem. Soc.* 1976.
Braude, E. A., and Stern, E. S. (1948b). *J. Chem. Soc.* 1982.
Bruckenstein, S., and Kolthoff, I. M. (1956). *J. Amer. Chem. Soc.* **78**, 10.
Bunnett, J. F. (1961). *J. Amer. Chem. Soc.* **83**, 4968.
Bunnett, J. F., and Olsen, F. B. (1966). *Canad. J. Chem.* **44**, 1917.
Bunton, C. A., and Hadwick, T. (1957). *J. Chem. Soc.* 3043.
Bunton, C. A., and Hendy, B. N. (1962). *J. Chem. Soc.* 2562.
Bunton, C. A., Ley, J. B., Rhind-Tutt, A. J., and Vernon, C. A. (1957). *J. Chem. Soc.* 2327.
Bunton, C. A., James, D. H., and Senior, J. B. (1960). *J. Chem. Soc.* 3364.
Campbell, H. J., and Edward, J. T. (1960). *Canad. J. Chem.* 38, 2109.
Carlin, R. B., Nelb, R. G., and Odioso, R. C. (1951). *J. Amer. Chem. Soc.* **73**, 1002.
Cieuciuch, R. F. W., and Westheimer, F. H. (1963). *J. Amer. Chem. Soc.* **85**, 2591.
Cook, A. G., and Mason, G. W. (1966). *J. Inorg. Nuclear Chem.* **28**, 2579.
Corsaro, G. (1964). *Chem. and Ind.* 75.
Cox, B. G., and McTigue, P. T. (1967). *Aust. J. Chem.* **20**, 1815.

Dahn, H., and Gold, H. (1963). *Helv. Chim. Acta*, **46**, 983.
Dahn, H., Loewe, L., and Rotzler, G. (1960). *Chem. Ber.* **93**, 1572.
Dallinga, G., and ter Maten, G. (1960). *Rec. Trav. Chim.* **79**, 737.
Dayagi, S. (1961). *Bull. Res. Council. Israel A*, **10**, 152.
D'Cruz, W. A. (1964). *Diss. Abs.* **25**, 829.
de Ligny, C. L., and Alfenaar, M. (1967). *Rec. Trav. Chim.* **86**, 1182.
de Ligny, C. L., Loriaux, H., and Ruiter, A. (1961). *Rec. Trav. Chim.* **80**, 725.
Deno, N. C., Jaruzelski, J. J., and Schriesheim, A. (1955). *J. Amer. Chem. Soc.* **77**, 3044.
de Ruyter de Steveninck, A. W. (1958). *J. Chem. Soc.* 2066.
Dolman, D., and Stewart, R. (1967). *J. Amer. Chem. Soc.* **45**, 903.
Eaborn, C. (1953). *J. Chem. Soc.* 3148.
Entelis, S. G., and Chirkov, N. M. (1957). *Zhur. fiz. Khim.* **31**, 1311.
Entelis, S. G., Eckel, G. V., and Chirkov, N. M. (1960a). *Doklady Akad. Nauk. S.S.S.R.* **130**, 826.
Entelis, S. G., Korovina, G. V., and Chirkov, N. M. (1960b). *Doklady Akad. Nauk. S.S.S.R.* **134**, 856.
Entelis, S. G., Korovina, G. V., and Chirkov, N. M. (1960c). *Izvest. Akad. Nauk. S.S.S.R., Otdel. Khim. Nauk.* 2050.
Franks, F., and Ives, D. J. G. (1966). *Quart. Rev.* **20**, 1.
Furuya, Y., and Ogata, Y. (1963). *Bull. Chem. Soc. Japan*, **36**, 419.
Gaboriaud, R. (1967). *Compt. rend.* **265**, 425.
Gebhart, H. J., and Adams, K. H. (1954). *J. Amer. Chem. Soc.* **76**, 3925.
Gold, V., and Hawes, B. W. V. (1951). *J. Chem. Soc.* 2102.
Gramstad, T. (1959). *Tidsskr. Kjemi Bergv. Metall.* **19**, 62.
Grinter, R., and Mason, S. F. (1964). *Trans. Faraday Soc.* **60**, 882.
Grunwald, E., and Berkowitz, B. J. (1951). *J. Amer. Chem. Soc.* **73**, 4939.
Gudmundsen, C. H., and McEwen, W. E. (1957). *J. Amer. Chem. Soc.* **79**, 329.
Gutbezahl, B., and Grunwald, E. (1953). *J. Amer. Chem. Soc.* **75**, 559, 565.
Hall, N. F., and Spengeman, W. F. (1940). *J. Amer. Chem. Soc.* **62**, 2487.
Hamer, D., and Leslie, J. (1960). *J. Chem. Soc.* 4198.
Hammett, L. P., and Deyrup, A. J. (1932). *J. Amer. Chem. Soc.* **54**, 4239.
Hawke, J. G., and Stimson, V. R. (1956). *J. Chem. Soc.* 4676.
Hogeveen, H., and Drenth, W. (1963). *Rec. Trav. Chim.* **82**, 410.
Homer, R. B., and Moodie, R. B. (1965). *J. Chem. Soc.* 4399.
Homer, R. B., Moodie, R. B., and Rydon, H. N. (1965). *J. Chem. Soc.* 4403.
Hyman, H. H., and Garber, R. A. (1959). *J. Amer. Chem. Soc.* **81**, 1847.
Jaffé, H. H., and Gardner, R. W. (1958). *J. Amer. Chem. Soc.* **80**, 319.
Juillard, J., and Simonot, N. (1968). *Bull. Soc. Chim. France*, 1883.
Kaeding, W. W., and Andrews, L. J. (1952). *J. Amer. Chem. Soc.* **74**, 6189.
Kalidas, Ch., and Palit, S. R. (1961). *J. Chem. Soc.* 3998.
Knessl, O., Roček, J., and Marek, M. (1955). *Coll. Czech. Chem. Comm.* **20**, 631.
Kolthoff, I. M., and Bruckenstein, S. (1956). *J. Amer. Chem. Soc.* **78**, 1.
Kolthoff, I. M., and Willman, A. (1934). *J. Amer. Chem. Soc.* **56**, 1007.
Kreevoy, M. M. (1956). *J. Amer. Chem. Soc.* **78**, 4236.
Kwart, H., and Herbig, J. (1963). *J. Amer. Chem. Soc.* **85**, 226.
Kwart, H., and Weisfeld, L. B. (1958). *J. Amer. Chem. Soc.* **80**, 4670.
Long, F. A., and Paul, M. A. (1957). *Chem. Rev.* **57**, 935.
Ludwig, F. J., and Adams, K. H. (1954). *J. Amer. Chem. Soc.* **76**, 3853.
Mackenzie, H. A. E., and Winter, E. R. S. (1944). *Trans. Faraday Soc.* **44**, 159, 171, 243.

Mackor, E. L., Smit, P. J., and van der Waals, J. H. (1957). *Trans. Faraday Soc.* **53**, 1309.
Manelis, G. B., Vinnik, M. I., and Chirkov, N. M. (1959). *Zhur. fiz. Khim.* **33**, 1030.
Martin, R. J. L., and Reece, I. H. (1959). *Aust. J. Chem.* **12**, 524.
Mörikofer, A., Simon, W., and Heilbronner, E. (1959). *Helv. Chem. Acta*, **42**, 1737.
Nachbaur, E. (1960). *Monatsh. Chem.* **91**, 749.
Nahlovsky, B., and Chvalovsky, V. (1968). *Coll. Czech. Chem. Comm.* **33**, 3122.
Novatskii, G. N., Ionin, B. I., Bagal, L. I., and Golod, E. L. (1968). *Russ. J. Phys. Chem.* **42**, 1578.
Noyce, D. S., and Castelfranco, P. (1951). *J. Amer. Chem. Soc.* **73**, 4482.
Noyce, D. S., and Jorgenson, M. J. (1961). *J. Amer. Chem. Soc.* **83**, 2525.
Noyce, D. S., and Jorgenson, M. J. (1962). *J. Amer. Chem. Soc.* **84**, 4312.
Noyce, D. S., and Jorgenson, M. J. (1963). *J. Org. Chem.* **28**, 3208.
Noyce, D. S., and Pryor, W. A. (1955). *J. Amer. Chem. Soc.* **77**, 1397.
Noyce, D. S., and Snyder, L. R. (1958). *J. Amer. Chem. Soc.* **80**, 4324.
Noyce, D. S., and Snyder, L. R. (1959). *J. Amer. Chem. Soc.* **81**, 620.
Ogata, Y., and Okano, M. (1956). *J. Amer. Chem. Soc.* **78**, 5423.
Ogata, Y., and Tabushi, I. (1959). *Bull. Chem. Soc. Japan*, **32**, 108.
O'Reilly, D. E., and Leftin, H. P. (1960). *J. Phys. Chem.* **64**, 1555.
Palm, V. A. (1958). *Zhur. fiz. Khim.* **32**, 380.
Patai, S., and Dayagi, S. (1962). *J. Chem. Soc.* 726.
Paul, M. A. (1950). *J. Amer. Chem. Soc.* **72**, 3813.
Paul, M. A. (1954). *J. Amer. Chem. Soc.* **76**, 3236.
Paul, M. A., and Hammett, L. P. (1936). *J. Amer. Chem. Soc.* **58**, 2182.
Paul, M. A., and Long, F. A. (1957). *Chem. Rev.* **57**, 1.
Pritchard, J. G., and Long, F. A. (1958). *J. Amer. Chem. Soc.* **80**, 4162.
Rangaiah, V., and Brahmaji Rao, S. (1967). *Indian J. Chem.* **5**, 452.
Reynaud, R. (1968). *Compt. rend.* **267C**, 989.
Reynaud, R. (1969). *Bull. Soc. Chim. France*, 699.
Roček, J. (1957). *Coll. Czech. Chem. Comm.* **22**, 1.
Roček, J. (1960). *Coll. Czech. Chem. Comm.* **25**, 375.
Rosenthal, D., Hetzer, H. B., and Bates, R. G. (1964). *J. Amer. Chem. Soc.* **86**, 549.
Sadek, H., Abu Elamayem, M. S., and Sidahmed, I. M. (1966). *Suomen Kemi*, **39B**, 225.
Salomaa, P. (1957a). *Acta Chem. Scand.* **11**, 125.
Salomaa, P. (1957b). *Acta Chem. Scand.* **11**, 141.
Salomaa, P. (1957c). *Acta Chem. Scand.* **11**, 235.
Salomaa, P. (1957d). *Acta Chem. Scand.* **11**, 239.
Salomaa, P. (1957e). *Acta Chem. Scand.* **11**, 461.
Sanders, W. N., and Berger, J. E. (1967). *Analyt. Chem.* **39**, 1473.
Satchell, D. P. N. (1956). *J. Chem. Soc.* 3911.
Satchell, D. P. N. (1957a), *J. Chem. Soc.* 2878.
Satchell, D. P. N. (1957b). *J. Chem. Soc.* 3524.
Satchell, D. P. N. (1958a). *J. Chem. Soc.* 1919.
Satchell, D. P. N. (1958b). *J. Chem. Soc.* 1927.
Satchell, D. P. N. (1958c). *J. Chem. Soc.* 3910.
Schwarzenbach, G., and Stensby, P. (1959). *Helv. Chim. Acta*, **42**, 2342.
Smith, L. C., and Hammett, L. P. (1945). *J. Amer. Chem. Soc.* **67**, 23.
Smith, T. L., and Elliott, J. H. (1953). *J. Amer. Chem. Soc.* **75**, 3566.
Smyth, C. P., and Rogers, H. E. (1930). *J. Amer. Chem. Soc.* **52**, 1824.

Stamhuis, E. J., and Drenth, W. (1963). *Rec. Trav. Chim.* **82**, 385.
Stewart, R., and Mathews, T. (1960). *Canad. J. Chem.* **38**, 602.
Stimson, V. R. (1955a). *J. Chem. Soc.* 2010.
Stimson, V. R. (1955b). *J. Chem. Soc.* 2673.
Stimson, V. R. (1955c). *J. Chem. Soc.* 4020.
Stimson, V. R., and Watson, E. J. (1954). *J. Chem. Soc.* 2848.
Swain, C. G., and Spalding, E. T. (1960). *J. Amer. Chem. Soc.* **82**, 6104.
Tillett, J. G., and Young, R. C. (1968). *J. Chem. Soc. B*, 209.
Torck, B., Hellin, M., and Coussemant, F. (1960). *Actes Congr. Intern. Catalyse*, 2^{e}, *Paris*, **1**, 213.
Torck, B., Hellin, M., and Coussemant, F. (1962a). *Bull. Soc. Chim. France*, 1657.
Torck, B., Hellin, M., and Coussemant, F. (1962b). *Bull. Soc. Chim. France*, 1664.
Utyanskaya, E. Z., and Vinnik, M. I. (1968). *Russ. J. Phys. Chem.* **42**, 204.
Valenta, J. G. (1964). *Diss. Abs.* **24**, 3996.
Van Looy, H., and Hammett, L. P. (1959). *J. Amer. Chem. Soc.* **81**, 3872.
Vinnik, M. I., Kruglov, R. N., and Chirkov. N. M. (1957). *Zhur. fiz. Khim.* **31**, 832.
Vinnik, M. I., Manelis, G. B., and Chirkov, N. M. (1960). *Probemy Kinetiki i Kataliza Akad. Nauk S.S.S.R.* **10**, 285.
Wells, C. F. (1965). *Trans. Faraday Soc.* **61**, 2194.
Wells, C. F. (1966). *Trans. Faraday Soc.* **62**, 2815.
Wells, C. F. (1968). "Hydrogen-bonded Solvent Systems" (A. K. Covington, and P. Jones, eds.), p. 323. Taylor and Francis, London.
Whalley, E. (1964). *Adv. Phys. Org. Chem.* **2**, 93.
White, W. N., and Stout, C. A. (1962). *J. Org. Chem.* **27**, 2915.
Wiberg, K. B., and Evans, R. J. (1958). *J. Amer. Chem. Soc.* **80**, 3019.
Wichterle, O., and Cefelin, P. (1957). *Coll. Czech. Chem. Comm.* **22**, 1083.
Wichterle, O., Laita, Z., and Pazlar, M. (1955). *Chem. Listy*, **49**, 1612.
Winstein, S., and Fainberg, A. H. (1957). *J. Amer. Chem. Soc.* **79**, 5937.
Wynne-Jones, W. F. K. (1968). "Hydrogen-bonded Solvent Systems" (A. K. Covington, and P. Jones, eds.), p. 246. Taylor and Francis, London.
Yeh, S. J., and Jaffe, H. H. (1959). *J. Amer. Chem. Soc.* **81**, 3274.
Zajac, W. W., and Nowicki, R. B. (1965). *J. Phys. Chem.* **69**, 2649.

CHAPTER 7

Acidity Functions for Concentrated Solutions of Bases

Far fewer acidity function studies for concentrated base solutions have been made than for concentrated acid solutions (Bowden, 1966; Rochester, 1966a). The available results are reviewed in this chapter. They show that the rules and problems associated with the interpretation of acidity function behaviour for reactions in strongly basic solutions are essentially similar to those for strongly acidic solutions.

7.1. Aqueous Solutions of Bases

7.1.1. *The H_- Acidity Function: Experimental Scales*

The definition of the H_- acidity function is given by equation (3.24). The measurement of an H_- scale involves the determination of pK_{SH} and the variation of $\log_{10}(C_{S^-}/C_{SH})$ with increasing hydroxide ion concentration for a weakly acidic neutral indicator SH. The H_- scale is effectively a measure of the ability of the strongly basic solution to abstract a proton from the weakly acidic neutral solute SH. The ratio (C_{S^-}/C_{SH}) is measured by electronic absorption spectrophotometry (equation 1.24) with the proviso that the neutral acid SH and its conjugate base S^- must have measurably different absorption spectra. Two methods are available for the determination of pK_{SH}. Equation (7.1) relates the extent of ionization of SH according to equilibrium (7.2) to the

$$pK_w - pK_{SH} = \log_{10}(C_{S^-}/C_{SH}) - \log_{10} C_{OH^-} + \log_{10}(y_{S^-}\, a_w / y_{SH}\, y_{OH^-}) \quad (7.1)$$

$$SH + OH^- \rightleftharpoons S^- + H_2O \quad (7.2)$$

hydroxide ion concentration. For strong bases MOH extrapolation of a graph of $\log_{10}(C_{S^-}/C_{SH}) - \log_{10} C_{MOH}$ against C_{MOH} to $C_{MOH}=0$ (where a_w and all solute activity coefficients are unity by definition) gives a value for $pK_w - pK_{SH}$ and hence for pK_{SH} (More O'Ferrall and Ridd, 1963a). For the more acidic indicators these plots are linear and the extrapolation is reliable. However, for the less acidic indicators the plots are curved and a stepwise comparison method (Section 1.4) must then be used for the evaluation of pK_{SH}. The relationship between the ionization ratios of an indicator SH and a more weakly acidic indicator RH in an aqueous solution is given by equation (7.3).

$$\log_{10}(C_{S^-}/C_{SH}) - \log_{10}(C_{R^-}/C_{RH}) = pK_{RH} - pK_{SH} + \log_{10}(y_{R^-}y_{SH}/y_{S^-}y_{RH}) \tag{7.3}$$

$$\log_{10}(y_{R^-}y_{SH}/y_{S^-}y_{RH}) = \text{ca. } 1 \tag{7.4}$$

$$pK_{RH} = pK_{SH} + \log_{10}(C_{S^-}/C_{SH}) - \log_{10}(C_{R^-}/C_{RH}) \tag{7.5}$$

Equation (7.4) is applicable providing H_- and $d\log_{10}(C_{S^-}/C_{SH})/dC_{MOH}$ for a particular C_{MOH} are independent of indicator. Hence pK_{RH} may be deduced via equation (7.5) by stepwise comparison of the ionization behaviour of the two indicators. pK_{SH} for a series of progressively weaker acids may be deduced by this method. The values are referred to standard state pure water via the use of equation (7.1) for the more acidic indicators. Substitution in equation (3.24) of pK_{SH} and $\log_{10}(C_{S^-}/C_{SH})$ as a function of concentration of base for each indicator defines the H_- scale.

The definition of the H_- function requires that the observed changes in spectra for a particular weak acid SH arise through a simple proton loss (equation 7.2) and not through base addition or through interaction of more than one hydroxide ion with the weak acid. It is often difficult to establish the exact mode of ionization of weak acids in strongly basic solutions although p.m.r. spectra are sometimes helpful in distinguishing between base addition and proton loss (Crampton and Gold, 1964, 1966; Servis, 1965; Norris and Osmundsen, 1965; Buncel *et al.* 1968).

Schwarzenbach and Sulzberger (1944) measured H_- scales for aqueous NaOH and KOH solutions at 20°C using a series of indigo derivatives as weak acids. Because these compounds were insoluble in water they were added to the aqueous alkali in an immiscible solvent and colorimetric analysis was carried out after equilibrium was established. This method required the assumption that the partition coefficients of the neutral indicators between the two phases were independent of the electrolyte concentration in the aqueous phase. The scales were also based on the assumption that $H_- = 14{\cdot}00$ when $C_{MOH} = 1$ mole litre^{-1}. The acidity function for aqueous NaOH correlates closely with the ionization behaviour of thioacetamide in aqueous NaOH up to a concentration of ca. 5 mole litre^{-1} (Edward and Wang, 1962).

Yagil (1967a) has used a series of indoles for the determination of H_- scales for aqueous KOH, NaOH and LiOH. Fig. 7.1 shows that plots of $\log_{10}(C_{S^-}/C_{SH})$ against C_{KOH} are accurately parallel for indoles which ionize in overlapping ranges of KOH concentration. The only exception is 4-fluoroindole for which there appears to be a specific interaction influencing the results. Yagil's data show that substituted indoles are a series of structurally related weak acids which are admirably suited for the definition of H_- acidity functions for concentrated solutions of bases.

The acidity functions for aqueous NaOH and KOH defined by the ionization of indoles compare favourably (Table 7.1) with the scales of Schwarzenbach and Sulzberger (1944) at low hydroxide concentrations but deviate appreciably as the hydroxide concentration is increased. The deviations are too great to be

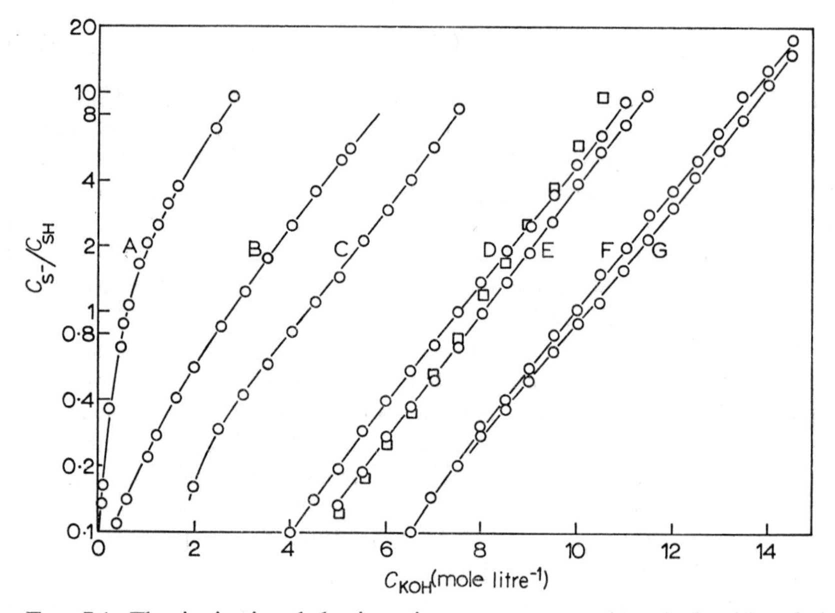

FIG. 7.1. The ionization behaviour in aqueous potassium hydroxide of A, Indazole; B, 5-nitroindole; C, 5-cyanoindole; D, 5-bromoindole; E, 5-fluoroindole; F, L-tryptophan; G, tryptophanol; □ 4-fluoroindole. Reproduced by permission from Yagil (1967a).

solely a temperature effect. They may arise either because the structurally different weak acids used by Yagil (1967a) and by Schwarzenbach and Sulzberger (1944) give rise to different acidity function behaviour or because the measurements using indigo derivatives were carried out in two phase systems. Yagil's scale for aqueous lithium hydroxide (Table 7.1) approximately parallels but is ca. 0·55 units more positive than the corresponding scale determined using substituted anilines and diphenylamines as indicators (Stewart and O'Donnell, 1964a). However, the latter measurements are unreliable because the primary and secondary amines were too insoluble in aqueous LiOH for accurate measurements to be made.

The ionization of 4-nitroaniline in aqueous NaOH has been studied by Rochester (1963a). The recorded spectral shifts observed for 4-nitroaniline in

TABLE 7.1

H_- acidity function scales for aqueous sodium hydroxide, aqueous potassium hydroxide and aqueous lithium hydroxide solutions

C_{MOH} (mole litre^{-1})	$H_-(25°)$† (NaOH)	$H_-(20°)$‡ (NaOH)	$H_-(25°)$† (KOH)	$H_-(20°)$‡ (KOH)	$H_-(25°)$† (LiOH)
0·1	12·99		13·00		
0·2	13·30		13·32		
0·5	13·71		13·75		13·68
0·8	13·92		14·00		
1·0	14·02	14·00§	14·11	14·00§	13·96
1·5	14·20		14·33		14·11
2·0	14·37	14·36	14·51	14·49	14·26
2·5	14·54		14·69		14·36
3·0	14·65	14·64	14·85	14·81	14·45
3·5	14·81		15·00		14·53
4·0	14·95	14·93	15·15	15·12	14·58
4·5	15·08		15·28		14·65
5·0	15·20	15·14	15·44	15·45	
6·0	15·40	15·37	15·72	15·74	
7·0	15·62	15·58	16·00	16·07	
8·0	15·75	15·76	16·33	16·39	
9·0	15·97	16·00	16·58	16·72	
10·0	16·20	16·17	16·90	17·22	
11·0	16·42	16·46	17·14	17·60	
12·0	16·58	16·72	17·39	18·12	
13·0	16·76	17·04	17·66	18·60	
14·0	16·93	17·37	17·95	18·90	
15·0	17·10	17·74	18·23	19·25	
16·0	17.30	18·13			

† Yagil (1967a).
‡ Schwarzenbach and Sulbzerger (1944).
§ Assumed values $H_- = 14$ when $C_{MOH} = 1$ mole litre^{-1}.

$C_{NaOH} < 4·5$ mole litre^{-1} arise entirely because of medium effects on the spectrum of the neutral 4-nitroaniline molecule (Dolman and Stewart, 1967). The shifts observed for higher NaOH concentrations are due to a 1:1 interaction (probably proton loss) between 4-nitroaniline and hydroxide ions. However, the logarithms of the ionization ratios do not parallel either the indole or the indigo H_- scales. Different structural types of indicators are apparently giving different acidity function behaviour in concentrated solutions of strong bases. 2,4-Dinitroaniline and 2,4,6-trinitroaniline are unsuitable as indicators for the definition of H_- acidity functions for aqueous alkali–metal hydroxide solutions because the neutral molecules in part undergo interaction with more than one hydroxide ion and irreversible substitution reactions occur (Rochester, 1963c; Gold and Rochester, 1964h). Competition between hydroxide addition to, and proton loss from, the primary amines also occurs.

Stewart and O'Donnell (1964a) used substituted anilines and diphenylamines to measure an H_- scale for benzyltrimethylammonium hydroxide in water. The H_- values ranged from 11·98 to 16·20 for concentrations of base from 0·01 to 2·38 mole litre^{-1} respectively. Similar measurements have been made for aqueous solutions of hydrazine (Deno, 1952; Schaal and Favier, 1959; Favier and Schaal, 1959), ethanolamine (Masure and Schaal, 1956), diethylaminoethanol (Mouronval *et al.* 1962) and ethylenediamine (Schaal, 1954a, 1955; Vermesse-Jacquinot, 1965b, 1965c). H_- acidity function scales for aqueous hydrazine, ethylenediamine and ethanolamine have been summarized by Stewart and O'Donnell (1964a) and by Bowden (1966).

Indicators used to establish the acidity functions for aqueous solutions of amines include carbazoles, primary and secondary amines, 4-nitrobenzyl cyanide, ketones and phenylhydrazones. In some of these studies not enough care has been taken to test the validity of the assumptions made in the acidity function approach that H_- and $d\log_{10}(C_{S^-}/C_{SH})/dC_{base}$ are independent of indicator for a particular base concentration if the defined H_- acidity functions are to be meaningful. It would be interesting to compare acidity functions defined by different sets of structurally similar indicators in order to see whether the dependence of acidity function on indicator structure found for acidic solutions (Fig. 3.2) is also applicable for strongly basic solutions. Plots such as those presented by Stewart and O'Donnell (1964a) of $\log_{10}(C_{A^-}/C_{HA})$ against H_- for structurally dissimilar indicators do suggest that acidity functions for concentrated base solutions are less sensitive to changing indicator structure than are acidity functions for concentrated acid solutions. However, there appears no clear reason why this should be so and care should be taken to confirm the acidity function postulates are applicable for the particular weak acids being studied. The most useful tabulations of pK_a values for weak acids which are suitable as indicators for measurement of H_- acidity functions are those of Yagil (1967a, 1967c) for pyrrole and indole derivatives, and of Stewart and O'Donnell (1964b) and Stewart and Dolman (1967) for substituted anilines and diphenylamines.

7.1.2. *Theoretical Calculations of the H_- Acidity Function*

Theoretical calculations of H_0 acidity functions for aqueous solutions of strong acids were discussed above (Section 2.3.1). Analogous calculations of H_- for aqueous NaOH and KOH have been attempted and are equally successful for moderate ($<$ ca. 8 mole litre^{-1}) concentrations of electrolyte. Equation (7.6) follows from the definition of the H_- acidity function.

$$H_- = pK_w + \log_{10} C_{OH^-} - \log_{10} a_w + \log_{10}(y_{SH}\, y_{OH^-}/y_{S^-}) \qquad (7.6)$$

For $C_{OH^-} <$ ca. 1 mole litre^{-1} the activity term $(y_{SH}y_{OH^-}/y_{S^-}a_w)$ in this equation approximates to 1 and therefore $H_- \approx pK_w + \log_{10} C_{OH^-}$. However, for concentrations of NaOH or KOH greater than 1 mole litre^{-1} the H_- values deviate from $pK_w + \log_{10} C_{MOH}$ by an amount which increases rapidly with increasing concentration of base (Table 7.1). Substitution of water activities (Stokes, 1945; Robinson and Stokes, 1949) in equation (7.6) shows that the $\log_{10} a_w$ term is insufficient to account for the deviation. The activity coefficient term is therefore making an important contribution to the observed H_- values. Unfortunately no information is available about the variation of y_{SH} and y_{S^-} with changing electrolyte concentration. However, following the approach developed by Bell and Bascombe (1957) it is profitable to consider the change in solvation in passing from the left to the right-hand side of equilibrium (7.2).

Equilibrium (7.2) may be rewritten as equation (7.7) in which p, q, and r are the hydration numbers of SH, OH^- and S^- respectively and $n = (p + q - r)$.

$$SH(pH_2O) + (OH,qH_2O)^- \rightleftharpoons (S,rH_2O)^- + (n+1)H_2O \quad (7.7)$$

Equation (7.6) now becomes

$$H_- = pK_w + \log_{10} C_{OH^-} - (n+1)\log_{10} a_w + \log_{10}(y_{SH}y_{OH^-}/y_{S^-}) \quad (7.8)$$

Edward and Wang (1962) tested this equation in relation to their results for the ionization of thioacetamide in aqueous sodium hydroxide solutions. They assumed that the activity coefficient term in equation (7.8) would vary with solvent activity according to equation (7.9) which leads to equation (7.10) for

$$\log_{10}(y_{SH}y_{OH^-}/y_{S^-}) \approx x \log_{10} a_w \quad (7.9)$$

$$H_- \approx pK_w + \log_{10} C_{OH^-} - (n+1-x)\log_{10} a_w \quad (7.10)$$

H_-. A plot of $H_- - \log_{10} C_{NaOH}$ against $\log_{10} a_w$ was linear for concentrations of NaOH up to ca. 5 mole litre^{-1}. Substituting the slope, $(n+1-x) = 3{\cdot}2$, into equation (7.10) enabled the calculation of H_- values which were in good agreement with the experimental H_- scale.

Yagil and Anbar (1963) made similar calculations of H_- via equation (7.11) in which C_w is the concentration of "free water" in the solution ($C_w = 1$ at infinite dilution). This implies the concentration of that water which is not bound to OH^- in hydration and was calculated from equation (7.12) in which

$$H_- = pK_w + \log_{10} C_{OH^-} - (n+1)\log_{10} C_w \quad (7.11)$$

$$C_w = d - 0{\cdot}001(M + 18{\cdot}0n)\,C_{OH^-} \quad (7.12)$$

d is the density of the solutions and M is the molecular weight of NaOH or KOH. This approach takes n as the hydration number of the hydroxide ion and therefore assumes that the numbers p and r of water molecules hydrating SH and S^- respectively (equation 7.7) are the same. With $n = 3$ a combination of equations (7.11) and (7.12) gave figures for H_- which agreed fairly well with the experimental results of Schwarzenbach and Sulzberger (1944) for $C_{MOH} <$ ca. 4 mole litre^{-1}. The agreement between the calculated H_- scale and the H_- acidity function defined by the ionization of indoles is considerably better (Yagil, 1967a). Thus the theoretical and experimental values are in accord with each other for concentrations of NaOH and KOH up to 7 mole litre^{-1} and 9 mole litre^{-1} respectively.

It is not surprising that these theories break down at higher concentrations in view of the drastic assumptions inherent in the calculations. Furthermore the factor n may be expected to decrease with increasing electrolyte concentration (Freeguard *et al.* 1965) whether it is regarded as the change in hydration in equilibrium (7.7) (Edward and Wang, 1962) or solely the hydration number of the hydroxide ion (Yagil, 1967a). The former interpretation seems preferable. In the calculations the concentration of hydroxide ions has been taken as equal to the stoichiometric concentration of alkali–metal hydroxide. Ion association is ignored. There is much evidence (Darken and Meier, 1942; Bell and Prue, 1949; Gutowsky and Saika, 1953; Jones, 1968b) that the extent of ion association for a particular concentration of alkali–metal hydroxide is in the order KOH < NaOH < LiOH. This led Rochester (1966a) to suggest that the observed order of basicity H_-(LiOH) < H_-(NaOH) < H_-(KOH) (Table 7.1) may at least in part arise because of the effects of ion association. This point was tested quantitatively by Jones (1968a) who showed that the deviation between the H_- scales (Schwarzenbach and Sulzberger, 1944; Edward and Wang, 1962) for aqueous NaOH and KOH was consistent with the greater extent of ion association of hydroxide ions with Na^+ than with K^+. The results for LiOH were less satisfactory but are improved if Yagil's (1967a) H_- scale for aqueous LiOH is used in the correlation. However, the agreement for the indole H_- scales of NaOH and KOH are not so good. Jones's approach on its own is unlikely to give a satisfactory account of the difference between the LiOH, NaOH and KOH H_- acidity functions. Preliminary calculations for aqueous NaOH and KOH have indicated that it may be better to evaluate the concentrations of unassociated hydroxide ions in the solutions and to substitute these for C_{OH^-} in equation (7.10) or in equations (7.11) and (7.12) (Rochester, 1964). However, the many assumptions inherent in such a treatment would render it more empirical than quantitative. This approach has been proved satisfactory for aqueous LiOH solutions (Yagil, 1967b). In general it may be concluded that hydration and ion association effects both play significant roles in determining the basicity of aqueous solutions of alkali–metal hydroxides.

The addition of potassium fluoride to aqueous KOH causes a considerable rise in the H_- basicity of the solution (Yagil, 1967b). Potassium chloride has a much smaller effect and potassium bromide has a negligible effect on the basicity. The changes in basicity are quantitatively explicable on the assumption that the numbers of water molecules hydrating each salt molecule are 3·5 for KF, 0·9 for KCl and 0 for KBr. Solvation of the salts reduces the free water concentration C_w (equation 7.11) in the solutions and therefore increases H_-.

7.1.3. *Correlation of H_- with Reaction Rates*

The Zucker–Hammett or Bunnett treatments (Chapter 4) may be considered in relation to the acidity function dependence of reaction rates in strongly basic solutions. For a reaction such as the base catalysed hydrolysis of chloroform the mechanism may be represented by equations (7.13) and (7.14).

$$PH + OH^- \underset{}{\overset{k'}{\rightleftharpoons}} P^- + H_2O \qquad (\text{fast}, C_{PH} \gg C_{P^-}) \qquad (7.13)$$

$$P^- \xrightarrow{k} \text{products} \qquad (\text{slow}) \qquad (7.14)$$

The experimental rate constant k_1 is given by equation (7.15) and is therefore related to H_- by equation (7.16) in which SH represents the weakly acidic indicators used to establish the H_- scale. K_{PH} is the acid ionization constant of PH.

$$k_1 = -(1/C_{PH})(dC_{PH}/dt) = k(C_{P^-}/C_{PH})/y_{P^-}/y^{+}) \qquad (7.15)$$

$$\log_{10} k_1 = \log_{10}(kK_{PH}) + H_- + \log_{10}(y_{PH}\, y_{S^-}/y^{+}\, y_{SH}) \qquad (7.16)$$

On the Zucker–Hammett approach the activity coefficient term may be approximated to zero and therefore $\log_{10} k_1$ should be a linear function of H_- with unit slope. Alternatively on the Bunnett approach, combination of equations (7.6) and (7.16) leads to equation (7.17) which by analogy to equation (7.8) may be written as equation (7.18). An extreme interpretation of t^*

$$\log_{10}(k_1/C_{OH^-}) = \log_{10}(kK_{PH}/K_w) - \log_{10} a_w + \log_{10}(y_{PH}\, y_{OH^-}/y^{+}) \qquad (7.17)$$

$$\log_{10}(k_1/C_{OH^-}) = \log_{10}(kK_{PH}/K_w) - (t^* + 1)\log_{10} a_w + \log_{10}(y_{PH}\, y_{OH^-}/y^{*}) \qquad (7.18)$$

(cf. Bunnett's w^*) would be that it represents the change in hydration in passing from the reactants ($PH + OH^-$) to the transition state in the slow step of the reaction.

An alternative mechanism for base catalysis is that in which the abstraction of a proton from the reacting substrate (equation 7.13) is the slow rate-determining step. Equation (7.19) relates the experimental rate constant to the H_- acidity function. It is illogical here to assume the identity of the ratios (y_{S^-}/y_{SH})

$$\log_{10} k_1 = \log_{10}(k' K_w) + H_- + \log_{10} a_w + \log_{10}(y_{PH}\, y_{S^-}/y^{+}\, y_{SH}) \qquad (7.19)$$

and (y^{+}/y_{PH}) because the transition state consists of the ion P^- plus a molecule of water. Anbar *et al.* (1963) considered the modified form of equation (7.19)

$$\log_{10} k_1 = \log_{10}(k' K_w) + H_- + \log_{10} C_w + \log_{10}(y_{PH}\, y_{S^-}\, y_w/y^{+}\, y_{SH}) \qquad (7.20)$$

in which C_w (equation 7.12) is the concentration of "free water" and y_w is given by $a_w = y_w C_w$. The activity coefficient term in this equation was considered less likely to vary with changing concentration of base. Thus a graph of $\log_{10} k_1$ against $H_- + \log_{10} C_w$ was linear with slope 0·98 for the elimination reaction of DL-serine phosphate to form pyruvate, phosphate and ammonia and catalysed by aqueous sodium hydroxide.

The base catalysed formation of hydrazine from chloramine and ammonia involves a rapid pre-equilibrium proton transfer (7.21) followed by a slow bimolecular attack (7.22) of ammonia on the $NHCl^-$ anion (Yagil and Anbar, 1962).

$$NH_2Cl + OH^- \rightleftharpoons NHCl^- + H_2O \quad \text{(fast)} \tag{7.21}$$

$$NHCl^- + NH_3 \xrightarrow{k} N_2H_4 + Cl^- \quad \text{(slow)} \tag{7.22}$$

Providing $C_{PH} \ll C_{P^-}$ (where PH is NH_2Cl) in equilibrium (7.21) then equation (7.23) for the experimental rate constant k_2 leads to equation (7.24) which relates k_2 with H_-. Anbar *et al.* (1963) proposed that equation (7.25) was

$$k_2 = -(1/C_{PH}\,C_{NH_3})(dC_{PH}/dt) \tag{7.23}$$

$$\log_{10} k_2 = \log_{10}(kK_{PH}) + H_- + \log_{10}(y_{NH_3}\,y_{PH}\,y_{S^-}/y_{SH}\,y^{\ddagger}) \tag{7.24}$$

$$(y_{NH_3}\,y_{PH}/y^{\ddagger}) = (y_{SH}/y_{S^-}) \tag{7.25}$$

reasonable and therefore $\log_{10} k_2$ should be a linear function of H_- with unit slope. Data for both NaOH and KOH as catalysing bases conformed to the same linear graph which had a slope of 0·90 (Yagil and Anbar, 1962).

A possible mechanism for base catalysis is one in which a pre-equilibrium proton abstraction from the reacting substrate is followed by a slow bimolecular interaction between the conjugate base so formed and water. Anber *et al.* (1963) deduced that $\log_{10} k_1$ should be a linear function with unit slope of $H_- + \log_{10} C_w$. Unfortunately there are no data available for reactions known to proceed by this mechanism.

Of the four mechanisms discussed above it has been predicted that $\log_{10} k_1$ should parallel H_- for two of the mechanisms and $H_- + \log_{10} C_w$ for the other two. Anbar *et al.* (1963) proposed that k_1 should be proportional to C_{OH^-}/C_w^4 for the former two and C_{OH^-}/C_w^3 for the latter two. This proposal involves the assumption that n in equation (7.11) and the corresponding change in solvation in passing from $PH + OH^-$ to the transition state for any reaction is equal to the hydration number of the hydroxide ion which is taken as 3 (Yagil and Anbar, 1963; Yagil, 1967a). Two precise alternatives are therefore envisaged for the dependence of reaction rate on solution basicity. The evidence (Bunnett, 1961) for reactions in concentrated acid solutions would suggest that this is an unlikely contingency. The acidity function dependence of reaction rates will probably not only depend on the mechanism of the reaction and the hydration of the hydroxide ion but also on the hydration characteristics and activity coefficient behaviour of the particular reacting substrate and transition state in the reaction. By analogy with the Bunnett w (equation 4.35) and w^* (equation 4.38) factors it would be interesting to consider whether the definition of v (equation 7.26) and v^* (equation 7.27) factors for reactions in strongly

$$\log_{10} k_1 - H_- = v \log_{10} a_w + \text{constant} \qquad (7.26)$$

$$\log_{10} k_1 - \log_{10} C_{OH^-} = v^* \log_{10} a_w + \text{constant} \qquad (7.27)$$

basic media would provide useful criteria for the mechanism of such reactions. This possibility cannot be assessed until many more experimental results are available for consideration.

7.1.4. *The J_- Acidity Function*

The J_- acidity function is defined by equation (7.28) in which K is the

$$J_- = -\log_{10}(a_{H^+} y_{ROH^-}/y_R a_w) = p(KK_w) + \log_{10}(C_{ROH^-}/C_R) \qquad (7.28)$$

equilibrium constant for the addition of hydroxide ions to a electrically neutral indicator molecule R (Gold and Hawes, 1951; Rochester, 1963a; Gold and Rochester, 1964g). Equation (7.29) follows from equation (7.28).

$$J_- = pK_w + \log_{10} C_{OH^-} + \log_{10}(y_{OH^-} y_R/y_{ROH^-}) \qquad (7.29)$$

Hence (Rochester, 1963a) the relationship between H_- (equation 7.6) and J_- is given by equation (7.30).

$$J_- = H_- + \log_{10} a_w + \log_{10}(y_R y_{S^-}/y_{SH} y_{ROH^-}) \qquad (7.30)$$

This is similar to equation (3.5) for the relationship between J_0 and H_0. By analogy to equation (7.8) it may be more appropriate to consider equation (7.29) in the modified form (7.31) in which m is the difference in hydration number between ROH^- and $R + OH^-$ (Rochester, 1966a). Comparison of

$$J_- = pK_w + \log_{10} C_{OH^-} - m \log_{10} a_w + \log_{10}(y_{OH^-} y_R/y_{ROH^-}) \qquad (7.31)$$

equations (7.8) and (7.31) leads to equation (7.32) for the modified relationship between J_- and H_-.

$$J_- = H_- + (n - m + 1) \log_{10} a_w + \log_{10}(y_R y_{S^-}/y_{ROH^-} y_{SH}) \qquad (7.32)$$

Although correlation with the H_- acidity function of the behaviour of equilibria involving hydroxide addition to neutral solutes has been attempted (Schaal, 1954b) the simple hydroxide addition equilibrium is often obscured by other effects (Rochester, 1963c; Gold and Rochester, 1964e, 1964f, 1964g, 1964h). Unfortunately no studies of equilibria involving hydroxide addition have yet provided results which are suitable for the definition of a J_- acidity function. The only J_- scale at present available is based on kinetic measurements (Rochester, 1967).

The experimental rate constant k_1 for a reaction which conforms with the generalized mechanism (7.33) is given by equation (7.34) in which $k_0 = k_1 k_2/(k_{-1}+k_2)$ providing an appreciable concentration of ROH^- never builds

$$R + OH^- \underset{k_{-1}}{\overset{k_1}{\rightleftharpoons}} ROH^- \xrightarrow{k_2} \text{products} \tag{7.33}$$

$$k_1 = -(1/C_R)(dC_R/dt) = k_0\, C_{OH^-}\, y_R\, y_{OH^-}/y^{\ddagger} \tag{7.34}$$

up during the reaction. Hence equation (7.35) follows from the definition of J_- (Rochester, 1963b, 1966a, 1967). In the absence of equilibrium measurements of a J_- scale Rochester (1967) defined a "kinetic" $J_-(k)$ acidity function

$$\log_{10} k_1 = \log_{10}(k_0 K_w) + J_- + \log_{10}(y_{ROH^-}/y^{\ddagger}) \tag{7.35}$$

for aqueous NaOH via equation (7.36). The aromatic S_N2 reactions between

$$J_-(k) = \log_{10}(k_0 K_w) - \log_{10} k_1 = -\log_{10}(a_{H^+}\, y^{\ddagger}/a_w\, y_R) \tag{7.36}$$

hydroxide ions and 2,4-dinitroanisole, 2,4-dinitrophenetole, and 1-chloro-2,4-dinitrobenzene conform to the mechanism (7.33) and therefore the variation of their experimental rate constants with sodium hydroxide concentration was used to define the $J_-(k)$ scale. The three reactions gave results which were consistent with the same acidity function.

Other reactions which conform with the generalized mechanism (7.33) are the S_N2 hydrolyses of chloramine, dimethylchloramine (Anbar, and Yagil, 1962), carbon disulphide (Lazarev *et al.* 1965; Lazarev and Moiseev, 1965) and ethyl iodide (Anbar *et al.* 1963). Plots of $\log_{10} k_1$ for the NaOH catalysed hydrolyses of chloramine and carbon disulphide against Rochester's (1967) $J_-(k)$ acidity function were linear and parallel. However, the slopes were 0·87 showing that the S_N2 hydrolyses of chloramine and carbon disulphide give a distinctly different acidity function dependence from the corresponding S_N2 hydrolysis of 1-substituted 2,4-dinitrobenzenes. For the former two reactions the rate-determining step is probably the attack of hydroxide ions on the neutral chloramine or carbon disulphide whereas for the latter reactions a rate-determining unimolecular decomposition of the ROH^- addition complex occurs. The difference in acidity function behaviour may therefore have some mechanistic significance. Furthermore the difference disproves the proposal that all reactions conforming with mechanism (7.33) should give a linear correlation with unit slope of $\log_{10} k_1$ against $H_- + \log_{10} C_w$ (Anbar *et al.* 1963). Although this correlation is applicable for the hydrolyses of chloramine, ethyl iodide and carbon disulphide it does not hold for the S_N2 reactions of 1-substituted 2,4-dinitrobenzenes.

The rates of the S_N2 reactions of 2,4-dinitroanisole, 2,4-dinitrophenetole and 1-chloro-2,4-dinitrobenzene in aqueous NaOH are consistent with equations (7.37) and (7.38) (Rochester, 1967). Plots of $\log_{10}(k_1/C_{OH^-})$ against

$-\log_{10} a_w$ (equation 7.37) were linear and parallel with slopes of ca. 7 which is rather large if h is to be taken as $(h_R + h_{OH^-} - h_\pm)$ where h_i is the hydration

$$\log_{10}(k_1/C_{OH^-}) = \log_{10} k_0 - h \log_{10} a_w + \log_{10}(y_R\, y_{OH^-}/y^{+}) \qquad (7.37)$$

$$\log_{10}(k_1/C_{OH^-}) = \log_{10} k_0 - h \log_{10} C_w + \log_{10}(y_R\, y_{OH^-}/y^{+}\, y_w{}^{h}) \qquad (7.38)$$

number of species i. For 2,4-dinitroanisole a plot of $\log_{10}(k_1/C_{OH^-})$ against $-\log_{10} C_w$ (equation 7.38; C_w calculated via equation 7.12 with $n = h_{OH^-}$) had a slope of $h = 5{\cdot}0$ when h_{OH^-} was taken as 3 and a slope of $h = 3{\cdot}5$ with $h_{OH^-} = 4$. Rochester (1967) concluded that the latter was the more acceptable result because it is unlikely that $h > h_{OH^-}$ which would require $h_\pm < h_R$. There is some evidence for a solvation number of 4 for the hydroxide ion (Glueckauf, 1955; van Panthaleon van Eck *et al.* 1957). In view of the available evidence for $h_{OH^-} = 3$ (Yagil and Anbar, 1963; Yagil 1967a) it may well be that the values of h deduced from these plots are empirical constants of no real significance. Their values may, however, provide useful criteria for the mechanism of reactions in strongly basic solutions.

An alternative estimate of h uses McTigue's (1964) approach in which the activity coefficients in equation (7.34) are calculated via the expression deduced by Glueckauf (1955, 1959) for the activity coefficients of ions in concentrated electrolyte solutions. For the nucleophilic attack of hydroxide ions on 1-substituted 2,4-dinitrobenzenes the calculations gave $h =$ ca. $2{\cdot}5$ (Rochester, 1967) which is a reasonable value. Further tests of the applicability of this approach to acidity function behaviour would be interesting.

The rates of formation of 4-nitrophenol from *p*-dinitrobenzene and hydroxide ions give a linear plot of $\log_{10} k_1$ against $H_- + \log_{10} a_w$ for the reaction in aqueous ethylenediamine solutions (Jacquinot-Vermesse and Schaal, 1964). This is in accord with the generalized mechanism (7.33) providing J_- (equation 7.35) is assumed equal to $H_- + \log_{10} a_w$ (equation 7.30). In contrast the rates of reaction of 1-chloro-2,4-dinitrobenzene in aqueous ethylenediamine are approximately proportional to the base concentration (Jacquinot-Vermesse and Schaal, 1964). This arises because predominantly substitution of the chloro group by the $NH_2CH_2CH_2NH$— group occurs in this case (Vermesse-Jacquinot, 1965b).

7.1.5. *The H_{2-} Acidity Function*

The ability of an aqueous strongly basic solution to abstract a proton from a mononegatively charged weakly acidic solute HA^- according to equilibrium (7.39) is given by the H_{2-} acidity function which is defined by equation (7.40).

$$HA^- + OH^- \rightleftharpoons A^{2-} + H_2O \qquad (7.39)$$

$$H_{2-} = -\log_{10}(a_{H^+}\, y_{A^{2-}}/y_{HA^-}) = pK_{HA^-} + \log_{10}(C_{A^{2-}}/C_{HA}) \qquad (7.40)$$

An H_{2-} scale has been measured for aqueous benzyltrimethylammonium hydroxide using substituted diphenylamine-carboxylates and aniline-carboxylates as indicators (Bowden *et al.* 1966), for aqueous NaOH and LiOH using sulphonate and carboxylate derivatives of aniline and diphenylamine as indicators (Hallé *et al.* 1968), and for aqueous KOH using substituted indole-carboxylates as indicators (Yagil, 1967a, 1967c). In all cases H_{2-} increases more rapidly with increasing concentration of base than does the corresponding H_- scale.

Combination of equations (3.24) and (7.40) leads to equation (7.41). Yagil (1967a) has considered to what extent the electrostatic terms (Glueckauf,

$$H_{2-} - H_- = -\log_{10}(y_{SH}\, y_{A^{2-}}/y_{S^-}\, y_{HA^-}) \qquad (7.41)$$

1955, 1959) of the activity coefficients in equation (7.41) contribute to the observed deviation between H_- and H_{2-} for aqueous KOH. Assuming that the ion size parameters a are the same for S^- (the conjugate base of an indole), HA^- (an indole-carboxylate anion) and A^{2-} it follows that the electrostatic

$$-\log_{10}(y_{SH}\, y_{A^{2-}}/y_{S^-}\, y_{HA^-})_{el} = 2AC_{KOH}^{1/2}/(1 + BaC_{KOH}^{1/2}) \qquad (7.42)$$

contribution to $H_{2-} - H_-$ is given by equation (7.42) in which A and B are the usual Debye–Hückel constants. Values based on $a = 6$ Å and $a = 4$ Å were calculated and proved that the difference between H_{2-} and H_- is in part accounted for in this way. However, the calculations are only rough and would not be expected to give a complete account of the acidity function behaviour.

7.2. Solutions of Bases in Methanol

7.2.1. *The H_M Acidity Function*

The H_M acidity function is defined by equation (7.43) in which pK_{SH}^{M} is the ionization constant of SH in methanol and the activity coefficients y refer to methanol as standard state (Lambert and Schaal, 1962b). H_M scales for

$$H_M = -\log_{10}(a_{H^+}\, y_{S^-}/y_{SH}) = pK_{SH}^{M} + \log_{10}(C_{S^-}/C_{SH}) \qquad (7.43)$$

solutions of lithium, sodium and potassium methoxides in methanol are given in Table 7.2 (Schaal and Lambert, 1962b, 1962c; Peuré and Schaal, 1963; Terrier and Schaal, 1965, 1966). The acidity functions are based on measurements using substituted anilines and diphenylamines as indicators. The values for pK_{SH}^{M} substituted in equation (7.43) were based on $pK_{MeOH} = 16{\cdot}92$ where K_{MeOH} is the autoprotolysis constant of methanol (Schaal and Lambert, 1962a). This figure is for K_{MeOH} in mole^2 litre^{-2} units at 25°C (Koskikallio, 1957a). More O'Ferrall and Ridd (1963a; Ridd, 1957) also determined on H_M scale for methanolic sodium methoxide using substituted anilines and diphenylamines as indicators. When their scale is corrected in accord with $pK_{MeOH} =$

16·92 rather than $pK_{MeOH} = 16{\cdot}71$ (which is for K_{MeOH} in molal[2] units at 25°) it agrees satisfactorily with the acidity function given in Table 7.2. Strictly speaking it would be better to use pK_{MeOH} at 20° (Koskikallio, 1957b) as the experimental measurements of indicator ratios were made at this temperature.

TABLE 7.2

H_M acidity function scales for three alkali–metal methoxides in methanol at ca. 20° (Schaal and Lambert, 1962b, 1962c; Peuré and Schaal, 1963; Terrier and Schaal, 1965, 1966)

C_{Base} (mole litre^{-1})	H_M (KOMe)	H_M (NaOMe)	H_M (LiOMe)
0·2	16·35	16·37	16·19
0·4	16·73	16·69	16·49
0·6	16.99	16·92	16·69
1·0	17·54	17·34	17·00
1·4	18·05	17·72	17·24
1·8	18·53	18·12	17·52
2·2	19·01	18·44	17·78
2·6	19·51	18·87	18·04
3·0	19·99	19·30	18·30
3·6	20·73	19·96	18·68
4·0	21·23	20·36	
4·5	21·84	20·86	
5·0	22·43		
5·5	23·06		

Bowden (1966) has retabulated the H_M acidity functions for alkali–metal methoxides in methanol as H_- acidity functions defined by equation (3.24) in which pK_{SH} and the activity coefficients are referred to water as standard state. The success of this approach depends not only upon the normal acidity function requirements (Section 1.4) but also on the identity of $pK_{SH}^{M} - pK_{SH}$ for all the weak bases used to define the acidity function (Rochester, 1966a). This identity is not maintained for the substituted diphenylamine and aniline indicators (More O'Ferrall and Ridd, 1963a; Vermesse-Jacquinot, 1965a; Schaal and Lambert, 1962a; Lambert and Schaal, 1962). It must be concluded that the H_M acidity functions are considerably more meaningful and less ambiguous than the H_- acidity functions for strongly basic solutions in methanol solvent. The suggestion (Bowden, 1966) that H_M scales are superfluous is unacceptable.

The H_M acidity function scale for methanolic sodium methoxide depends upon the chemical structure of the weak acids used to measure the scale. Thus use of 2-t-butyl substituted phenols, 2,6-di-t-butyl substituted phenols

(Rochester, 1965a, 1965c) and pentamethylphenol (Rochester, 1966b) leads to three H_M scales which differ slightly from each other and which differ significantly (Fig. 7.2) from the amine H_M acidity function. By analogy to equation

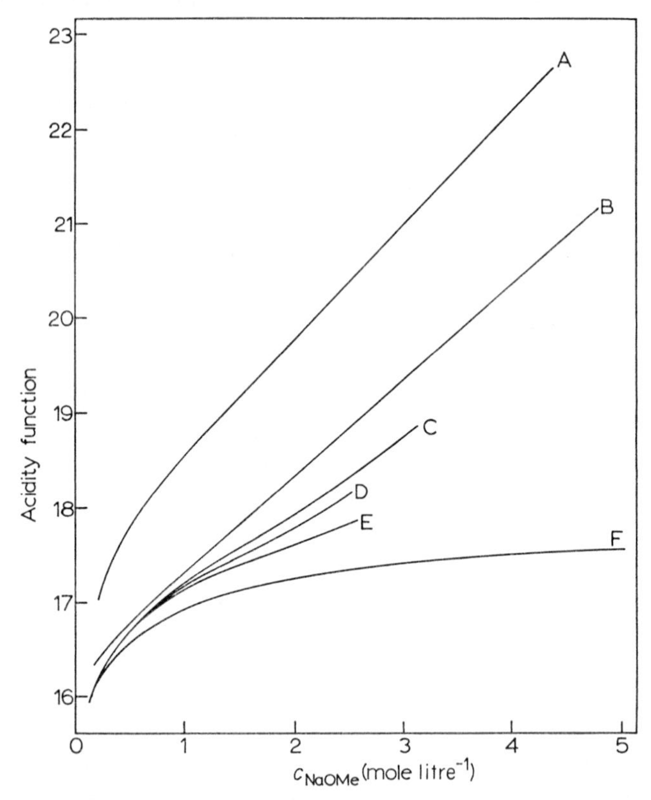

FIG. 7.2. Comparison of acidity function scales for methanolic sodium methoxide. A, H_M^{2-}; B, H_M (amines as indicators); C, H_M (2,6-di-t-butyl phenols); D, H_M (2-t, butyl phenols); E, H_M (pentamethylphenol); F, $pK_s + \log_{10} C_{NaOMe}$.

(7.8) for the H_- scale in aqueous solution it follows that H_M may be written as equation (7.44) in which a_{MeOH} is the activity of the solvent and p may represent

$$H_M = pK_{MeOH} + \log_{10} C_{OMe^-} - (p+1)\log_{10} a_{MeOH} + \log_{10}(y_{SH}\, y_{OMe^-}/y_{S^-}) \tag{7.44}$$

the difference in solvation between $SH + OMe^-$ and S^-. Plots (Rochester, 1966b) of $(H_M - \log C_{NaOMe})$ against $-\log_{10} a_{MeOH}$ (Freeguard *et al.* 1965; Terrier, 1967) for the four H_M scales decreased in slope with increasing sodium methoxide concentration in accord with the proposal (Freeguard *et al.* 1965) that p will decrease as the concentration of solvent is decreased. At low concentrations of base the deduced values of p were 3 for pentamethylphenol, 4·1 for the 2-t-butyl phenols and 5·6 for the 2,6-di-t-butyl phenols. If these figures are related to the changes in solvation for the indicator acid–base equilibria then they indicate that the phenol anions have solvation numbers in the order pentamethylphenoxide > 2-t-butylphenoxide > 2,6-di-t-butyl-phenoxide. This order is consistent with the steric effect of 2-t-butyl substituents in reducing the ability of methanol molecules to hydrogen bond with the phenoxide charge centre in the phenol anions. However, the values of p (particularly for the amine H_M scale) are too large to be equated entirely with the proposed change in solvation and therefore it must be concluded that the high basicity of alkali–metal methoxide solutions is incompletely understood. The observed order (Table 7.2) $H_M(\text{KOMe}) > H_M(\text{NaOMe}) > H_M(\text{LiOMe})$ for a particular stoichiometric molarity of base may be in part due to the effects of ion association on basicity.

Terrier *et al.* (1967) have measured H_M^{2-} acidity functions for methanolic sodium methoxide and potassium methoxide using a series of substituted diphenylamine-carboxylates as weak acids. This acidity function is defined with reference to methanol standard state by an equation similar to equation (7.40) for aqueous solution. The H_M^{2-} acidity functions increased more rapidly with increasing base concentration than the corresponding H_M scales (Fig. 7.2). Analogous results have been obtained for aqueous solutions of strong bases (Section 7.1.5).

7.2.2. *Correlation of H_M with Reaction Rates*

The methanolysis of chloroform in methanolic sodium methoxide has been studied by More O'Ferrall and Ridd (1963b; Ridd, 1957; Allison *et al.* 1958). The mechanism may be contracted to reactions (7.45) and (7.46) (B-1 methanolysis).

$$CHCl_3 + OMe^- \rightleftharpoons CCl_3^- + MeOH \quad \text{(fast)} \qquad (7.45)$$

$$CCl_3^- \rightarrow \text{products} \quad \text{(slow, B-1)} \qquad (7.46)$$

By analogy with arguments discussed in Section 7.1.3 $\log_{10} k_1$ for the methanolysis of chloroform might be expected to parallel the H_M acidity function (cf. equation 7.16). The plot of $\log_{10} k_1$ against H_M was curved with an average slope of 0·8 for 1 mole litre^{-1} < C_{NaOMe} < 3 mole litre^{-1}. Barbaud *et al.* (1965) obtained a linear plot of $\log_{10} k_1$ against H_M for the reaction in methanolic sodium methoxide and lithium methoxide solutions although the slope (ca. 0·75) was much the same as for the earlier result. The linear correlation breaks down, however, at high concentrations of methoxide (Georgoulis *et al.* 1969). This appears to be the only reaction studied for which a direct correlation between $\log_{10} k_1$ and H_M might be expected on simple theory. The low value of the slope of the plot may in part arise because chloroform is a carbon acid whereas nitrogen acids (amines) were used to establish the H_M

scale. Carbon bases and nitrogen bases show appreciably different acidity function behaviour in concentrated acid solutions (Fig. 3.2; see also Table 4.2).

The acidity function dependence of the methanolysis of chloroform is apparently independent of temperature in the range $20° < t < 80°$ (Barbaud *et al.* 1965). A similar result was obtained by Rochester (1967) for the S_N2 reaction between hydroxide ions and 2,4-dinitroanisole in aqueous NaOH.

More O'Ferrall and Ridd (1963b) found that the experimental rate constant for the E-2 elimination reaction of phenethyl chloride in methanolic sodium methoxide was directly proportional to the sodium methoxide concentration in dilute solutions, but increased to about third order when $C_{NaOMe} =$ ca. 4 mole litre^{-1}. This acceleration in rate with increasing concentration of base is not sufficient to give a correlation between $\log_{10} k_1$ and H_M. The mechanism involves the slow attack of phenethyl chloride by methoxide ions which will only have been partially neutralized at the transition state. The change in solvation in passing from reactants to transition state will probably therefore be less than for the methanolysis of chloroform in which the methoxide ion is completely neutralized before the transition state. The methanol activity is lowered as the NaOMe concentration is increased and this will have the largest effect on the rates of reactions for which there is a large solvation change in reaching the transition states. The different behaviour for the reactions of chloroform and phenethyl chloride may tentatively be explained in this way.

7.2.3. *The J_M Acidity Function*

Three sets of measurements of a J_M acidity function for methanolic sodium methoxide have been made and are compared in Fig. 7.3. This function is defined by equation (7.47) in which K is the equilibrium constant for reaction (7.48) in methanol and R is a neutral solute species.

$$J_M = -\log_{10}(a_{H^+} y_{ROMe^-}/y_R a_{MeOH}) = p(KK_{MeOH}) + \log_{10}(C_{ROMe^-}/C_R) \quad (7.47)$$

$$R + OMe^- \rightleftharpoons ROMe^- \quad (7.48)$$

The acidity function scale is thus, as for H_M, referred to pure methanol as standard state.

Kroeger and Stewart (1967) used a series of substituted α-cyanostilbenes (Stewart and Kroeger, 1967) as indicators R and hence deduced a J_- acidity function (equation 7.28) for methanolic sodium methoxide referred to water as reference state. The J_M scale plotted in Fig. 7.3 has been interpolated from their data by adding $(pK_{MeOH} - pK_w) = 2{\cdot}92$ to the J_- values. The true J_M scale will differ from Kroeger and Stewart's J_- scale by a constant amount equal to the equilibrium constant for reaction (7.49) in methanol where A is

$$A + MeOH \rightleftharpoons AOMe^- + H^+ \quad (7.49)$$

α-cyano-4,4′-dinitrostilbene minus 14·42 which was the value taken for the equilibrium constant referred to water as standard state.

Several polynitrobenzene derivatives undergo methoxide addition in methanolic alkali–metal methoxide solutions. Rochester (1965b, 1966a) defined a J_M scale for sodium methoxide solutions based on measurements with 2,4-dinitroanisole as indicator. The scale increases more rapidly with increasing

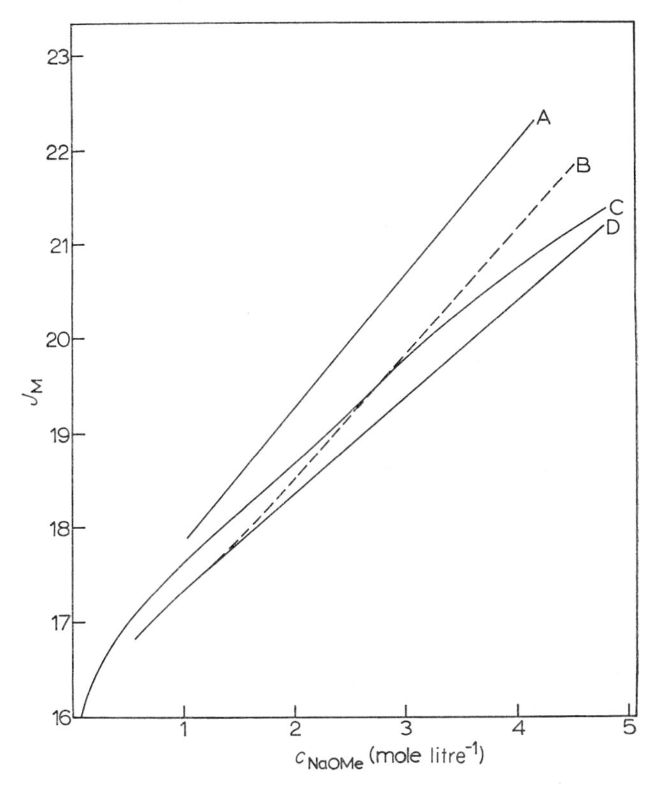

FIG. 7.3. The experimental J_M scales for sodium methoxide in methanol compared with the amine H_M acidity function (curve D). Indicators were: A, polynitrobenzene derivatives (Rochester, 1965b); B, polynitrobenzene derivatives (Terrier and Schaal, 1966); C, α-cyanostilbenes (Kroeger and Stewart, 1967).

concentration of base than does the α-cyanostilbene J_M acidity function (Fig. 7.3). Similar studies of methoxide addition to 2,4-dinitroanisole and 2,6-dinitroanisole (Terrier and Schaal, 1966) led to nearly parallel results to those of Rochester (1965b) although the two scales were displaced ca. 0·7–0·8 units from each other (Fig. 7.3). This primarily arises because of a difference of 0·79 units in the equilibrium constant K (equation 7.45) for 2,4-dinitroanisole deduced in the two studies. It must be noted that the pK values listed by Rochester (1965b) should all be 0·22 units higher in accord with the value of p$K_{\text{MeOH}} = 16{\cdot}92$ where K_{MeOH} is in units of mole^2 litre^{-2}. In general polynitrobenzene derivatives often interact with alkoxide ions in overlapping equilibria of different stoichiometry and also give irreversible substitution reactions which are sometimes photochemical (Gold and Rochester, 1964a, 1964b, 1964c, 1964d; Rochester, 1965b). Considerable care must be exercised if acidity function scales defined by measurements using these compounds are to be reliable and unambiguous.

Rochester (1965b) has studied the addition of methoxide ions to the conjugate base of 2,4-dinitroaniline, the picrate anion, and the 1:1 adducts of methoxide ions with both 2,4-dinitroanisole and 2,4,6-trinitroanisole. The four methoxide addition equilibria gave acidity function behaviour consistent with the same J_{M}^{2-} acidity function for methanolic sodium methoxide. Furthermore the J_{M}^{2-} scale apparently coincides with the J_{M} acidity function defined by the addition of methoxide ions to 2,4-dinitroanisole.

7.2.4. *Correlation of J_{M} with Reaction Rates*

Schaal and Peuré (1963a, 1963b) showed that $\log_{10} k_1$ was a linear function of the H_{M} acidity function for the S_N2 reaction of methoxide ions with 1,4-dinitrobenzene in methanol. Use of lithium, sodium or potassium methoxides as the source of OMe^- led to the same plot of $\log_{10} k_1$ against H_{M}. Similar results were obtained by Schaal and Latour (1964) for the corresponding reaction of 1,2-dinitrobenzene. The S_N2 reactions of methoxide ions with 2,5-dinitroanisole, 3,4-dinitroanisole, 1,4-dinitronaphthalene, 4-chloronitrobenzene, 2-chloronitrobenzene and 2-chloro-3-nitropyridine have also been studied by Terrier and Schaal (1967; Terrier, 1965). In all cases plots of $\log_{10} k_1$ against H_{M} were linear although the slopes of the plots were significantly different for some of the reactions.

The generalized mechanism for the S_N2 substitution of OMe^- for X in a neutral molecule RX may be represented by equations (7.50) and (7.51)

$$\text{RX} + \text{OMe}^- \rightleftharpoons \left[\text{R}\begin{matrix}\diagup \text{X} \\ \diagdown \text{OMe}\end{matrix}\right]^- \tag{7.50}$$

$$\left[\text{R}\begin{matrix}\diagup \text{X} \\ \diagdown \text{OMe}\end{matrix}\right]^- \rightarrow \text{ROMe} + \text{X}^- \tag{7.51}$$

In the reactions being considered here the addition (7.50) of methoxide to RX is a rapid pre-equilibrium lying predominantly on the $\text{RX} + \text{OMe}^-$ side. Reaction (7.51) is the rate-controlling step. Combination of equation (7.52)

$$k_1 = -(1/C_{\text{RX}})(dC_{\text{RX}}/dt) = k_0\, C_{\text{OMe}^-}\, y_{\text{RX}}\, y_{\text{OMe}^-}/y^{\ddagger} \tag{7.52}$$

for the experimental rate constant k_1 with the definition (7.47) of J_M leads to equation (7.53) for the relationship between k_1 and J_M.

$$\log_{10} k_1 = \log_{10}(k_0 K_{MeOH}) + J_M + \log_{10}(y_{RX}\, y_{ROMe^-}/y_R\, y^{\ddagger}) \qquad (7.53)$$

Thus $\log_{10} k_1$ might be expected to be a linear function of J_M with unit slope providing the activity coefficient term is independent of electrolyte concentration. This is clearly not so for all the reactions studied by Terrier and Schaal (1967). They have therefore adopted an approach which is rather analogous to that of Bunnett (1963) for reactions in concentrated acid solutions.

Combination of equation (7.52) with the definition (7.43) of H_M leads to equation (7.54). Following the theoretical interpretation of acidity function behaviour in terms of solvation effects this may be rewritten as equation (7.55)

$$\log_{10} k_1 = \log_{10}(k_0 K_{MeOH}) + H_M + \log_{10} a_{MeOH} + \log_{10}(y_{RX}\, y_{S^-}/y_{SH}\, y^{\ddagger}) \qquad (7.54)$$

$$\log_{10} k_1 = \log_{10}(k_0 K_{MeOH}) + H_M + (n+1)\log_{10} a_{MeOH} + \log_{10}(y_{RX}\, y_{S^-}/y_{SH}\, y^{\ddagger}) \qquad (7.55)$$

in which n is the difference in solvation between (SH + ‡) and (S^- + RX). Terrier and Schaal (1967) deduced values of n for which plots of $\log_{10} k_1$ against $H_M + (n+1)\log_{10} C_{MeOH}$ were linear with unit slope. Concentrations C_{MeOH} of "free" methanol, calculated assuming a solvation number of 3 for OMe^-, were used in these correlations rather than the methanol activities. Values of n for the S_N2 substitutions were −1 for 1,4-dinitronaphthalene and 2-chloro-3-nitropyridine, +1 for 2,5-dinitroanisole, 3,4-dinitroanisole and 2-chloronitrobenzene, and 3 for 4-chloronitrobenzene. Values of −1 and 0 may be deduced from earlier results for 1,4-dinitrobenzene (Schaal and Peuré, 1963a, 1963b) and 1,2-dinitrobenzene (Schaal and Latour, 1964) respectively. Clearly here, as for reactions in concentrated acid solutions (Bunnett, 1963) the acidity function dependencies for reactions of similar mechanism are influenced appreciably by the structures and solvation requirements of the reacting substrate and transition state in a particular reaction.

7.3. Non-aqueous Solvents other than Methanol

7.3.1. *t-Butyl Alcohol*

Bethell and Cockerill (1966a; Bethell, 1963) have studied the ionization of seven primary aromatic amine indicators in solutions of sodium t-butoxide, potassium t-butoxide and benzyltrimethylammonium hydroxide in t-butyl alcohol solvent at 30°C. Rigorous H_- acidity functions could not be defined for these systems but Bethell and Cockerill (1966a) deduced empirical acidity function scales designated H_r^{BuOH} which give a satisfactory indication of the variation of basicity of the solutions with changing concentration of base.

Potassium t-butoxide solutions are about twice as basic as sodium t-butoxide solutions at the same concentration in the range 0·0005 mole litre^{-1} < C_{M+OBu^-} < 0·06 mole litre^{-1}. In this range plots of H_r^{BuOH} against $\log_{10} C_{M+OBu^-}$ were linear but had slopes of ca. 1·13 for potassium t-butoxide and ca. 1·25 for sodium t-butoxide. Solutions of benzyltrimethylammonium hydroxide appear to be at least 1000 times more basic than those of potassium t-butoxide. Clearly the measured basicity is influenced to a large extent by the effects of ion association. The dielectric constant of t-butyl alcohol is 11·2 at 30°. The dielectric constant decreases with increasing temperature and a corresponding decrease in the ionization ratios for 4-chloro-2-nitroaniline was observed. Raising the temperature from 30° to 50° approximately halved the basicity of potassium t-butoxide solutions.

Correlations between the rates of several reactions and the H_r^{BuOH} acidity functions have been studied by Bethell and co-workers. The mechanism of the formation of bifluorenylidene from substituted 9-bromofluorenes either involves a rate-determining unimolecular decomposition of the conjugate base and another 9-bromofluorene molecule. In either case $\log_{10} k_{obs}$ (k_{obs} is the experimental rate constant) was a linear function of H_r^{BuOH} with ca. unit slope (Bethell, 1963; Bethell and Cockerill, 1964). In contrast plots of $\log_{10} k_{obs}$ against H_r^{BuOH} had ca. 0·71 slope for the E-2 β-elimination of hydrogen bromide from 9-bromo-9,9′-bifluorenyl (Bethell and Cockerill, 1966b). This difference in acidity function behaviour confirmed that the conversion of 9-bromo-9,9′-bifluorenyl into bifluorenylidene could not be the rate-determining step in the mechanism for the formation of bifluoroenylidene from 9-bromofluorene. The rates of all these reactions were halved if sodium t-butoxide was substituted for potassium t-butoxide as the base with the values of H_r^{BuOH} the same. This effect is explicable if the unimolecular and bimolecular rate-determining steps do not involve reaction of the separate 9-bromofluorenyl carbanions (the conjugate bases of substituted 9-bromofluorenes) but rather involve reaction of the ion-pairs formed by association of the carbanions with sodium or potassium ions (Bethell *et al.* 1967). A similar effect is apparent from a comparison of the rates of autoxidation of fluorene promoted by potassium t-butoxide or by benzyltrimethylammonium hydroxide in t-butyl alcohol (Bethell and Talbot, 1968).

The rates of proton abstraction from the α-carbon atom of 4-nitrodiphenylmethyl chloride are greater in solutions of potassium t-butoxide in t-butyl alcohol than for the same concentrations of sodium t-butoxide in t-butyl alcohol (Bethell and Cockerill, 1966c). However, a graph of $\log_{10} k_{obs}$ against H_r^{BuOH} leads to the same linear plot (slope 0·82) for the two systems. The rates of deuteron removal from 8,9-dideuteriofluorene in solutions of equal H_r^{BuOH} are also independent of the particular cation present in solution (Bethell and Talbot, 1968). Bethell and Cockerill (1966c) have suggested that the dependence of rates on H_r^{BuOH} alone is characteristic of reactions in which the formation of a carbanion is rate determining. If, on the other hand, the rate-determining step involves reaction of a carbanion then the observed reaction rate will depend not only on H_r^{BuOH} but also on the particular cations which are associated in ion aggregates with the carbanions. Sensitivity of reaction rate to changes in cation may provide a useful mechanistic criterion in studies of reactions in strongly basic solutions for which the solvent has a low dielectric constant.

7.3.2. *Other Alcohols*

The majority of acidity function studies for solutions of bases in anhydrous alcohols have been concerned with methanol or t-butyl alcohol as solvent. A few equilibrium measurements of the ionization of amine indicators in solutions of alkali–metal alkoxides in the corresponding alcohols have also been made for ethanol (Stearns and Wheland, 1947; Schaal and Gadet, 1961), isopropanol (Hine and Hine, 1952; Vermesse-Jacquinot *et al.* 1960), and t-pentanol (Jacquinot-Vermesse *et al.* 1960; Vermesse-Jacquinot, 1965d). A comparison of H_- values for dilute ($< 0{\cdot}1$ mole litre^{-1}) solutions of alkali–metal alkoxides in methanol, ethanol, isopropanol, t-butanol and t-pentanol has been given by Bowden (1965, 1966).

An H_- acidity function for solutions of sodium aminoethoxide in β-amino-ethanol has been measured by Jacquinot-Vermesse and Schaal (1962; Vermesse-Jacquinot, 1965a). Comparison of the H_- value (17·96) for a solution of unit activity of $NH_2CH_2CH_2ONa$ with the H_- value (12·82) for a solution of unit activity of β-aminoethanol hydrochloride in β-aminoethanol enabled a figure of $7{\cdot}28 \times 10^{-6}$ mole^2 litre^{-2} to be deduced for the ionic product of β-aminoethanol at 20°C. However, the logarithms of the ionization ratios of primary and secondary amines in solutions of $NH_2CH_2CH_2ONa$ in $NH_2CH_2CH_2OH$ are not parallel functions of the concentration of base (Vermesse-Jacquinot, 1965a). An H_- scale for sodium β-diethylaminoethoxide in β-diethylaminoethanol has also been determined using phenylhydrazones as indicators (Gaboriaud *et al.* 1961). The ionization of substituted diphenylamines was consistent with a different H_- scale. This may in part arise because of the effects of ion association in β-diethylaminoethanol for which the dielectric constant is 9·1 at 20°.

7.4. Solutions of Bases in Mixed Solvent Systems

7.4.1. *Water/Dimethyl Sulphoxide Mixtures*

The basicity of aqueous solutions of strong bases is considerably increased by the addition of dimethyl sulphoxide. An H_- acidity function (Table 7.3) has been measured for dimethyl sulphoxide/water mixtures containing 0·011

TABLE 7.3

The H_- (Dolman and Stewart, 1967) and H_{2-} (Bowden *et al.* 1966) acidity functions for water/dimethyl sulphoxide mixtures containing 0·011 mole litre^{-1} tetramethylammonium hydroxide at 25°C

Mole % Me_2SO	H_-	Mole % Me_2SO	H_-	Mole % Me_2SO	H_{2-}
10·32	13·17	73·69	19·90	1·0	12 04
15·20	13·88	76·12	20·14	5·0	12·32
20·18	14·49	78·36	20·38	9·9	12·70
23·57	14·86	80·78	20·68	14·9	13·24
26·95	15·22	83·14	20·97	19·8	13·85
30·11	15·54	85·46	21·27	24·7	14·34
33·42	15·87	88·79	21·61	29·6	14·84
36·79	16·17	90·07	21·98	34·5	15·39
39·86	16·48	92·47	22·45	39·3	15·77
43·27	16·83	94·74	23·01	44·1	16·18
46·54	17·12	95·77	23·32	48·8	16·59
49·59	17·42	96·21	23·48	53·7	16·96
52·55	17·73	97·13	23·88	58·6	17·32
55·95	18·08	97·89	24·25	63·5	17·70
58·56	18·34	98·29	24·50	68·3	18·18
62·27	18·72	98·71	24·84	73·2	18·48
64·20	18·92	99·14	25·30	78·0	18·88
69·09	19·41	99·59	26·19	82·8	19·30
71·35	19·65			87·5	19·84
				90·9	20·27
				92·3	20·71
				94·6	21·31
				96·9	21·92

mole litre^{-1} tetramethylammonium hydroxide (Stewart and O'Donnell, 1962, 1964a; Dolman and Stewart, 1967). Substituted anilines and diphenylamines (Stewart and O'Donnell, 1964b; Stewart and Dolman, 1967), were used as indicators as both the primary and secondary amines conformed to the same H_- scale. The H_- values show that there is a 10^{14} increase in basicity in going from pure water to 95% dimethyl sulphoxide as solvent. The ionization of a weakly acidic indicator SH in a solution containing hydroxide ions may be written as equilibrium (7.7) consideration of which leads to equation (7.8) for H_-. Adding dimethyl sulphoxide to aqueous solutions decreases the water activity a_w both by a dilution effect and also via the ability of dimethyl sulphoxide to form strong hydrogen bonds with water. However, this is not sufficient to account completely for the steep rise in H_- with decreasing water concentration (Dolman and Stewart, 1967). The activity of the hydroxide ion also increases rapidly as dimethyl sulphoxide replaces water in the solvent. The dimethyl sulphoxide and hydroxide ions compete for the solvent water. Increasing the dimethyl sulphoxide concentration therefore leads to a less hydrogen bonded more reactive hydroxide species.

Bowden and Cockerill (1967) have established an H_- scale for 0·0471 mole litre^{-1} tetramethylammonium hydroxide in 40–95 mole % aqueous dimethyl sulphoxide using a series of substituted fluorenes as indicators. The scale increases more rapidly with changing solvent composition than the amine scale. It has been shown that different structural classes of weak acids give different acidity function behaviour in water/dimethyl sulphoxide solutions (Steiner and Starkey, 1967; Bowden and Cockerill, 1967). This point is further emphasized by comparison of the differences ΔpK_a between pK_a in dimethyl sulphoxide and pK_a in water for structurally dissimilar weak acids. Thus ΔpK_a varies appreciably for different acids (Kolthoff and Reddy, 1962; Steiner and Gilbert, 1965; Ritchie and Uschold, 1967). If the correlation of ionization ratios with an acidity function H_- is to be used as a means of measuring the ionization constant of a weak acid SH in water then $\log_{10}(C_{S^-}/C_{SH})$ for the acid must be a linear function with unit slope of the H_- scale. Furthermore, the H_- scale must be correctly referred to standard state water. Only if these requirements are applicable can measurements of acid–base equilibria in water–dimethyl sulphoxide mixtures give reliable pK_a values referred to water standard state. This does provide an important method for the determination of pK_a values as high as ca. 25 units.

An H_{2-} scale has been measured for 0·011mole litre^{-1} tetramethylammonium hydroxide in water–dimethyl sulphoxide mixtures by Bowden *et al.* (1966) using series of substituted aminobenzoates and diphenylamine-carboxylates as indicators (Table 7.3). When $C_{Me_2SO} = 0$, $H_- = H_{2-} = 12$ but over the rest of the range of dimethyl sulphoxide concentrations $H_- > H_{2-}$. However, the deviation between the two scales (equation 7.41) is never very large.

The rates of the sodium hydroxide promoted elimination reaction of dimethyl-2-phenylethylsulphonium bromide in water–dimethyl sulphoxide mixtures have been measured by Cockerill (1967). In accord with the E-2 mechanism for this reaction a plot of $\log_{10} k_{obs}$ against $H_- + \log_{10} C_w$ (compare equation 7.20) was linear with a slope of 0·98. The water concentrations were calculated as the molar concentrations of water divided by 55·5. Because $C_w \gg C_{OH^-}$ it was not necessary to allow for the water molecules which are used up in solvating hydroxide ions. Other reactions promoted by hydroxide ions in water–dimethyl sulphoxide mixtures and which show a marked acceleration in rate with increasing dimethyl sulphoxide concentration include the hydrolyses of esters (Tommila and Murto, 1963; Tommila and Palenius, 1963; Roberts, 1964, 1965), fluoronitrobenzenes (Murto and Huro, 1964), methyl iodide (Murto, 1961) and benzyl chloride (Tommila and Pitkanen,

1966). Jones (1967b) has obtained linear plots of $\log_{10} k_{obs}$ against H_- for many of these reactions although the slopes of the lines are less than unity. A slope of 0·5 was similarly obtained for the detritiation of [α-^{3}H] acetophenone in aqueous NaOH/dimethyl sulphoxide solutions (Jones and Stewart, 1967).

7.4.2. *Mixtures of Alcohols with Dimethyl Sulphoxide*

H_- and J_- scales have been measured for solutions of sodium methoxide in methanol/dimethyl sulphoxide mixtures and for sodium ethoxide in ethanol/dimethyl sulphoxide mixtures (Table 7.4). The H_- acidity function for methanol

TABLE 7.4

H_- and J_- acidity function scales (referred to standard state water) for solutions of sodium alkoxides in alcoholic dimethyl sulphoxide

Mole % Me_2SO	H_-† (EtOH/EtONa)	J_-‡ (EtOH/EtONa)	H_-§ (MeOH/MeONa)	J_-‡ (MeOH/MeONa)
0	13·99		12·23	
1	14·07	14·01	12·32	11·89
5	14·25	14·34	12·63	12·31
10	14·45	14·73	12·97	12·85
20	14·92	15·45	13·67	13·85
30	15·40	16·15	14·30	14·70
35	15·68	16·47	14·56	15·10
40	16·11	16·82	14·84	15·54
45	16·61	17·19	15·15	15·95
50	17·03	17·55	15·43	16·40
55	17·37	17·94	15·77	16·81
60	17·75	18·34	16·16	17·27
65	18·13	18·70	16·47	17·77
70	18·45	19·11	16·80	18·23
75	18·69	19·59	17·21	18·75
80	18·97	20·08	17·64	19·22
85	19·28	20·60	18·06	19·77
90	19·68	21·15	18·51	20·41
92·5	20·05	21·56	18·85	20·80
95	20·68		19·37	21·27

† $C_{RONa} = 0·01$ mole litre^{-1}; indicators were carbon acids (Bowden and Stewart, 1965).

‡ $C_{RONa} = 0·01$ mole litre^{-1}; indicators were α-cyanostilbenes (Kroeger and Stewart, 1967).

§ $C_{RONa} = 0·025$ mole litre^{-1}; indicators were amines (Stewart *et al.* 1962).

solutions was based on the ionization behaviour of substituted anilines and diphenylamines as indicators (Stewart *et al.* 1962), whereas that for ethanol solutions was based on measurements with a series of carbon acids (Bowden and Stewart, 1964). The two scales were approximately parallel functions of the concentration of dimethyl sulphoxide. The J_- acidity functions were measured using α-cyanostilbenes as indicators (Kroeger and Stewart, 1967; Stewart and Kroeger, 1967). J_- increased more steeply with increasing dimethyl sulphoxide concentration than did H_-. The theoretical relationship between J_- and H_- is given by equation (7.32). The J_- scales in Table 7.4 are based on an arbitrary pK (equation 7.28) of 0·42 for α-cyano-4,4'-dinitrostilbene.

The change in the H_r^{BuOH} acidity function (Section 7.3.1) on adding up to 2·17 mole litre^{-1} dimethyl sulphoxide to 0·0163 mole litre^{-1} potassium t-butoxide in t-butyl alcohol has been determined using 4-(4-nitrophenylazo)-aniline as weak acid (Bethell and Cockerill, 1966a). As for aqueous solutions the enormous increase in basicity on adding dimethyl sulphoxide to solutions of alkoxides in alcohols is explicable in terms of the marked reduction in alcohol activity and the increase in alkoxide ion activity which occurs as the concentration of dimethyl sulphoxide is increased. Steiner and Gilbert (1963) have suggested that $H_- =$ ca. 33 for a 0·02% mole ratio of alcohol in dimethyl sulphoxide.

A plot of $\log_{10} k_1$ against H_- was linear with slope 0·87 for the racemization of (+)-2-methyl-3-phenylpropionitrile in sodium methoxide (0·025 mole litre^{-1})/methanol/dimethyl sulphoxide solutions (Stewart *et al.* 1962). The two possible mechanisms are represented by reactions (7.56) in which k_2 and

$$\underset{\text{optically active}}{AH + OMe^-} \underset{-1}{\overset{1}{\rightleftharpoons}} \underset{\text{optically active}}{A^- \ldots HOMe} \xrightarrow{2} \underset{\text{racemic}}{A^- \ldots HOMe} \xrightarrow{-1} \underset{\text{racemic}}{AH + OMe^-} \qquad (7.56)$$

$k_{-1} \gg k_1$ and either (mechanism 1) $k_{-1} \gg k_2$ or (mechanism 2) $k_2 \gg k_{-1}$. The acidity function dependence is in accord with mechanism 1 in which the racemization step (k_2) is rate controlling. This conclusion is consistent with the acidity function dependence for the methanolysis of chloroform in methanolic sodium methoxide (More O'Ferrall and Ridd, 1963b).

Plots of $\log_{10} k_1$ against H_- for the detritiation of [α-^{3}H] acetophenone in mixtures of dimethyl sulphoxide with ethanolic sodium ethoxide, methanolic sodium methoxide and aqueous tetramethylammonium hydroxide were linear with slopes of 0·40, 0·31 and (slightly curved) 0·42–0·54 respectively (Jones and Stewart, 1967). These low slopes are consistent with a mechanism in which the rate-determining step is the abstraction of a triton by the appropriate base. The acidity function correlation is analogous to the result for the E-2 elimination reaction of β-phenethyl chloride in methanolic sodium methoxide for which the slow step is the attack of the substrate by methoxide ions (More O'Ferrall and Ridd, 1963b).

In general the rates of reactions involving alkoxide ions correlate with the appropriate acidity functions if the rate-determining step is the unimolecular decomposition of a conjugate base (H_- correlation) or OR^- addition complex (J_- correlation) of the reacting substrate. Plots of $\log_{10} k_1$ against the acidity function will have ca. unit slope although some variation arises because of the different solvation requirements of the reactants and transition state compared with those of the indicators used to measure the acidity functions and their anions. Reactions for which bimolecular attack (either proton abstraction or OR^- addition) of OR^- on the substrate is rate determining are sometimes much less sensitive to medium effects. Thus plots of $\log_{10} k_1$ against acidity function are often either non-linear or if linear give slopes which are much less than 1.

The rate of conversion of 9-bromo-2-methoxyfluorene into 2,2′-dimethoxy-bifluorenylidene by potassium t-butoxide in t-butyl alcohol is enhanced by addition of dimethyl sulphoxide although a change of mechanism also occurs (Bethell *et al.* 1967). Similar results are obtained if tetrahydrothiophen *S*,*S*-dioxide or pyridine *N*-oxide are added in place of dimethyl sulphoxide. In contrast the detritiation of [α-^{3}H] acetophenone (Jones and Stewart, 1967) and the reaction of 2-arylethyl bromide (Cockerill *et al.* 1967) with alkoxide ions show no apparent change in mechanism as dimethyl sulphoxide is added to the solvent (Jones, 1967a). Changes in mechanism should destroy the linearity of plots of $\log_{10} k_1$ against acidity function.

7.4.3. *Miscellaneous Mixed Solvent Systems*

An H_- acidity function for solutions of phenyltrimethylammonium hydroxide in aqueous sulpholane has been determined by Langford and Burwell (1960; Bowden, 1966) using amines and carbon acids as indicators. Stewart and O'Donnell (1962, 1964a) made similar measurements for 0·011 mole litre^{-1} tetramethylammonium hydroxide in aqueous sulpholane and confirmed that the addition of sulpholane increased the basicity of the solutions. Thus $H_- = 19{\cdot}28$ for 0·011 mole litre^{-1} aqueous Me_4NOH in 95 mole % sulpholane. However, the corresponding value of $H_- = 22{\cdot}5$ for 95 mole % dimethyl sulphoxide confirms the superior ability of dimethyl sulphoxide to enhance solution basicity.

Stewart and O'Donnell (1962, 1964a) have investigated the basicity of solutions of benzyltrimethylammonium hydroxide in 50 mole % and 30 mole % aqueous pyridine and of 0·001 mole litre^{-1} tetramethylammonium hydroxide in water–pyridine mixtures. H_- scales were deduced and tabulated. Increasing the pyridine concentration increases the extent of ionization of weak amine acids in these solutions. However, the effect of pyridine is less marked than that of sulpholane and much less marked than that of dimethyl sulphoxide. The extent of ionization of 4-(4-nitrophenylazo)aniline in 0·0163 mole litre^{-1}

potassium t-butoxide in t-butyl alcohol is increased by the addition of tetrahydrothiophen *S,S*-dioxide or pyridine *N*-oxide to the solvent (Bethell and Cockerill, 1966a). The former has much the larger effect.

H_{2-} scales for 0·011 mole litre^{-1} tetramethylammonium hydroxide in pyridine/water and sulpholane/water mixtures were deduced by Bowden *et al.* (1966) who used substituted aminobenzoates and diphenylamine-carboxylates as indicators. In both cases the H_{2-} scale increased less steeply than the corresponding H_- scale with increasing concentration of the non-aqueous solvent component. However, the deviation between the H_- and H_{2-} scales for a particular solvent system was never more than ca. 0·7 units.

A study of the ionization of amine indicators in alcohol/water mixtures containing either 0·001 mole litre^{-1} or 0·005 mole litre^{-1} sodium alkoxide has enabled Bowden (1965) to list H_- values for these concentrations of base in mixtures of water with methanol, ethanol, isopropanol, t-butanol and t-pentanol. In view of the specificity of acid–base behaviour (Wynne-Jones, 1968) and the complexity of solvent structure (Franks and Ives, 1966) in these systems it seems unlikely that the acidity function approach will be useful when changes in solvent composition are being considered. This conclusion is borne out by the results for solutions of strong acids in alcohol/water mixtures (Section 6.4). Studies involving changes in concentration of base in a particular solvent composition might be more useful.

7.5. Conclusions

There have been considerably fewer studies involving acidity functions for strongly basic solutions than for strongly acidic solutions. However, from the limited amount of data available for the former it is clear that the results for solutions at the extremes of high and low acidity are essentially similar. The general usefulness and limitations of the acidity function approach which have been investigated in some detail for strongly acidic solutions, are apparently equally applicable for strongly basic solutions.

REFERENCES

Allison, M. F. L., Bamford, C., and Ridd, J. H. (1958). *Chem. and Ind.* 718.
Anbar, M., and Yagil, G. (1962). *J. Amer. Chem. Soc.* **84**, 1790.
Anbar, M., Bobtelsky, M., Samuel, D., Silver, B., and Yagil, G. (1963). *J. Amer. Chem. Soc.* **85**, 2380.
Barbaud, J., Georgoulis, C., and Schaal, R. (1965). *Compt. rend.* **260C**, 2533.
Bell, R. P., and Bascombe, K. N. (1957). *Discuss. Faraday Soc.* **24**, 158.
Bell, R. P., and Prue, J. E. (1959). *J. Chem. Soc.* 362.
Bethell, D. (1963). *J. Chem. Soc.* 666.
Bethell, D., and Cockerill, A. F. (1964). *Proc. Chem. Soc.* 283.

Bethell, D., and Cockerill, A. F. (1966a). *J. Chem. Soc. B*, 913.
Bethell, D., and Cockerill, A. F. (1966b). *J. Chem. Soc. B*, 917.
Bethell, D., and Cockerill, A. F. (1966c). *J. Chem. Soc. B*, 920.
Bethell, D., and Talbot, R. J. E. (1968). *J. Chem. Soc. B*, 638.
Bethell, D., Cockerill, A. F., and Frankham, B. (1967). *J. Chem. Soc. B*, 1287.
Bowden, K. (1965). *Canad. J. Chem.* **43**, 2624.
Bowden, K. (1966). *Chem. Rev.* **66**, 119.
Bowden, K., and Cockerill, A. F. (1967). *Chem. Comm.* 989.
Bowden, K., and Stewart, R. (1964). *Tetrahedron*, **21**, 261.
Bowden, K., Buckley, A., and Stewart, R. (1966). *J. Amer. Chem. Soc.* **88**, 947.
Buncel, E., Norris, A. R., and Russell, K. E. (1968). *Quart. Rev.* **22**, 123.
Bunnett, J. F. (1961). *J. Amer. Chem. Soc.* **83**, 4956, 4968, 4973, 4978.
Cockerill, A. F. (1967). *J. Chem. Soc. B*, 964.
Cockerill, A. F., Rottschaefer, S., and Saunders, W. H. (1967). *J. Amer. Chem. Soc.* **89**, 901.
Crampton, M. R., and Gold, V. (1964). *Proc. Chem. Soc.* 298.
Crampton, M. R., and Gold, V. (1966). *J. Chem. Soc. B*, 893.
Darken, L. S., and Meier, H. F. (1942). *J. Amer. Chem. Soc.* **64**, 621.
Deno, N. C. (1952). *J. Amer. Chem. Soc.* **74**, 2039.
Dolman, D., and Stewart, R. (1967). *Canad. J. Chem.* **45**, 911.
Edward, J. T., and Wang, I. C. (1962). *Canad. J. Chem.* **40**, 399.
Favier, P., and Schaal, R. (1959). *Compt. rend.* **249**, 1231.
Franks, F., and Ives, D. J. C. (1966). *Quart. Rev.* **20**, 1.
Freeguard, G. F., Moodie, R. B., and Smith, D. J. G. (1965). *J. Appl. Chem.* **15**, 179.
Gaboriaud, R., Monnaye, B., and Schaal, R. (1961). *Compt. rend.* **254C**, 4027.
Georgoulis, C., Pataillot, J., Vial, M., and Valéry, J-M. (1969). *Compt. rend.* **268C**, 761.
Glueckauf, E. (1955). *Trans. Faraday Soc.* **51**, 1235.
Glueckauf, E. (1959). *In* "The Structure of Electrolytic Solutions" (W. J. Hamer, ed.), p. 97. Wiley, New York.
Gold, V., and Hawes, B. W. V. (1951). *J. Chem. Soc.* 2102.
Gold, V., and Rochester, C. H. (1964a). *J. Chem. Soc.* 1687.
Gold, V., and Rochester, C. H. (1964b). *J. Chem. Soc.* 1692.
Gold, V., and Rochester, C. H. (1964c). *J. Chem. Soc.* 1697,
Gold, V., and Rochester, C. H. (1964d). *J. Chem. Soc.* 1704.
Gold, V., and Rochester, C. H. (1964e). *J. Chem. Soc.* 1710.
Gold, V., and Rochester, C. H. (1964f). *J. Chem. Soc.* 1717.
Gold, V., and Rochester, C. H. (1964g). *J. Chem. Soc.* 1722.
Gold, V., and Rochester, C. H. (1964h). *J. Chem. Soc.* 1727.
Gutowsky, H. S., and Saika, A. (1953). *J. Chem. Phys.* **21**, 1688.
Hallé, J-C., Terrier, F., and Schaal, R. (1968). *Compt. rend.* **267C**, 29.
Hine, J., and Hine, M. (1952). *J. Amer. Chem. Soc.* **74**, 5266.
Jacquinot-Vermesse, C., and Schaal, R. (1962). *Compt. Rend.* **254C**, 3679.
Jacquinot-Vermesse, C., and Schaal, R. (1964). *Compt. rend.* **258C**, 2334.
Jacquinot-Vermesse, C., Schaal, R., and Souchay, P. (1960). *Bull. Soc. Chim. France*, 141.
Jones, J. R. (1967a). *Chem. Comm.* 710.
Jones, J. R. (1967b). Personal communication.
Jones, J. R. (1968a). *Chem. Comm.* 513.
Jones, J. R. (1968b). *Trans. Faraday Soc.* **64**, 440.

Jones, J. R., and Stewart, R. (1967). *J. Chem. Soc. B*, 1173.
Kolthoff, I. M., and Reddy, T. B. (1962). *Inorg. Chem.* **1**, 189.
Koskikallio, J. (1957a). *Suomen Kemi B*, **30**, 111.
Koskikallio, J. (1957b). *Suomen Kemi B*, **30**, 155.
Kroeger, D. J., and Stewart, R. (1967). *Canad. J. Chem.* **45**, 2163.
Lambert, G., and Schaal, R. (1962). *J. Chim. phys.* 1170.
Langford, C. H., and Burwell, R. L. (1960). *J. Amer. Chem. Soc.* **82**, 1503.
Lazarev, V. I., and Moiseev, Y. V. (1965). *Russ. J. Phys. Chem.* **39**, 231.
Lazarev, V. I., Moiseev, Y. V., and Golyand, S. M. (1965). *Russ. J. Chem.* **39**, 193.
Masure, F., and Schaal, R. (1956). *Bull. Soc. chim. France*, 1138, 1141.
McTigue, P. T. (1964). *Trans. Faraday Soc.* **60**, 127.
More O'Ferrall, R. A., and Ridd, J. H. (1963a). *J. Chem. Soc.* 5030.
More O'Ferrall, R. A., and Ridd, J. H. (1963b). *J. Chem. Soc.* 5035.
Mouronval, S., Gaboriaud, R., and Schaal, R. (1962). *Compt. rend.* **255C**, 2605.
Murto, J. (1961). *Suomen Kemi B*, **34**, 92.
Murto, J., and Huro, H. M. (1964). *Suomen Kemi B*, **37**, 177.
Norris, W. P., and Osmundsen, P. (1965). *J. Org. Chem.* **30**, 2407.
Peuré, F., and Schaal, R. (1963). *Bull. Soc. chim. France*, 2636.
Ridd, J. H. (1957). *Chem. and Ind.* 1268.
Ritchie, C. D., and Uschold, R. E. (1967). *J. Amer. Chem. Soc.* **89**, 2752.
Roberts, D. D. (1964). *J. Org. Chem.* **29**, 2039, 2714.
Roberts, D. D. (1965). *J. Org. Chem.* **30**, 3516.
Robinson, R. A., and Stokes, R. H. (1949). *Trans. Faraday Soc.* **45**, 612.
Rochester, C. H. (1963a). *Trans. Faraday Soc.* **59**, 2820.
Rochester, C. H. (1963b). *Trans. Faraday Soc.* **59**, 2826.
Rochester, C. H. (1963c). *Trans. Faraday Soc.* **59**, 2829.
Rochester, C. H. (1964). Unpublished results.
Rochester, C. H. (1965a). *J. Chem. Soc.* 676.
Rochester, C. H. (1965b). *J. Chem. Soc.* 2404.
Rochester, C. H. (1965c). *J. Chem. Soc.* 4603.
Rochester, C. H. (1966a). *Quart. Rev.* **20**, 511.
Rochester, C. H. (1966b). *J. Chem. Soc. B*, 121.
Rochester, C. H. (1967). *J. Chem. Soc. B*, 1076.
Schaal, R. (1954a). *Compt. rend.* **238**, 2156.
Schaal, R. (1954b). *Compt. rend.* **239**, 1036.
Schaal, R. (1955). *J. Chim. phys.* **52**, 784, 796.
Schaal, R., and Favier, P. (1959). *Bull. Soc. chim. France*, 2011.
Schaal, R., and Gadet, C. (1961). *Bull. Soc. chim. France*, 2154.
Schaal, R., and Lambert, G. (1962a). *J. Chim. phys.* 1151.
Schaal, R., and Lambert, G. (1962b). *J. Chim. phys.* 1164.
Schaal, R., and Lambert, G. (1962c). *Compt. rend.* **255**, 2256.
Schaal, R., and Latour, J-C. (1964). *Bull. Soc. chim. France*, 2177.
Schaal, R., and Peuré, F. (1963a). *Bull. Soc. chim. France*, 2638.
Schaal, R., and Peuré, F. (1963b). *Compt. rend.* **256**, 4020.
Schwarzenbach, G., and Sulzberger, R. (1944). *Helv. chim. Acta*, **27**, 348.
Servis, K. L. (1965). *J. Amer. Chem. Soc.* **87**, 5495.
Stearns, R. S., and Wheland, G. W. (1947). *J. Amer. Chem. Soc.* **69**, 2025.
Steiner, E. C., and Gilbert, J. M. (1963). *J. Amer. Chem. Soc.* **85**, 3054.
Steiner, E. C., and Gilbert, J. M. (1965). *J. Amer. Chem. Soc.* **87**, 382.
Steiner, E. C., and Starkey, J. D. (1967). *J. Amer. Chem. Soc.* **89**, 2751.

Stewart, R., and Dolman, D. (1967). *Canad. J. Chem.* **45**, 925.
Stewart, R., and Kroeger, D. J. (1967). *Canad. J. Chem.* **45**, 2173.
Stewart, R., and O'Donnell, J. P. (1962). *J. Amer. Chem. Soc.* **84**, 493.
Stewart, R., and O'Donnell, J. P. (1964a). *Canad. J. Chem.* **42**, 1681.
Stewart, R., and O'Donnell, J. P. (1964b). *Canad. J. Chem.* **42**, 1694.
Stewart, R., O'Donnell, J. P., Cram, D. J., and Rickborn, B. (1962). *Tetrahedron*, **18**, 917.
Stokes, R. H. (1945). *J. Amer. Chem. Soc.* **67**, 1689.
Terrier, F. (1965). *Compt. rend.* **261C**, 1001.
Terrier, F. (1967). *Compt. rend.* **265C**, 1433.
Terrier, F., Hallé, J-C., and Schaal, R. (1967). *Compt. rend.* **264C**, 2126.
Terrier, F., and Schaal, R. (1965). *Compt. rend.* **260C**, 5567.
Terrier, F., and Schaal, R. (1966). *Compt. rend.* **263C**, 476.
Terrier, F., and Schaal, R. (1967). *Compt. rend.* **264C**, 465.
Tommila, E., and Murto, M. L. (1963). *Acta Chem. Scand.* **17**, 1947.
Tommila, E., and Palenius, I. (1963). *Acta Chem. Scand.* **17**, 1980.
Tommila, E., and Pitkanen, I. P. (1966). *Acta Chem. Scand.* **20**, 937.
van Panthaleon van Eck, C. L. Mendel, H., and Boog, W. (1957). *Disc. Faraday Soc.* **24**, 200.
Vermesse-Jacquinot, C. (1965a). *J. Chim. phys.* 185.
Vermesse-Jacquinot, C. (1965b). *J. Chim. phys.* 235.
Vermesse-Jacquinot, C. (1965c). *J. Chim. phys.* 366.
Vermesse-Jacquinot, C. (1965d). *J. Chim. phys.* 198.
Vermesse-Jacquinot, C., Schaal, R., and Rumpf, P. (1960). *Bull. Soc. chim. France*, 2030.
Wynne-Jones, W. F. K. (1968). "Hydrogen-bonded Solvent Systems" (A. K. Covington and P. Jones, eds.), p. 246. Taylor and Francis, London.
Yagil, G. (1967a), *J. Phys. Chem.* **71**, 1034.
Yagil, G. (1967b). *J. Phys. Chem.* **71**, 1045.
Yagil, G. (1967c). *Tetrahedron*, **23**, 2855.
Yagil, G., and Anbar, M. (1962). *J. Amer. Chem. Soc.* **84**, 1797.
Yagil, G., and Anbar, M. (1963). *J. Amer. Chem. Soc.* **85**, 2376.

CHAPTER 8

Appendix

Since the manuscript for this book was handed over to the publishers several relevant papers have appeared in the chemical literature. The more important of these are given here and are mentioned under the numbers of the appropriate sections in the book to which they make a contribution. A detailed general review on acidity functions has been written by Boyd (1969).

Section 1.2

Staples and Bates (1969) have proposed two new primary standards, to be added to the five given in Table 1.1, for the National Bureau of Standards pH scale. The variation of pH with temperature for the two standards is given in Table 8.1.

TABLE 8.1

Variation of pH with temperature for two further standards for the NBS pH scale (Staples and Bates, 1969)

t °C	pH†	pH‡
0	3·863	10·317
5	3·840	10·245
10	3·820	10·179
15	3·802	10·118
20	3·788	10·062
25	3·776	10·012
30	3·766	9·966
35	3·759	9·925
40	3·753	9·889
45	3·750	9·856
50	3·749	9·828

† 0·05 *m* solution of potassium dihydrogen citrate.

‡ Solution of sodium bicarbonate (0·025 *m*) and sodium carbonate (0·025 *m*).

Section 2.2.7

Akitt *et al.* (1969) have re-emphasized their conclusion from p.m.r. and ^{35}Cl magnetic resonance studies that perchloric acid is completely ionized in water up to ca. 6 mole litre^{-1}. Dawber's conclusion (page 67) that incomplete ionization occurs in 3–4 mole litre^{-1} $HClO_4$ is probably less reliable. Wai and Yates (1969) have measured water activities for 40·03 to 74·78 w/w % $HClO_4$ in water (see also Section 2.3.1).

Section 3.2.1

Proudlock and Rosenthal (1969) have compared the activity coefficient and acid–base equilibrium behaviour of primary, secondary, and tertiary amines in concentrated aqueous LiCl solutions. The results are consistent with the assumption that the number of water molecules solvating an anilinium cation is approximately proportional to the number of hydrogen atoms bonded to the amine nitrogen atom (cf. also equations (3.11) and (3.12), page 78).

Section 3.2.2.

The protonation of phenyl carbamate and phenyl *N,N*-dimethylcarbamate in aqueous H_2SO_4 has been correlated with the amide H_A acidity function (Armstrong and Moodie, 1970).

The ionization ratios for the carbamates were difficult to estimate accurately because of pronounced medium effects on the ultraviolet spectra.

The protonation equilibria of sulphoxides in aqueous $HClO_4$ or H_2SO_4 also correlate with the H_A acidity scale (Landini *et al.* 1969). This necessitates a reappraisal of the interpretation of the results for the acid catalysed reduction and racemization of 4-tolyl methyl sulphoxide (pages 192–3).

Section 3.2.5

An acidity scale H_C based on the protonation equilibria of eleven carbon bases has been measured for 0·5 mole litre^{-1} $\leqslant C_{H_2SO_4} \leqslant$ 14·0 mole litre^{-1} and for 0·5 mole litre^{-1} $\leqslant C_{HClO_4} \leqslant$ 9·5 mole litre^{-1} in water (Reagan, 1969).

The indicators used were substituted 1,1-diarylethylenes, azulenes and aromatic polyethers. For sulphuric acid the H_C scale is 0·2–0·9 units less negative than the H'_R $(= J_0 - \log_{10} a_w)$ scale. Equation (3.21) is more rigorously written

$$J_0 - \log_{10} a_w = pK_{BH^+} - \log_{10}(C_{BH^+}/C_B) + \log_{10}(y_{R^+} y_B / y_{ROH} y_{BH^+}) \quad (8.1)$$

as equation (8.1) in which B is a carbon base and ROH is a triarylcarbinol (Section 3.1). Equation (3.20) is not exactly obeyed.

The rates of acid-catalysed reactions (e.g. the hydration of olefins or aromatic hydrogen exchange) involving protonation of a carbon base should be correlated with the H_C acidity function rather than with H_0 which is appropriate to the protonation of amines.

Section 3.5.3

Reeves and Kaiser (1969) have deduced the ionization constants of six azo dye sulphonic acids by measurement of the solubility of the corresponding sulphonates in aqueous H_2SO_4 solutions of increasing concentration. The pK_a values were in the range –0·7 to –2·0 and the azo sulphonic acids are therefore weaker acids than simple arylsulphonic acids (pages 98–9).

Section 5.1

Hopkinson (1969) has provided added evidence for the proposal of Jaques (page 131) that simple esters hydrolyse by an A-1 mechanism in concentrated aqueous H_2SO_4. The methyl esters of several aliphatic acids, except those containing halogen, hydrolysed by an $A_{AC}1$ mechanism in 95·90 % H_2SO_4. In 76·88 % H_2SO_4 the $A_{AC}2$ mechanism was maintained.

Section 5.5

The acid-catalysed hydrolysis of mixed aryl alkyl acetals of benzaldehyde exhibit general acid catalysis (Anderson and Capon, 1969). A concerted A-S_E2 mechanism has been proposed.

Section 5.13

The entropy and volume of activation for the acid catalysed hydration of crotonic acid at 83·0°C are –34·2 e.u. and –17·9 cm^3 $mole^{-1}$ respectively (Bhattacharyya and Purohit, 1969). This suggests an A-2 mechanism for the reaction in which at least one molecule of water is involved in the activated complex.

Section 5.17

A book on mechanistic aspects of sulphonation and desulphonation has recently appeared (Cerfontain, 1968).

Section 5.18

A further example has been reported (Katritzky *et al.* 1970) of the use of the Moodie–Schofield method for determining whether the nitration of weak bases involves nitration of the free base or nitration of its conjugate acid.

Section 5.19

Adsetts and Gold (1969) have measured the rates of tritiation of mesitylene in aqueous HCl (0·1 mole $litre^{-1} \leqslant C_{HCl} \leqslant 9{\cdot}16$ mole $litre^{-1}$). The experimental rate constants were correlated with the H_0 acidity function via equation (8.2)

$$\log_{10} k_1 + H_0 + \log_{10}(y_S/y_B) = AC_{HCl} + \text{constant} \tag{8.2}$$

in which it has been assumed that the ratio ($y_{BH^+}/y^{\ddagger}$) obeys the Setschenow equation. A value for A of less than 0·1 was compatible with the experimental results. For 3 mole litre^{-1} < C_{HCl} < 7 mole litre^{-1} aqueous HCl a plot of $\log_{10}k_1$ against $-H_0$ had ca. 1·7 slope. It is interesting to note that the H_C acidity scale (Reagan, 1969) plotted aginst H_0 had slopes of 1·6 for H_2SO_4 and 1·7 for $HClO_4$. $\text{Log}_{10}k_1$ for the tritiation of mesitylene would probably correlate closely with the H_C acidity function for aqueous HCl.

Section 5.20

The advantage of selecting the appropriate acidity function scale to correlate the rates of reaction of a particular organic substrate has been demonstrated by Bell and De Maria (1969). For the bromination of primary amines in aqueous $HClO_4$ the inverse acidity dependence parallels H_0 whereas for 4-bromo-*N*,*N*-dimethylaniline and 2,*N*,*N*-trimethylaniline (Bell and Ninkov, 1966) the inverse acidity dependence parallels the tertiary amine H_0''' acidity function.

Section 6.2.1.

The usefulness of a glass–acetous fibre calomel electrode system for the potentiometric determination of the acidity of solutions in acetic acid has been established (Medwick *et al.* 1969). $R_0(H)$ (see page 96) and a related acidity function for acetic acid solutions have been determined using the ferrocene–ferricinium cation redox system (Pendin and Brodskaya, 1969).

Section 6.4.1

Bates (1969a, b) has discussed the best approach to the experimental measurement of pH for mixed aqueous solvents. A pa_H scale which refers to water as the standard state for all solvent systems is not possible. The definition by equation (8.3) (cf. equation 1.8) of an operational pH* scale for a particular solvent system

$$\text{pH}^*(\text{X}) - \text{pH}^*(\text{S}) = \frac{F[E(\text{X}) - E(\text{S})]}{2{\cdot}303RT} \tag{8.3}$$

is recommended. The pH* scale refers to the particular solvent as standard state provided pH*(S) is identified with $pa_H{}^*$, referred to the same solvent standard state, for the standard solution.

Vetešník *et al.* (1969) have measured an acidity function scale for perchloric acid in 50% v/v $EtOH/H_2O$ solvent using 3-nitroaniline, 2-nitroaniline and a series of substituted 2-arylazo-4-*t*-butylphenols as indicators. The scale was more negative than the H_0 scale for aqueous $HClO_4$ when compared at the same molar concentration of $HClO_4$.

Sections 7.2.1 *and* 7.2.3

Terrier (1969a) has summarized the amine H_M acidity function scales for solutions of CH_3OK, CH_3ONa, and CH_3OLi in methanol (cf. Table 7.2). The deviations between the scales for the three alkali metal cations are discussed in terms of ion association effects. The J_M acidity functions for CH_3OK and CH_3ONa in methanol are also tabulated. The H_M^{2-} scales for CH_3OK and CH_3ONa in methanol have been compared with the corresponding H_M acidity functions (Terrier, 1969b).

Terrier *et al.* (1969) have used a series of substituted indoles to measure H_{M_i} (subscript *i* for indoles) acidity functions for solutions of the methoxides of potassium, sodium and lithium in methanol. The scales (Table 8.2) deviate

TABLE 8.2

The H_{M_i} acidity function scales for three alkali–metal methoxides in methanol at 20°. Indicators were substituted indoles (Terrier *et al.* 1969).

C_{Base} (mole litre^{-1})	H_{M_i} (KOMe)	H_{M_i} (NaOMe)	H_{M_i} (LiOMe)
0·01	14·92	14·92	14·92
0·1	15·92	15·92	15·92
0·5	16·62	16·62	16·62
1	17·02	16·95	16·92
2	17·59	17·47	17·38
2·7	17·98	17·83	17·69
3	18·15	17·98	
4	18·70	18·50	
5	19·26	19·00	
5·5	19·54		

much less from $(pK_{MeOH} + \log_{10} C_{OMe^-})$ (equation 7.44) than the H_M scales (Table 7.2) based on the use of substituted anilines and diphenylamines as indicators. This further emphasizes the fact (Fig. 7.2) that acidity function scales for strong bases in methanol are a function of the particular class of weakly basic indicators used to define the scales.

Section 7.4.1

Yagil (1969) has presented a theoretical model to explain the rise in basicity (Table 7.3) which occurs with increasing dimethyl sulphoxide content for Me_2SO/water mixtures containing 0·011 mole litre^{-1} tetramethylammonium hydroxide. If the hydration number of the hydroxide ion is assumed to be 3 then the steep rise in H_- is consistent with 2 water molecules being bound to each dimethyl sulphoxide molecule. Bowden and Cockerill (1970) using a series of substituted fluorenes as weak bases, have measured an H_- scale for 40–95 mole % dimethyl sulphoxide in water containing 0·0471 mole litre^{-1} tetramethylammonium hydroxide. This scale is therefore based on carbon-base indicators, and parallels the amine indicator scale (Table 7.3) for 0·011 mole litre^{-1} tetramethylammonium hydroxide in the same solvent mixtures.

REFERENCES

Adsetts, J. R., and Gold, V. (1969). *J. Chem. Soc. B*, 950.
Akitt, J. W., Covington, A. K., Freeman, J. G., and Lilley, T. H. (1969). *Trans. Faraday Soc.* **65**, 2701.
Anderson, E., and Capon, B. (1969). *J. Chem. Soc. B*, 1033.
Armstrong, V. C., and Moodie, R. B. (1970). *J. Chem. Soc. B*, 934.
Bates, R. G. (1969a). *Pure Appl. Chem.* **18**, 419.
Bates, R. G. (1969b). *In* "Solute–Solvent Interactions" (J. F. Coetzee and C. D. Ritchie, eds.), p. 45. Marcel Dekker, New York.
Bell, R. P., and De Maria, P. (1969). *J. Chem. Soc. B*, 1057.
Bell, R. P., and Ninkov, B. (1966). *J. Chem. Soc. B*, 720.
Battacharyya, S. K., and Purohit, G. B. (1969). *J. Phys. Chem.* **73**, 3278.
Bowden, K., and Cockerill, A. F. (1970). *J. Chem. Soc. B*, 173.
Boyd, R. H. (1969). *In* "Solute–Solvent Interactions" (J. F. Coetzee and C. D. Ritchie, eds.), p. 97. Marcel Dekker, New York.
Cerfontain, H. (1968). "Mechanistic Aspects in Aromatic Sulphonation and Desulphonation." Interscience, New York.
Hopkinson, A. C. (1969). *J. Chem. Soc. B*, 861.
Katritzky, A. R., Tarhan, H. O. and Tarhan, S. (1970). *J. Chem. Soc. B*, 114.
Landini, D., Modena, G., Scorrano, G., and Taddei, F. (1969). To be published; Modena, G., Personal communication.
Medwick, T., Kaplan, G., and Weyer, L. G. (1969). *J. Pharm. Soc.* **58**, 308.
Pendin, A. A., and Brodskaya, Yu. S. (1969). *Zhur. fiz. Khim.* **43**, 1512.
Proudlock, W., and Rosenthal, D. (1969). *J. Phys. Chem.* **73**, 1695
Reagan, M. T. (1969). *J. Amer. Chem. Soc.* **91**, 5506
Reeves, R. L., and Kaiser, R. S. (1969). *J. Phys. Chem.* **73**, 2279.
Staples, B. R., and Bates, R. G. (1969). *J. Res. Natl. Bur. Std.* **73A**, 37.
Terrier, F. (1969a). *Ann. Chim.* **4**, 153.
Terrier, F. (1969b). *Bull. Soc. chim. France*, 1894.
Terrier, F., Millot, F., and Schaal, R. (1969). *Bull. Soc. chim. France*, 3002.
Vetešník, P., Rothschein, K., and Večeřa, M. (1969). *Coll. Czech. Chem. Comm.* **34**, 1087.
Wai, H., and Yates, K. (1969). *Canad. J. Chem.* **47**, 2326.
Yagil, G. (1969). *J. Phys. Chem.* **73**, 1610.

Author Index

Numbers in italics are those pages on which References are listed

10

C

H

N

O

P

Y

Z

Subject Index

Q

R

Chemical Compound Index

B

C

D

E

Q

U

V

X

Z